W9-BGF-318

ELECTROLESS AND OTHER NONELECTROLYTIC PLATING TECHNIQUES

Electroless and Other Nonelectrolytic Plating Techniques

Recent Developments

Edited by J.I. Duffy

NOYES DATA CORPORATION
Park Ridge, New Jersey, U.S.A.
1980

Library of Congress Catalog Card Number: 80-19494
ISBN: 0-8155-0818-2
Printed in the United States

Published in the United States of America by
Noyes Data Corporation
Noyes Building, Park Ridge, New Jersey 07656

Library of Congress Cataloging in Publication Data

Duffy, Joan Irene.
Electroless and other nonelectrolytic plating
techniques.

(Chemical technology review ; no. 171)
Supplement to John McDermott's [i. e. M. W.
Ranney's] Electroless plating and coating of
metals, 1972.
Includes indexes.
1. Electroless plating--Patents. I. Ranney,
Maurice William, 1934– Electroless plating
and coating of metals 1972. I. Title. II. Series.
TS662.R35 Suppl. 671.7'3'0272 80-19494
ISBN 0-8155-0818-2

FOREWORD

The detailed, descriptive information in this book is based on U.S. patents issued since July 1978 that deal with recent developments in electroless plating and related techniques. This title supersedes our previous title *Electroless Plating and Coating of Metals,* published in 1972.

This book is a data-based publication, providing information retrieved and made available from the U.S. patent literature. It thus serves a double purpose in that it supplies detailed technical information and can be used as a guide to the patent literature in this field. By indicating all the information that is significant, and eliminating legal jargon and juristic phraseology, this book presents an advanced commercially oriented review of recent developments in electroless plating.

The U.S. patent literature is the largest and most comprehensive collection of technical information in the world. There is more practical, commercial, timely process information assembled here than is available from any other source. The technical information obtained from a patent is extremely reliable and comprehensive; sufficient information must be included to avoid rejection for "insufficient disclosure." These patents include practically all of those issued on the subject in the United States during the period under review; there has been no bias in the selection of patents for inclusion.

The patent literature covers a substantial amount of information not available in the journal literature. The patent literature is a prime source of basic commercially useful information. This information is overlooked by those who rely primarily on the periodical journal literature. It is realized that there is a lag between a patent application on a new process development and the granting of a patent, but it is felt that this may roughly parallel or even anticipate the lag in putting that development into commercial practice.

Many of these patents are being utilized commercially. Whether used or not, they offer opportunities for technological transfer. Also, a major purpose of this book is to describe the number of technical possibilities available, which may open up profitable areas of research and development. The information contained in this book will allow you to establish a sound background before launching into research in this field.

Advanced composition and production methods developed by Noyes Data are employed to bring these durably bound books to you in a minimum of time. Special techniques are used to close the gap between "manuscript" and "completed book." Industrial technology is progressing so rapidly that time-honored, conventional typesetting, binding and shipping methods are no longer suitable. We have bypassed the delays in the conventional book publishing cycle and provide the user with an effective and convenient means of reviewing up-to-date information in depth.

The table of contents is organized in such a way as to serve as a subject index. Other indexes by company, inventor and patent number help in providing easy access to the information contained in this book.

16 Reasons Why the U.S. Patent Office Literature Is Important to You

1. The U.S. patent literature is the largest and most comprehensive collection of technical information in the world. There is more practical commercial process information assembled here than is available from any other source. Most important technological advances are described in the patent literature.
2. The technical information obtained from the patent literature is extremely comprehensive; sufficient information must be included to avoid rejection for "insufficient disclosure."
3. The patent literature is a prime source of basic commercially utilizable information. This information is overlooked by those who rely primarily on the periodical journal literature.
4. An important feature of the patent literature is that it can serve to avoid duplication of research and development.
5. Patents, unlike periodical literature, are bound by definition to contain new information, data and ideas.
6. It can serve as a source of new ideas in a different but related field, and may be outside the patent protection offered the original invention.
7. Since claims are narrowly defined, much valuable information is included that may be outside the legal protection afforded by the claims.
8. Patents discuss the difficulties associated with previous research, development or production techniques, and offer a specific method of overcoming problems. This gives clues to current process information that has not been published in periodicals or books.
9. Can aid in process design by providing a selection of alternate techniques. A powerful research and engineering tool.
10. Obtain licenses–many U.S. chemical patents have not been developed commercially.
11. Patents provide an excellent starting point for the next investigator.
12. Frequently, innovations derived from research are first disclosed in the patent literature, prior to coverage in the periodical literature.
13. Patents offer a most valuable method of keeping abreast of latest technologies, serving an individual's own "current awareness" program.
14. Identifying potential new competitors.
15. It is a creative source of ideas for those with imagination.
16. Scrutiny of the patent literature has important profit-making potential.

CONTENTS AND SUBJECT INDEX

INTRODUCTION

Thin film metallic coatings have been the focus of much interest in recent years. As the cost of metals soars, manufacturers are increasingly turning to more economical means of coating their products.

There are many different processes that can be considered under the heading of nonelectrolytic plating and coating. These include electroless plating (in which a metal compound is reduced to the metallic state by means of a chemical reducing agent in solution), hot dipping, plasma arc spraying, chemical vapor deposition, diffusion processes, vacuum coating, sputtering, and others. All these different methods have the goal of applying the desired thickness of metal onto a surface in the shortest period of time and at the lowest possible cost.

There has been a revolution in electrical circuitry manufacture with the advent of solid state components, and the large number of processes contained in this book attest to that fact. Cumbersome wiring and vacuum tubes have been replaced by printed circuits and transistors, and the industry has discovered new and better ways of producing electrically conductive coatings. Better ways of adhering these metals to plastic and ceramic substrates have also received much attention.

The computer industry has also benefitted from recent advances in the area of magnetic coatings, which are used to produce memory tapes and discs. New processes have produced coatings which are more oxidation-resistant and which can contain a larger amount of information using less space.

New processes involving photosensitive coatings are also contained here, such as in television screens, photographic and photocopy uses. Less traditional applications include solar cell technology.

Finally there are the more familiar uses of metallic coatings, for example, decorative and anticorrosion purposes. Here also are numerous new methods of applying these metallic layers in a more efficient manner with better results.

This book contains 225 patents which are representative of the current state-of-the-art in this dynamic field.

PRECIOUS METAL COATINGS

GOLD

Cyanide-Free Plating Bath

A.R. Burke, W.V. Hough and G.T. Hefferan; U.S. Patent 4,142,902; March 6, 1979; assigned to Mine Safety Appliances Company describe an electroless gold plating bath of excellent stability that does not require the use of cyanide ion, such a bath consisting essentially of a solution of a gold(III) salt and a tertiary amine borane reducing agent.

The process is based on the discovery that an aqueous solution of a water-soluble gold(III) salt and a water-soluble ether-substituted tertiary amine borane, maintained at a pH above about 12, preferably above 13, is stable and autocatalytically plates gold on a metallic substrate or nonmetallic substrates having catalytically activated surface. Practical plating rates are obtained at room temperature and increased plating rates are obtained at higher temperatures, suitably up to about 70°C, the bath retaining its excellent stability at such temperatures.

Gold is provided to the baths as a soluble gold(III) salt, such as, for example, halides, acetates, nitrates or sulfates of gold and corresponding alkali metal gold salts, most suitably alkali metal gold halides such as $KAuCl_4$. Gold salts having anions deleterious to the plating process should be avoided and, of course, gold cyanide salts, although operable, are undesirable in order to obtain the benefits of a bath free of cyanide ion.

The mild reducing agents used in the bath are water-soluble tertiary amine boranes of the formula $RO(C_aH_{2a}O)_xC_bH_{2b}NR'R''\cdot BH_3$ in which a is zero or an integer, x is zero or an integer, b is an integer, and R, R' and R'' are alkyl groups.

The most readily available amines and ethers have lower alkyl groups containing up to five carbon atoms, and, in the case of polyethers, up to five ether oxygen atoms, usually polymethylene, polyethylene or polypropylene glycol dialkyl ethers, from which are derived the preferred tertiary amine boranes in which a is 0 or an integer from 1 to 3; x is 0 or an integer from 1 to 4; b is an integer from 1 to 5; and R, R' and R'' are alkyl groups containing up to 5 carbon atoms. Of these, tertiary amines having a polyether substituent, (where x is an integer) are most preferred as they are more soluble in water than those tertiary amines having a simple ether substituent (where x is 0).

Example: The pH of 100 ml of a 3.0 g/ℓ solution of $KAuCl_4$ was raised to 12.5 by the addition of NaOH. In a separate beaker, the pH of 100 ml of a 2.00 g/ℓ solution of $CH_3OCH_2CH_2N(CH_3)_2 \cdot BH_3$ was also raised to 12.5 by the addition of NaOH. These two solutions were mixed with stirring and the resulting solution remained clear and stable. The surface of a nickel plate substrate was activated by (a) dipping for 20 seconds in an aqua regia to clear-etch the surface, (b) rinsing with distilled water and then dipping in a 0.5% $PdCl_2$ solution, and (c) rinsing with distilled water and then dipping in an amine borane solution for 2 minutes, following with a thorough rinse with distilled water. The activated substrate of nickel plate was immersed in the stirred bath for one hour at room temperature. Plating was noted immediately and after one hour a weight increase of 0.34 mg/cm^2 was measured.

Inorganic Fluoride Compound Binding Films

S. Singh, L.G. Van Uitert and G.J. Zydzik; U.S. Patent 4,146,309; March 27, 1979; assigned to Bell Telephone Laboratories, Incorporated describe a procedure for producing articles or devices with evaporated gold film.

The essential feature of the process is the use of certain inorganic fluoride compounds (lead fluoride, cadmium fluoride, and tin fluoride) as binding films in the bonding of evaporated gold films to certain inorganic fluoride surfaces. Such inorganic fluoride surfaces are useful in a large variety of devices including electrochromic displays, electrochromic switches, and certain capacitors and resistors containing inorganic fluorides.

The process pertains particularly to inorganic fluorides which are solids at ambient temperatures and in which gold adhesion is often a problem. More particularly it applies preferably to inorganic fluorides with cations other than posttransition metal ions since the posttransition metal ions generally have good adhesion to gold. Specifically these cations are the Group I-A elements (lithium, sodium, potassium, rubidium, cesium), the Group II-A elements (beryllium, magnesium, calcium, strontium, and barium) and the Group III-A elements (scandium, the transition metal ions, atomic numbers 22 to 28, 40 to 46, 72 to 78, yttrium, lanthanum and the rare earth elements, atomic numbers 57 to 71 and 89 to 102).

Particularly useful are surfaces of LiF, MgF_2 and LaF_3 because of their use in various type devices including electrochromic devices, solid-state switches, etc. Rare earth fluorides with atomic numbers 57 to 71 are also preferred because of their similarity to the lanthanum trifluorides. It is believed that these compounds occlude H_2O and thereby act as sources of protons (and possibly lithium ions in the case of LiF) for injection into electrochromic materials such as WO_3. Other fluorides may be useful but often such substances do not have physical properties suitable for use as a surface.

In general the binding film may be lead fluoride, cadmium fluoride, or tin fluoride or mixtures of these substances. Lead fluoride is preferred because it gives exceptionally good adherence, excellent electrical contact properties and resists electrical migration under the influence of an electrical field. One of the advantages of the process is the relative simplicity of the procedure. The binding film may be interposed between the inorganic fluoride and gold layer in a variety of ways. Generally it is most convenient to evaporate a relatively thin layer of binding film onto the inorganic fluoride surface prior to gold evaporation. Thickness of the binding film generally is not critical but might depend on the type of device performance required. Thicknesses of 25 to 2000 Å are convenient and result in good adhesion and electrical properties.

Evaporation pressure also is not generally critical but typical pressures are about 10^{-3} to 10^{-6} torr. Often it is advantageous to flush the vacuum system with an inert gas such as argon before evaporation takes place. In the fluoride evaporation, temperature including substrate temperature is not usually a critical parameter other than avoiding temperatures destructive to substrate fluoride, etc. For example, it is preferred that the substrate temperature remain below the melting point of the fluoride used in the binding film and inorganic fluoride surface. In the fluoride evaporation it is most preferred for convenience that the temperature be approximately ambient temperature.

Gold evaporation may also be carried out over a wide range of temperatures. However, it is preferred that the substrate be in a temperature range between 80° and 180°C with 140° to 160°C most preferred.

SILVER

Alditols as Reducing Agents

H. Bahls; U.S. Patent 4,102,702; July 25, 1978; assigned to Peacock Industries, Inc. has found that striking improvement in reduction efficiency during the electroless deposition of silver can be obtained by the provision of a reducer solution in the form of an aqueous solution of a polyhydric alcohol of the formula $CH_2OH(CHOH)_nCH_2OH$, where n is an integer from 1 to 6. The sharp improvement of efficiency is reflected in a considerable saving of valuable silver in the usual commercial processing plant, wherein the unreacted (unreduced) silver which remains in solution overflows into a trough or other collecting chamber, which may then be reclaimed.

The process of reclamation is complicated by the presence of sludges and other undesirable side reaction products, to such an extent that in many silvering establishments the waste overflow solution and sludge is kept in a holding tank for a very substantial number of months before reprocessing is attempted. This results in the storage of waste materials containing silver of very substantial value, which is of no use whatsoever during the entire period of storage.

The polyhydric alcohols in accordance with this process are also known as alditols, and are acyclic, polyhydric alcohols that are normally derived from aldoses or ketoses by the known reduction of the corresponding carbonyl group. For example, glucose may be reduced in the presence of a nickel catalyst to form dextro-glucitol, which is a hexitol and is a member of the alditol series.

This process is applicable universally to systems using hydraulic pumps for the delivery and projection of the solutions, and to systems utilizing air pressure for actuating the sprays. It is highly desirable in accordance with this process to provide at least three solutions which are concurrently applied to the substrate. These are conveniently referred to as solutions (A), (B) and (C). However, there are some circumstances when it is preferable to mix two of these solutions together.

With respect to solution (A), this solution contains a silver salt, desirably silver nitrate, mixed with ammonia to provide a soluble silver ammine compound, having a formula of the general type $Ag_2O \cdot NH_3$ or $[Ag(NH_3)_2]NO_3$, etc., a colorless solution. The quantities of silver nitrate per gallon may vary within the general range of about 30 to 36 ounces avoirdupois per gallon, and the amount of ammonia may also vary, for example between 45 to 75 fluid ounces of ammonium hydroxide (28° Bé) per gallon.

With respect to solution (B), which might be referred to as an activator solution, this solution desirably contains an alkali and ammonia. For example, suitable alkalis include sodium hydroxide and potassium hydroxide, and it is preferred to use a concentration of about 1 to 1¾ pounds of sodium hydroxide per gallon of activator solution, together with about 0 to 20 fluid ounces of 28° Bé ammonium hydroxide per gallon of solution. Desirable specific compositions include 1.70 lb/gal of sodium hydroxide and 13 fluid ounces of ammonium hydroxide (28° Bé).

Regarding solution (C), which might be referred to as a reducer solution, this solution contains about ½ to 2 lb/gal of sorbitol (100% sorbitol) when sorbitol is the polyhydric alcohol selected for use in accordance with the process, and about 15 to 25 ml of formaldehyde (40% solution) per gallon of concentrated solution. Ideal compositions include about 2.5 lb/gal of 70% sorbitol and 19 ml of formaldehyde (40% solution), per gallon.

In commercial use, solutions (A), (B) and (C) may be made up separately and held in separate solution tanks (A), (B) and (C). These solutions may be separately conducted to separate guns and separately sprayed upon the surface of the substrate in which the mixing occurs just prior to reaching the substrate. On the other hand, it is sometimes considered desirable to mix solutions (A) and (B) with one another, and to conduct them jointly through one gun while conducting solution (C) through a separate gun for separate spraying upon the substrate. Other modifications and forms of procedures and apparatus may be substituted, provided however that the reducer solution (C) should preferably be kept separate from the other solutions until substantially the time of actual contact with the substrate itself.

It is highly preferable to maintain the pH of the reacting solutions at a value of about 12½ or higher, preferably 12.6 or higher, in order to obtain optimum efficiency.

Example: A plant run was conducted, using an air-atomized system and guns mounted on transverse arms moving back and forth over a moving glass plate which was running (downstream) on a continuous conveyor. Several sets of spray guns were utilized, in series, spaced about one foot apart, each set spraying just downstream of its predecessor. In this run, solutions were used as follows.

Solution (A): $AgNO_3$, 33 oz avoir/gal + 70 fl oz NH_4OH, 28° Bé

Solution (B) (activator): 1.70 lb/gal NaOH; 13 fl oz NH_4OH, 28° Bé

Solution (C) (reducer): 2.5 lb/gal sorbitol; 19 ml (40% solution) preservative

The rate of application of these solutions was varied by varying the pump setting to control and regulate the speed of the pump providing the concentrated solutions, utilizing a constant rate of coating in terms of square feet of glass coated per minute. After establishing steady state conditions, silver deposition of 70 mg/ft^2 was obtained using solutions (A), (B) and (C) separately, having the compositions defined above, and this produced a deposition rate of 44 ft^2/min utilizing a pump setting providing 33 ml/min of concentrated solution.

In contrast, a solution was prepared containing the same amount of sodium glucoheptonate instead of d-glucitol. A pump setting of 44 ml/min of concentrated solution was required to produce a coating rate of 44 ft^2/min, with silver deposition of 70 ml/ft^2.

The resulting silver film was very bright and evenly applied. The pump settings used for the sorbitol solution were lower than the pump settings used with sodium glucoheptonate. The silver film was heavier than that produced by the glucoheptonate reducers, and much brighter.

Salt spray tests were conducted based on a modified ASTM test method No. B 117-72 on mirror samples taken from these runs. The film was protected in the usual manner with standard mirror backing paint and the results showed excellent adherence of silver to glass. There was no leafing of the silver from the glass.

The samples of these mirrors were placed in salt spray. After 637 hours of continuous salt spray exposure (in accordance with a modified version of ASTM method B 117-72 for mirrors), the silver still adhered well with no signs of peeling from the glass.

Silvering of Mirrors

Silver is applied to mirrors generally in the form of three separate solutions, a silver solution, a reducer solution and a caustic solution. Generally, these solutions have been applied separately from three spray heads directed onto the glass to be silvered at a common point, so that the three solutions meet on the glass. This was done for the reason that when the solutions meet there is an immediate reaction with silver being reduced which tends to clog lines and sprays.

Generally these spray heads are assembled in multiple arrays, e.g., five sets of three heads in succession on a traversing assembly which moves back and forth transversely across a conveyor carrying the glass sheet to be silvered. Such arrays of sprays are very heavy and the glass must be at an elevated temperature in order to attain the maximum efficiency. An alternative practice has been to premix the silver and caustic and to introduce the mixture through one spray nozzle and the reducer through another. This reduces the number of spray nozzles but all other problems remain the same. One of the major problems in silvering mirrors is that the efficiency of deposition of silver on the glass is related to the thoroughness of mixing of the three solutions.

In all prior art practices, the average efficiency of silver deposition is about 85%, with efficiencies generally running between 80 and 90%, depending upon a variety of factors of which temperature is an important factor. In prior art practices, a temperature of at least about 120°F is necessary in order to obtain efficiency of deposition. With energy sources as they are this has become a serious problem.

F.M. Workens; U.S. Patent 4,135,008; January 16, 1979; assigned to Falconer Plate Glass Corporation has discovered a method of silvering which produces much higher efficiencies of silver deposition using less expensive and complex apparatus, with less maintenance and at lower temperatures. It was found that with appropriate controls and equipment all three of the solutions can be premixed and applied through a single nozzle at lower temperatures with a marked increase in the efficiency of silver deposition.

Preferably a mixing manifold is provided having an outlet for mixed solution, an inlet for silver solution and an inlet for caustic solution on one side of the outlet so that silver solution and caustic solution are mixed together prior to reaching the outlet.

There is also an inlet for reducing solution on the side of the outlet opposite the silver solution inlet and means for delivering the reducer solution to the reducer inlet and to the outlet at a slightly higher volume or pressure than that of the silver and caustic solutions so that the mixture of silver and caustic solutions cannot enter the reducer solution inlet side of the manifold. There is provided a nozzle on the manifold outlet delivering a large volume spray onto the glass being silvered, the nozzle having an orifice sufficiently large to pass the total volume of mixture through a single hole in a defined pattern. Preferably a hollow conical spray is used but a solid conical spray may also be used, with the entire silvering solution being applied through a single large-volume unimix nozzle.

In Figures 1.1a through 1.1d an apparatus according to this process is shown using a mixing manifold **30** having an outlet **31** intermediate its ends, a reducer solution inlet **32** on one side of the outlet and silver solution inlet **33** and caustic solution inlet **34** on the other side of the outlet. The outlet is connected to a conical spray nozzle **35**. The nozzle and manifold are traversed back and forth across the glass as in the prior art but there is only a single nozzle involved and the weight is about 3.5 lb as compared with 35 lb.

Figure 1.1: Silvering of Mirrors

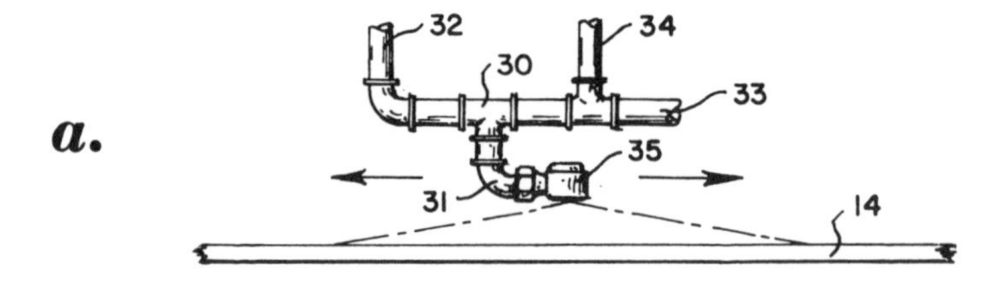

(continued)

Figure 1.1: (continued)

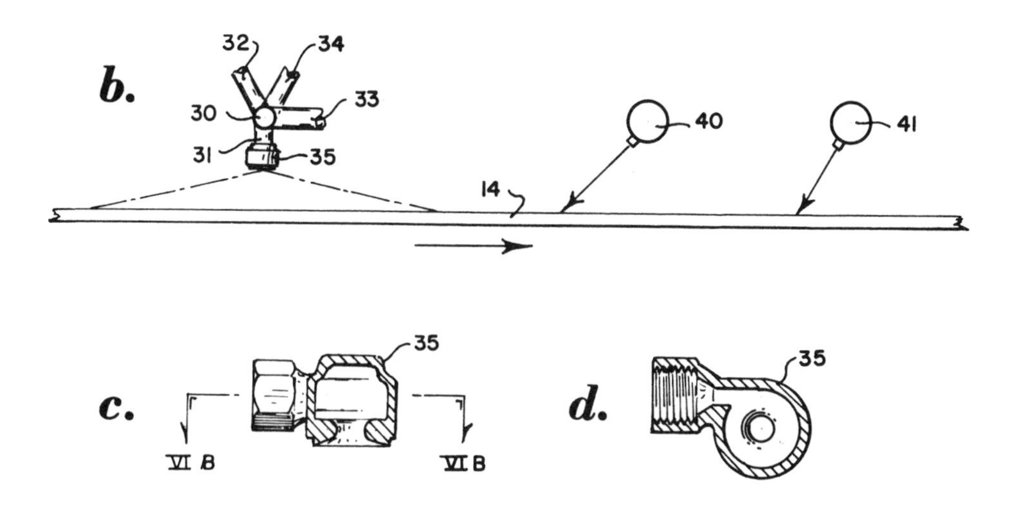

(a) Schematic end view of a silvering line
(b) Side elevational view of a silvering line of Figure 1.1a
(c)-(d) Sectional views of a spray nozzle

Source: U.S. Patent 4,135,008

In both cases an air knife blow-off **40** removes excess silvering fluid and is followed by a fresh water rinse **41**. In the apparatus and method of Figures 1.1a through 1.1d, the efficiency of silver deposition is in the range 90 to 99% with solution temperatures as low as 100°F. The reducer solution entering inlet **32** is delivered at a slightly higher volume than that of the two other solutions. This is preferably accomplished by means of flow meter controls in the delivery lines which regulate the flow of each solution to manifold **30**.

Supported Silver Catalysts

H. Diem, C. Dudeck, W. Simmler, S. Marquardt and W. Stingl; U.S. Patent 4,126,582; November 21, 1978; assigned to BASF AG, Germany have found that supported silver catalysts which are very suitable for oxidation reactions are obtained when metallic silver is deposited as a dense well-adhering layer from silver solutions onto the carrier, by a direct electroless method using reducing agents, and without prior activation or sensitization of the carrier surface.

Any materials which under the reaction conditions or at 800°C are still dimensionally stable and chemically inert may be used as the carriers. Rough-surfaced alumina, quartz moldings and commercial catalyst carriers based on α-alumina have proved particularly suitable for the manufacture of the catalyst.

Preferred reducing agents are hydrazine and hydrazine derivatives, glucose, and formalin hydroxylamine.

A particularly advantageous embodiment is, for example, one in which compounds of the general formula

$$NH_2RCOOH, \quad NH\begin{matrix} \diagup RCOOH \\ \diagdown R'COOH \end{matrix} \quad \text{or} \quad N\begin{matrix} \diagup RCOOH \\ -R'COOH \\ \diagdown R''COOH \end{matrix}$$

or salts thereof, where R, R' and R'' are identical or different hydrocarbon radicals, are added, individually or as mixtures, to the silver salt solution. These compounds are advantageously used in a ratio of from 0.02 to 0.3 mol/gram-atom of silver present in the solution. Silver is used in the form of water-soluble salts, e.g., silver nitrate.

The coating may in general be effected at from 0° to 100°C, preferably at from 10° to 30°C, at an alkaline, neutral or acid pH. The above compounds may already be added at the stage of preparing the silver salt solution.

To carry out the process it is possible, for example, to add a solution of hydrazine hydrate and distilled water in the ratio of 3:50 to the moldings to be used as carriers. The silver salt solution, comprising silver nitrate, nitrilotriacetic acid, ammonium hydroxide and distilled water in the weight ratio of, preferably, 5:0.2:5.2:95 or 25:1:26:475 is then added and the mixture is shaken gently. At reaction temperatures of from 10° to 30°C, the moldings are removed from the silver salt solution after reaction times of from 1 to 10 minutes, and are dried.

The advantages of the process are that the coating of the catalyst carrier is effected without prior impregnation, activation or sensitization of the surface, while such treatments are necessary in the conventional processes of metal coating. Furthermore, firmly adhering and electrically conductive silver layers of any desired thickness can be produced with silver yields of more than 80%, based on the silver employed. The process also avoids poisoning of the catalyst which may result from activation with foreign metals.

The example which follows describes the manufacture of a catalyst for the synthesis of formaldehyde from methanol.

Example: The carrier used consists of porcelain beads of size 1 to 2 mm, which have a tap density of 2.0 kg/ℓ and a conductivity of $<2.10^{-5}$ S.

50 ml of distilled water and 3 ml of analytical grade 100% strength hydrazine hydrate are added to 52.8637 g of these beads. To this mixture are added 100 ml of a silver solution comprising 50 g of $AgNO_3$, 1.5 g of nitrilotriacetic acid and 58 ml of 25% strength aqueous NH_4OH. The reaction is allowed to proceed for 3 minutes at 27°C, with gentle stirring. The supernatant liquid is then decanted and the silvered carrier is washed neutral with distilled water and then dried at 120°C. The weight of the silvered carrier is 55.7591 g; mg of Ag/g of carrier is 54.7; thickness of silver layer is 5 μ; surface area is 0.9 m^2/kg catalyst; and conductivity is 1.92 S.

The silver yield achieved in manufacturing the catalyst is 91%. For comparable conversions and yields, this catalyst makes it possible to lower the amount of silver used in the ratio of 1:50 as compared to solid silver catalysts in the manufacture of formaldehyde.

Germanium, Tin or Lead Promoter

N. Feldstein; U.S. Patent 4,144,361; March 13, 1979 describes a method for applying compositions which are admixed for electroless (aerosol) deposition of silver. The composition comprises a soluble silver salt, a complexing agent for the silver ions, a reducing agent for ions, a pH adjustor, and a soluble promoting compound. The presence of the promoter improves the efficiency of the deposited silver by reducing the amount of sludge formation.

The term "promoter" as used herein is intended to encompass compounds bearing the elements selected from the group consisting of germanium, tin and lead, and are preferably those inorganic compounds of such elements which are readily soluble in aqueous media. Accordingly, lead(II) is preferred. Also included are tungstate, vanadate, and similar compounds of other elements from the same groups and periods of the Periodic Table of the Elements. The incorporation of the promoter compound results in the increase of the deposition rate as well as the efficiency for the process, and diminishes the sludge formation.

PLATINUM

Plating of Titanium

S. Fujishiro and D. Eylon; U.S. Patent 4,137,370; January 30, 1979; assigned to U.S. Secretary of the Air Force describe an article of manufacture comprising a component fabricated from titanium or a titanium alloy, the component having a noble metal or noble metal alloy ion plated coating with a thickness of about 0.5 to 1.5 μ cohesively bonded to noble metal or noble metal alloy ion implanted surfaces of the component.

The ion-plated noble metal coating protects the underlying titanium or titanium alloy, which is highly reactive with oxygen, other atmospheric gases and chemicals. The mechanical properties of the titanium or titanium alloy, such as fatigue, creep and postcreep tensile ductility, are thereby increased both at room temperature and at elevated temperatures.

The method inherently involves a two-stage process. First, there is the impregnation stage in which the ionized metal atoms are forced, i.e., implanted, into the substrate. Second, the ions are plated out onto the ion-implanted substrate, forming a protective coating. The two stages of the process can be termed ion implantation and ion plating, respectively. It is the combination of ion implantation and ion plating which makes possible the cohesive coating which is resistant to oxidation and which is not subject to separation or spalling upon exposure to high temperatures for extended periods of time. The initial ion implantation is an important and essential step prior to the ion plating since the ion-implanted substrate provides a sound and stable base that is conducive to film formation by ion plating because of the atomistic cohesion between the implanted ions and the plated ions.

Procedures and apparatus described in the literature for conducting ion implantation can be used. As an ion implantation apparatus, a standard model equipment manufactured by Temescal can be conveniently employed. The bias potential between the molten noble metal or noble metal alloy source (anode) and the component substrate (cathode) usually ranges from about 2 to 4.5 kV.

The ion current generally falls in the range of about 30 to 90 mA. The vacuum chamber is evacuated to 1×10^{-6} torr and backfilled with argon up to 3 to 5 μ pressure. The two stages, i.e., ion implantation and ion plating, usually take from about 2 to 5 minutes.

As used here, noble metal includes ruthenium (Ru), rhodium (Rh), palladium (Pd), osmium (Os), iridium (Ir), gold (Au) and platinum (Pt). An example of a preferred alloy is 80% platinum and 20% rhodium. It is often preferred to employ platinum in the process.

Platinum and Palladium Catalysts

F.G.A. Stone, M. Green and J.L. Spencer; U.S. Patent 4,098,807; July 4, 1978; assigned to Air Products and Chemicals, Inc. describe methods of depositing a thin film of metallic platinum or palladium on a substrate, to form a catalyst useful in heterogeneous catalysis.

According to one aspect of the process, a method of preparing bis(cis,cis-cycloocta-1,5-diene) platinum or palladium is described in which a reducing agent is allowed to react with $[PtCl_2(1,5\text{-}C_8H_{12})]$ or $[PdCl_2(1,5\text{-}C_8H_{12})]$ and excess cis,cis-cycloocta-1,5-diene in the presence of a solvent not having an active hydrogen atom.

One preferred reducing agent is the lithium derivative of cycloocta-1,3,5,7-tetraene ($Li_2C_8H_8$), in which case diethyl ether is a preferred solvent. Examples of other possible reducing agents are alkali metal naphthalides, lithium metal in pyridine as solvent, and $NaH_2Al(OCH_2CH_2OMe)_2$.

Bis(cis,cis-cycloocta-1,5-diene) platinum and palladium are convenient starting materials for the preparation of other platinum and palladium compounds, such as $Pt(Q)_3$ or $Pd(Q)_3$, where Q is a compound which contains at least one olefinic or acetylenic double or triple bond, such as an olefin, allene, acetylene, or substituted olefin.

The cyclooctadiene molecules in $Pt(C_8H_{12})_2$ and $Pd(C_8H_{12})_2$ can be replaced by other ethylenically unsaturated ligands, or acetylenically unsaturated ligands, e.g., ethylene, butadiene or acetylene, to form the complexes $Pt(Q)_3$ or $Pd(Q)_3$ mentioned above, by reacting them with the appropriate unsaturated compound in a suitably inert solvent, e.g., petroleum ether.

These complexes $Pt(Q)_3$ and $Pd(Q)_3$ may also be prepared by reacting

$$[PtCl_2(1,5\text{-}C_8H_{12})] \quad \text{or} \quad [PdCl_2(1,5\text{-}C_8H_{12})]$$

with a reducing agent (e.g., $Li_2C_8H_8$) and the appropriate olefinically or ethylenically unsaturated compound Q. An example of such compounds which are suitable is bicyclo[2.2.1] heptene.

The complexes $Pt(Q)_3$ and $Pd(Q)_3$ are easily decomposed by heating and thus afford a relatively simple way of depositing a layer of pure platinum or palladium metal on a substrate. For example, trisethylene platinum decomposes to deposit pure platinum, the only other product being C_2H_4 which comes off as gas, leaving no impurities in the platinum layer. A convenient way to apply trisethylene platinum to a substrate is in soluble form, although as it is slightly volatile and its vapor stable in a binary mixture with ethylene gas, it may also be deposited from the vapor.

A mixture of ethylene and trisethylene platinum is passed over the substrate until the substrate has become coated with the trisethylene platinum. This is then decomposed, e.g., either by gentle heating or by passing over the substrate a gas other than ethylene. A platinum layer can also be deposited on a substrate by coating the substrate with cis,cis-cycloocta-1,5-diene platinum, e.g., in solution form and decomposing it. Platinum and palladium layers thus deposited have potentially valuable catalytic properties.

Example 1: *Preparation of cis,cis-Cycloocta-1,5-Diene Platinum* – A sample of the compound $[PtCl_2(1,5\text{-}C_8H_{12})]$ (3.7 g, 10 mmol) was finely powdered and suspended in freshly distilled cis,cis-cycloocta-1,5-diene (15 cm^3). The mixture was degassed and cooled to -40°C and a solution of the lithium derivative of cyclooctatetraene ($Li_2C_8H_8$) (10 mmol) in diethyl ether was added over 5 minutes. The resulting slurry was allowed to warm to -10°C (1 hr) and the solvent was evaporated at reduced pressure. Extraction of the residue with dry toluene (5 x 60 cm^3 portions) at 0°C gave a brown solution which was filtered through a short column (12 cm) of alumina. The volume of solvent was reduced in vacuo to ~20 cm^3 and the mother liquor decanted from the white crystalline product, cis,cis-cycloocta-1,5-diene platinum (yield 50%).

The $Li_2C_8H_8$ was prepared by suspending lithium foil (1 g) in dry diethyl ether (80 cm^3), cyclooctatetraene (3 cm^3) was added and the mixture stirred for 16 hours. The resulting solution was standardized by hydrolysis of a known volume and titration with standard aqueous hydrochloric acid.

Example 2: The following is the preparation of $Pt(Q)_3$ and $Pd(Q)_3$ compounds, i.e., trisethylene platinum, trisethylene palladium. Ethylene (at 1 atm, 18°C) displaces the cycloocta-1,5-diene from $Pt(C_8H_{12})_2$ in petroleum ether solution to give trisethylene platinum as a white crystalline solid (on cooling). This is unstable in solution except under an ethylene atmosphere. The preparation of the corresponding Pd compound is analogous but in this case it is important to ensure that the reaction temperature does not rise above -20°C.

MISCELLANEOUS PROCESSES

Encapsulated Metal Powder

O.N. Collier and S.J. Hackett; U.S. Patent 4,130,506; December 19, 1978; assigned to Johnson, Matthey & Co., Limited, England describe a metal powder which will withstand high temperatures, and means for its production. The metals for which the process will have special application are platinum, palladium, rhodium, ruthenium, iridium, osmium and gold and silver.

The method includes the steps of nucleating the metal, for example gold, or alloy onto the surface of the substrate particles and then growing further metal or alloy in bulk onto the resulting nuclei. The resultant metal or alloy film or coating forms an encapsulation around each particle of the substrate. It is generally necessary, in order to produce and to reproduce consistently a metal powder in which the particles are of the desired size and the encapsulation is of the desired thickness, to control closely the steps of nucleation and growth.

The metal constitutes between 50 and 95 wt %, e.g., 50 to 65 wt %, of the total weight of the powder.

The heat-resistant substrate may be made from any material from the range of naturally occurring and synthetic refractories. Examples of naturally occurring refractories are clays, silica, alumina, titania, zirconia and/or mixtures of these, and examples of synthetic refractories are silicon nitride, silicon carbide and/or mixtures of these.

A first stage of the process is to activate the substrate. "Activating" means ensuring that the maximum surface area of the substrate becomes available for nucleation and subsequent growth. Methods of activation vary according to the nature of the substrate. For example, activation of a clay or a naturally occurring mineral can be achieved by boiling it in water. Optionally, the water may also contain a solution of a strong reducing agent, such as hydrazine hydrate or sodium sulfite. An alternative activation procedure for a clay or a naturally occurring mineral is to boil it in dilute mineral acid.

On the other hand, an activation procedure for a synthetic refractory compound, if the inherent activity is too low, is to deposit active sites on the surface of the refractory. This may be carried out by using any or all of the methods of preparing ceramic and other heat-resistant substrates for catalytic purposes, methods which are well known to those skilled in the art.

A second stage of the process is to nucleate particles of metal onto the surface of the activated substrate particles and this may be achieved by adding a suspension of the substrate particles in the activating agent to an aqueous solution of a salt of the metal or applying an organo compound of the metal and subsequently decomposing the same.

By way of example and with particular reference to gold powders, nucleation may be achieved by dispersing the refractory substrate particles in a solution of an organic sulfur-containing gold compound in an organic solvent, evaporating the solvent and thermally decomposing the gold compound.

Nucleation is then initiated by reducing the metal, e.g., gold salt, with a strong reducing agent which may be that already optionally present in the activation solution. If no reducing agent is present in the activation solution, nucleation may be induced by adding a strong reducing agent to the suspension of substrate in the mixture of activating agent and metal salt solution. The chemical nature of the strong reducing agent added to initiate nucleation may be similar to that of the reducing agent added to the activation solution, for example hydrazine hydrate or sodium sulfite. Vigorous stirring is desirable at this stage to ensure adequate and uniform dispersion of the substrate particles in the metal salt solution.

It is preferable to add to the solution of metal salt, prior to the addition of the suspension of substrate particles in the activation solution, a colloidal protective agent. This agent controls nucleation and prevents agglomeration of the nucleated substrate particles. Examples of suitable colloidal protective agents are gum acacia, gelatin, egg albumin and dextrin, but in general the requirements of the colloidal protective agents are that they should have a high molecular weight and be capable of being adsorbed onto the surface of the nucleated substrate particles so that their agglomeration is physically prevented.

A third stage of the process is to grow further metal in bulk on the nuclei already present on the substrate surface. This is achieved by adding to the second-stage suspension a weak reducing agent, such as hydrogen peroxide or hydroquinone. It is preferred to add the reducing agent in portions and any foam generated may readily be suppressed by a spray of, for example, isopropanol. After all the weak reducing agent has been added, the suspension is stirred for some hours to complete the growth stage of the process. The resulting metal powder is then filtered off, washed and dried.

Example: 75.0 g of gold as gold ammonium chloride was dissolved in 2.4 ℓ of distilled water in a 5 ℓ beaker. 40 ml of a 10% gum acacia solution was added and the mixture stirred to ensure complete dissolution of the gold salt. Meanwhile, 25 g of china clay was activated by boiling in 100 ml of distilled water containing 10 drops (= 0.45 ml) of a 6% hydrazine hydrate aqueous solution. The china clay/hydrazine hydrate suspension was then added with vigorous stirring to the gold solution. Upon addition, the color of the mixture changed from yellow to yellow-green. After stirring for 10 minutes, 400 ml of 40 volume hydrogen peroxide was added, as a result of which the color changed from green to brown and foam was formed from reaction gases generated.

The foam was suppressed using the minimum quantity of isopropanol from a laboratory spray. After 10 minutes, a further 100 ml of 40 volume hydrogen peroxide was added, which caused further foaming, and the final 100 ml of hydrogen peroxide was added after a further 10 minutes. The reaction mixture was then stirred for 5 hours to complete the reaction, after which the clear supernatant liquor was decanted off and the powder was filtered off, washed and dried until constant weight was achieved.

The resulting gold powder was eminently suitable for use as a pigment in a burnish gold preparation for decorating pottery and porcelain and for firing at high temperatures. Firing schedules for decorations and in common use employ a peak temperature of about 800°C, but modern furnaces are designed to operate at temperatures up to about 1050°C (for chinaware) and up to about 1400°C (for porcelain). At these temperatures, using burnish gold preparations containing standard gold powders, breakdown of the film occurs due to the gold sintering and forming into agglomerates.

However, using a gold powder according to this process, the resulting film has surprisingly high cohesive properties at temperatures as high as 1400°C and the spatial configuration of the gold powder particles in the film is maintained. The resulting films are capable of being burnished to a continuous decorative film with good adhesion and no wrinkling.

Metal Leaf

In the handicraft method for producing a metal leaf, a piece of metal such as gold or silver is finely patted by a hammer to a thickness of about 100 mμ. Therefore, a skilled art is required for the production of metal leaves.

H. Narui, I. Akune and Y. Kobiki; U.S. Patent 4,100,317; July 11, 1978; assigned to Oike & Co., Ltd., Japan describe a process for producing metal leaf having the same characteristics as those of a conventional metal leaf by handicrafting.

This metal leaf can be produced (1) by applying an undercoating composition on a base film to give an undercoating layer having a thickness of 0.35 to 1 μ, (2) by depositing a metal under a vacuum on the undercoating layer to give a metal deposition layer having a thickness of 0.03 to 0.1 μ, and (3) by applying an overcoating composition on the metal deposition layer to give an overcoating layer having a thickness of 0.35 to 1 μ, to result in an integrated material of three layers (undercoating layer + metal deposition layer + overcoating layer), having a thickness of 0.73 to 2.1 μ and a tensile strength of 0.01 to 1.4 kg/mm^2, and then peeling the integrated material out of the base film.

The method comprises elongating the base film having the metal leaf thereon to mechanically peel the metal leaf and putting the peeled leaf on a supporting paper to recover it. As a result, the metal leaf can be continuously and mechanically peeled out of the base film without any cracking or breaking.

Examples of the base film are a film of polytetrafluoroethylene, polyethylene, polypropylene, polyethylene terephthalate, polyvinyl chloride, polyamide, cellulose acetate, regenerated cellulose, polycarbonate, water-resistant polyvinyl alcohol, and the like. The thickness of these films may be about 6 to 100 μ.

As an undercoating composition there may be employed those prepared from either thermoplastic resins or thermosetting resins. Examples of these resins are acrylic resins, thermosetting-type acrylic resins such as epoxy-modified acrylic resin, vinyl chloride-vinyl acetate copolymer, polycarbonate, polyvinyl butyral, rosin-modified maleic resin, urea resin, melamine resin, nitrocellulose, cellulose acetate, alkyd resin, urethane resin, rosin, shellac, and the like. These resins alone or together with each other are employed in a form of solvent solution or aqueous solution.

Examples of the solvent are halogenohydrocarbons such as trichloroethylene, ketones such as methyl ethyl ketone and methyl isobutyl ketone, lower alkyl acetates such as ethyl acetate and butyl acetate, lower alcohols such as methanol, ethanol, isopropanol and n-butanol, aromatic hydrocarbons such as xylene and toluene, Cellosolves such as methyl Cellosolve, ethyl Cellosolve and butyl Cellosolve, dioxane, and the like. If necessary, additives, e.g., plasticizer, viscosity adjusting agent, etc., may be added to the coating composition. Further, a colorant may also be added into the composition. Examples of the colorant are a solvent-soluble dyestuff and a clear lacquer consisting of resin and pigment. A thickness of the undercoating layer (the overcoating layer is also the same) is suitable in the range of about 0.35 to 1 μ.

Prior to the undercoating, a releasing agent such as silicone resins or waxes may be coated onto a base film. Then, the metal deposition is carried out on the undercoating layer. Examples of metals to be deposited are gold, silver, aluminum, copper, and the like.

The metal deposition may be carried out according to a conventional method, e.g., vacuum metallizing method or metal sputtering method. For instance, a pressure in the vacuum chamber is preferably about 10^{-3} to 10^{-6} torr, and a temperature of evaporation source is selected from the range of about 1200° to 2000°C in accordance with the kind of metal. In case of gold or copper it is preferably carried out at about 1500° to 2000°C and in case of silver or aluminum at about 1200° to 1600°C.

As an overcoating composition there may be employed the same composition as that in the undercoating. Preferably both compositions are the same, and thereby the curling of metal leaf can be prevented. However, a solvent having a large evaporation velocity is preferable for the overcoating composition.

The metal leaf on a long-size base film is continuously and mechanically peeled without any cracking or breaking by elongating the base film. It is then placed on a supporting paper, which is fed from a supplying roll, and recovered.

The method can be better understood with reference to the accompanying drawing which is the schematic representation of one specific embodiment.

In Figure 1.2, **1** is a roll on which a base film **2a** having metal leaf is wound up. While winding back the film having the metal leaf **2b**, the metal leaf is peeled out of the film by turning up only the base film at the position of guide bar **3** and rewound up onto a roll **4**. The rewinding onto the roll is carried out with elongation to peel the metal leaf mechanically out of the base film. An elongating ratio of the base film suitable for peeling is selected from the range of about 0.1 to 20%. If necessary, the elongation may be carried out while heating to assist elongation. **5** is an earthing. **6** is a supplying roll for supporting paper **7**. While winding back from the roll, the supporting paper is put on the metal leaf side of the base film and progressed, and thereafter wound up on a winding roll **8** without any cracking or breaking.

Figure 1.2: Production of Metal Leaf

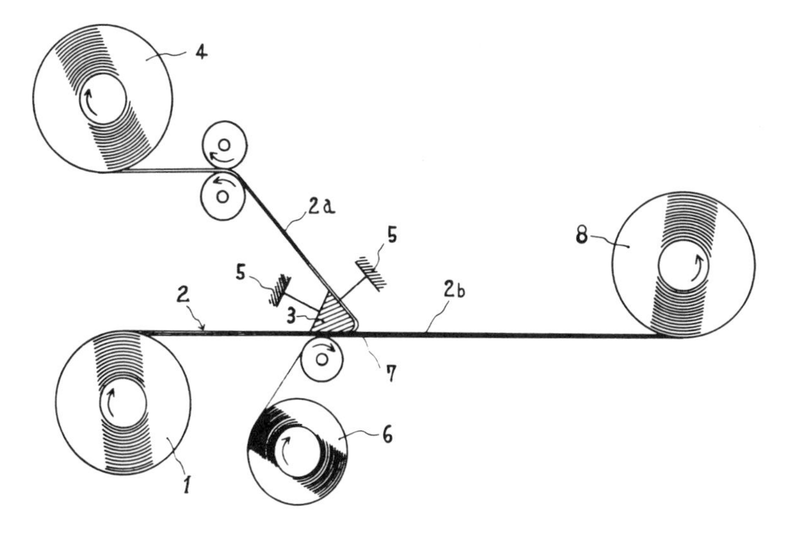

Source: U.S. Patent 4,100,317

As a supporting paper there may be employed pulp paper, rice paper or dermic paper, preferably so-called supporting paper for metal leaf. Alternatively, the metal leaf **2b** may be wound back after or instead of winding onto the roll **8** and optionally cut to give the desired product.

Example: An undercoating composition (consisting of 20 parts of vinyl chloride-vinyl acetate copolymer, 100 parts of methyl isobutyl ketone and 100 parts of ethyl acetate) was applied on a polypropylene film having a thickness of 30 μ by means of the so-called gravure method. Hereinafter all applications were carried out by the gravure method. The composition was then dried to give an undercoating layer having a thickness of 1 μ. On the undercoating layer was deposited gold under a vacuum of about 3 x 10^{-10} torr at an evaporation temperature of 1700°C to give a gold deposition layer having a thickness of about 75 mμ. Further, an overcoating composition (the same as the undercoating composition) was applied on the deposition layer and dried to give an overcoating layer having a thickness of 1 μ.

Using the apparatus as illustrated in Figure 1.2, the resultant was set so the polypropylene film side was in the upward direction, and only base film was turned up at the guide bar, elongated in an elongating ratio of about 10% to peel out of the gold leaf, and then wound up around the film-winding roll.

On the other hand, the gold leaf thus peeled out of the base film was put on a supporting paper fed from a supplying roll which was put on the gold leaf side of base film in the underside at the position of the guide bar, and progressed and recovered.

Thus-obtained gold leaf is graceful and there is neither cracking nor breaking. Besides, the gold leaf has the same appearance and hand touchness by comparison to a conventional gold leaf by handicrafting. A tensile strength of the resultant (test piece: 8 x 4 cm) was measured by using a tensile strength testing machine Autograph S-100 under conditions of a temperature of 20°C, a relative humidity of 60% and a tensile velocity of 100 mm/min to give a tensile strength of 1.3 kg/mm^2. On the contrary, a conventional handicrafted gold leaf has a thickness of 100 mμ and a tensile strength of 0.7 kg/mm^2.

BASE METAL COATINGS

COPPER

Control of Dissolved O_2 Content

W.A. Alpaugh and T.D. Zucconi; U.S. Patent 4,152,467; May 1, 1979; assigned to International Business Machines Corporation have found that improved products can be obtained by maintaining the level of dissolved oxygen in a copper electroless plating bath between 2 and 4 ppm, and preferably between 2.5 and 3.5 ppm, during the plating process. These amounts of dissolved oxygen are below the saturation level of dissolved oxygen obtained by using air. By maintaining the dissolved oxygen level in this range, the formation of voids in the coating and particularly in holes in the substrate, is significantly reduced if not completely eliminated, and the formation of nodules on the surface of the substrate is also significantly reduced.

In addition, the process makes it possible to use only one bath both to initiate and to provide for continuous plating of the desired substrates. The necessity for a strike- or flash-bath is eliminated. The process also results in short take times, e.g., times for plating of the initial thickness of coating in the holes. For instance, take time of many processes were minutes or less.

The proper level of oxygen is preferably maintained during the plating by introducing into the plating tank a mixture of oxygen and an inert gas. The amount of oxygen in the mixture is such that there will be less than the amount of oxygen dissolved in the bath than if the bath were aerated with air alone. For instance, if nitrogen is used as the inert gas, then the amount of oxygen in the mixture would be less than the amount normally found in air.

The oxygen portion of the mixture can be from pure oxygen or preferably from air primarily because of economics. Mixed with the air or oxygen is an inert gas such as hydrogen, nitrogen, argon, neon, krypton, and the like. The preferred inert gas is nitrogen. Mixtures of inert gases can be used. The relative amount of inert gas and oxygen will depend upon the relative solubilities of oxygen and the inert gas in water at the temperature of operation of the electroless bath in order to obtain the desired amount of oxygen in the bath. For instance, with nitrogen as the inert gas in admixture with air, at a plating temperature of 73±0.5°C, 14 scfm/1,000 gal of bath of air and between about 0.7 to 1 scfm/1,000 gal of bath of nitrogen provided O_2 levels between about 2.5 and 3.5 ppm. The saturation level of oxygen at 73°C in the bath due to aeration with only air would be about 5 to 5.5 ppm. The saturation level of oxygen decreases with increasing temperatures.

The inert gas is preferably premixed with the oxygen or air prior to introduction into the bath. However, the individual gases can be introduced into the bath separately if desired.

In addition to including the required level of dissolved oxygen, the copper electroless plating bath is generally an aqueous composition which includes a source of cupric ion, a reducing agent, a complexing agent for the cupric ion, and a pH adjuster. The plating baths also preferably include a cyanide ion source and a surface-active agent.

The cupric ion source generally used is a cupric sulfate or a cupric salt of the complexing agent to be used. When using cupric sulfate, it is preferred to use amounts from about 3 to 15 g/ℓ and, most preferably, from about 8 to 12 g/ℓ. The most common reducing agent used is formaldehyde which is used in amounts from about 0.7 to 7 g/ℓ and, most preferably, from about 0.7 to 2.2 g/ℓ. Examples of some other reducing agents include formaldehyde precursors or derivatives such as paraformaldehyde, trioxane, dimethylhydantoin, glyoxal; borohydrides such as alkali metal borohydrides (sodium and potassium borohydride) and substituted borohydrides such as sodium trimethoxyborohydride; boranes such as amine borane (isopropylamine borane and morpholine borane).

Examples of some suitable complexing agents include Rochelle salts, ethylenediaminetetraacetic acid, the sodium (mono-, di, tri- and tetrasodium) salts of ethylenediaminetetraacetic acid, nitrilotriacetic acid and its alkali salts, gluconic acid, gluconates, triethanolamine, glucono-γ-lactone, modified ethylenediamine acetates such as N-hydroxyethylethylenediaminetriacetate. The amount of complexing agent is dependent upon the amount of cupric ions present in the solution and is generally from about 20 to 50 g/ℓ, or in a 3 to 4 fold molar excess.

The plating bath can also contain a surfactant which assists in helping to wet the surface to be coated. A satisfactory surfactant is, for instance, an organic phosphate ester [Gafac (RE-610)]. Generally, the surfactant is present in amounts from about 0.02 to 0.3 g/ℓ. In addition, the pH of the bath is also generally controlled, for instance, by the addition of a basic compound such as sodium hydroxide or potassium hydroxide in the desired amount to achieve the desired pH. Preferably the pH of the electroless plating bath is between 11.6 and 11.8.

The plating bath also contains a cyanide ion and, most preferably, contains about 10 to 25 mg/ℓ to provide a cyanide ion concentration in the bath within the range of 0.0002 to 0.0004 molar. Examples of some cyanides which can be used are alkali metal, alkaline earth metal, and ammonium cyanides such as sodium, potassium, calcium, and ammonium cyanide. In addition, the plating baths can include other minor additives as is well-known in the prior art.

The preferred plating baths have a specific gravity within the range of 1.060 to 1.080. In addition, the temperature of the bath is maintained between 70° and 80°C and preferably between 70° and 75°C. The overall flow rate of the gases into the bath is generally from about 1 to 20 scfm/1,000 gal of bath and preferably from about 3 to 8 scfm/1,000 gal of bath. The plating rate achieved by the process is generally between about 0.06 to 0.15 mil per hour and the bath loading is between about 20 and 200 cm^2/ℓ.

Addition of Gallium Stabilizer

M. Gulla and C. Savas; U.S. Patent 4,124,399; November 7, 1978; assigned to Shipley Company Inc. have found that the addition of a small, but effective amount of a source of gallium to an electroless copper solution improves stability without substantially retarding the rate of deposition. Moreover, it has been found that the addition of a combination of gallium with another stabilizer results in substantially increased stability. Accordingly, the process provides an electroless copper deposition solution comprising (a) a source of copper ions; (b) a reducing agent, such as formaldehyde; (c) a pH adjuster; (d) a complexing agent for the copper ions sufficient to prevent their precipitation in solution; and (e) a stabilizer for the solution which may be a source of gallium alone or in combination with another stabilizer.

The solutions are used to deposit copper in conventional manner. The surface of a part to be plated should be free of grease and contaminating material. Next, the surface to receive the metal deposit is sensitized to render it catalytic to the reception of the electroless metal as by the well-known treatment of contact with a colloid of palladium having a protective stannic acid colloid. Thereafter, following known rinsing steps and the like, the part is immersed in the plating solution at a temperature dependent upon the solution used for a time sufficient to provide a deposit of desired thickness.

The process will be better understood by reference to the following examples where stability of solution was measured by the time a bath spontaneously decomposes (triggers) when plating catalyzed cloth at one-eighth square foot per liter. Catalyzed cloth is cloth immersed in Catalyst 6F (Shipley Company Inc.).

Examples 1 Through 8: Catalyzed cloth was plated with the following formulation at room temperature with gallium nitrate added in amounts and with results as set forth in the following table. The formulation comprised 8 g cupric sulfate pentahydrate, 7.5 g formaldehyde, 40 g sodium/potassium tartrate, 17 g sodium hydroxide and water to 1 liter.

Example	Gallium Nitrate (ppm)	Time (min)
1	0	45
2	5	45
3	25	45
4	50	45
5	100	45
6	250	90
7	500	50
8	1,000	85

The above results show an improvement with gallium. It should be noted that the results set forth above are approximate as they are based upon visual observation.

Boron Hydride and Copper Complexing Agent

Y. Arisato and H. Koriyama; U.S. Patent 4,138,267; February 6, 1979; assigned to Okuno Chemical Industry Company, Ltd., Japan describe a composition for chemical copper plating having a pH of 12 to 14 and containing a boron hydride compound serving as a reducing agent, a water-soluble copper compound and a copper complexing agent, the composition being characterized in that the copper complexing agent is at least one of hydroxyalkyl-substituted ethylenediamines represented by the formula

$$\begin{matrix} R_1 & & R_3 \\ & \diagdown\text{N}-(\text{CH}_2)_2-\text{N}\diagup & \\ R_2 & & R_4 \end{matrix}$$

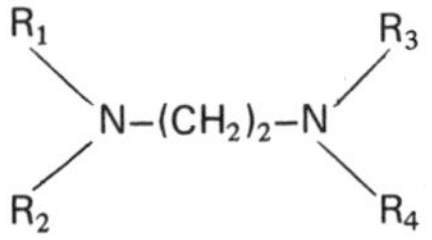

where R_1, R_2, R_3 and R_4 are the same or different and are each unsubstituted lower alkyl or hydroxyl- and/or carboxyl-substituted lower alkyl, and at least one of R_1, R_2, R_3 and R_4 is hydroxyl-substituted lower alkyl, and hydroxyalkyl-substituted diethylenetriamines represented by the formula

$$\begin{matrix} R_5 & & R_9 & & R_7 \\ & \diagdown\text{N}-(\text{CH}_2)_2- & \text{N} & -(\text{CH}_2)_2-\text{N}\diagup & \\ R_6 & & & & R_8 \end{matrix}$$

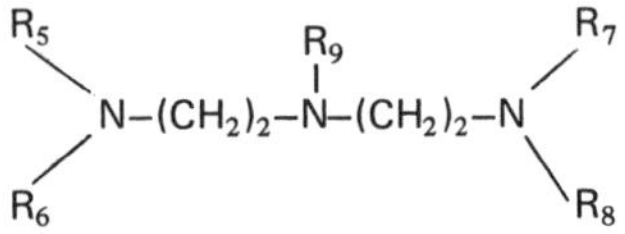

where R_5, R_6, R_7 and R_8 are the same or different and are each unsubstituted lower alkyl or hydroxyl- and/or carboxyl-substituted lower alkyl; R_9 is hydrogen or unsubstituted lower alkyl or hydroxyl- and/or carboxyl-substituted lower alkyl; and at least one of R_5, R_6, R_7, R_8 and R_9 is hydroxyl-substituted lower alkyl.

As compared with conventional chemical copper plating compositions containing a boron hydride compound as a reducing agent, the chemical copper plating compositions of this process have higher stability and give coatings having an outstanding gloss which is better than the gloss of the coatings formed from conventional chemical copper plating compositions of the copper-formaldehyde type. Furthermore, the compositions are free of pollution problems since they contain no formaldehyde.

Aromatic Nitro Compound Stabilizer

J.M. Jans; U.S. Patent 4,118,234; October 3, 1978; assigned to U.S. Phillips Corporation has found that simple, aromatic nitro compounds are particularly effective for stabilizing electroless copper plating solutions containing formaldehyde as a reducing agent. Furthermore, a considerable improvement of the selectivity of the patterns is obtained when intensified with the use of this bath as compared to a bath without these nitro compounds.

According to the process, an aqueous alkaline copper plating bath which contains cuprous ions, a compound which forms complexes with cuprous ions, alkali for adjusting the pH and formaldehyde or a compound which yields formaldehyde, is characterized in that it also contains an additional substance consisting of a substituted aromatic nitro compound having at least one substituent selected from aldehyde, alkyl, nitro, sulfonic acid, hydroxyalkyl, hydroxyketoalkyl ($COCH_2OH$) and amino.

It is advantageous if the copper plating bath also contains at least one polyalkylene oxide compound of at least 4 alkaline oxidic groups to improve the ductility of the depositing copper.

The process will be further explained with reference to the following examples. Various aromatic nitro compounds were added to the baths of the following two compositions.

Bath A	
$CuSO_4 \cdot 5H_2O$	7.5 g/ℓ
Sodium potassium tartrate (Rochelle salt)	85 g/ℓ
Na_2CO_3	15 g/ℓ
NaOH	12 g/ℓ
Formalin solution, 37% by weight	36 ml/ℓ
Working temperature	25 °C
Bath B	
$CuSO_4 \cdot 5H_2O$	7.5 g/ℓ
Tetrasodium salt of ethylenediaminetetraacetic acid	21 g/ℓ
NaOH	3 g/ℓ
Formalin solution, 37% by weight	7 ml/ℓ
Triton QS 44 (Rohm and Haas)*	2 g/ℓ
Working temperature	70 °C

* Alkylphenoxy polyethylene phosphate ester having a molecular weight of approximately 800 and approximately 8 ethoxy groups.

One of the following nitro-benzene derivatives and thereafter 10 ml/ℓ of a solution of 2 g/ℓ $PdCl_2$ were added to Baths A and B. The survey below specifies the time in minutes after which the bath became unstable.

Compound Added	Bath A	Bath B
None	0-2	0-2
2-Cl-4-nitroaniline	10-15	5-10
m-Nitrobenzaldehyde	15-20	15-20
p-Nitrotoluene	20-25	5-10
m-Nitrobenzenesulfonic acid	20-25	5-10
o-Nitrobenzaldehyde	25-30	25-30
1,3-Dinitrobenzol	30-35	25-30
p-Nitrobenzaldehyde	60-65	40-45

No improvement in stability was found when inter alia halogenized nitrobenzol, methoxylated nitrobenzol and unsubstituted nitrobenzol were added. The deposition rates of the copper, that is, 2 μm/hr for Bath A and 4 μm/hr for Bath B at the specified working temperatures were not influenced by the addition.

Hot-Dip Coating Method

A. Gierek, L. Bajka and M. Machnicka; U.S. Patent 4,142,011; February 27, 1979; assigned to Politechnika Slaska im. Wincentego Pstrowskiego, Poland describe a method of obtaining diffusion coatings of copper alloys in a hot-dip process on workpieces made of ferrous alloys, with simultaneous heat treatment of the products. The coatings are able to provide a notable increase in the corrosion-resistance of the products, especially in highly corrosive environments, mainly in water and sea environment, as well as in hot industrial waters containing certain contaminations, such as, for instance, chlorides or compounds of sulfur. The coating can also be applied on wear-resistant parts of bearings and on elements of other friction connections.

The method comprises dipping the workpieces to be coated into a bath of molten alloys of Cu with Sn, Si, Al, P, In, Ga, Be, at a temperature within the range of 700° to 1100°C, in a two-stage or single-stage continuous movement, and the workpieces are held therein for 15 seconds up to 60 minutes, whereafter they are taken out of the bath and cooled at any rate. The dipping of the workpieces into the bath is performed in a single stage, or in two stages, in which the workpiece is dipped into the bath and held beneath the surface, whereafter it is introduced into a deeper layer of the bath.

For instance, a coating of Cu-Si alloys can be obtained by dipping steel workpieces, with previously prepared surfaces, into a bath of fused metal containing 84% Cu and 16% Si at a speed of 5 m/min, holding the workpieces just below the surface of the bath for a time less than 1 minute, and then immersing the workpieces deeper into the bath, adjacent to the bottom of the crucible. After the products are held thereat for 10 minutes, the products are brought to the surface at a speed of 1 m/min and slowly cooled in the air. The temperature of the molten metal is 850°C.

IRON

Iron- or Copper-Coated Scrap Particles

S.M. Kaufman; U.S. Patent 4,129,443; December 12, 1978; assigned to Ford Motor Company describes a method of making sintered shapes from ferrous-based metal particles containing oxidizable ingredients, which method not only decreases the energy and cost requirements of the mode of comminution of the metal, but also improves the diffusion kinetics for sintering of the comminuted metal into a desired shape. Other method objects of this process comprise: (a) a method of making an intermediate powder useful in powder metallurgy techniques; (b) an intermediate powder composition made from scrap machine turnings; and (c) a method of making a cold, compacted shape which can be shipped as a commodity useful in subsequent sintering techniques to make a stable permanent metal part.

A preferred mode for carrying out the method is depicted in Figure 2.1 and is as follows: (1) Scrap metal and particularly machine turnings **10** are selected as the starting material. Machine turnings are defined herein to mean segments of ribbons of low alloy steel. They typically are shavings cut from an alloy bar.

(2) The selected scrap pieces **10** are then put into a suitable charging passage **11** leading to a ball-milling machine **12** or equivalent impacting device. Within the passage, means **13** for freezing such metal pieces is introduced, such as liquid nitrogen; it is sprayed directly onto the metal pieces. Mere contact of the liquid nitrogen with the scrap pieces will freeze them instantly. The application of the liquid nitrogen should be applied uniformly throughout its path to the point of impaction. The iron ball-milling elements **14** are motivated, preferably by rotation of the housing **17**, to contact and impact the frozen pieces **15** of

scrap metal causing them to fracture and be comminuted. Such impaction is carried out to apply sufficient fracturing force (less than 1 ft-lb) for a sufficient period of time and rate to reduce the scrap pieces to a powder form. The powder **16** will have both a coarse and a fine powder proportion. Both proportions will be comprised of particles which are flake or layered in configuration; each particle will be highly irregular in shape and dimension, none being spherical. A typical screen analysis for the powder after step (2) would be as follows (for a 100 g sample).

Mesh	No Milling (g)	After 72 Hours (g)
60	60.0	31.5
100	19.5	11.0
140	5.5	7.5
200	6.5	18.0
325	4.5	22.5
-325	4.0	9.5

Figure 2.1: Iron- or Copper-Coated Scrap Particles

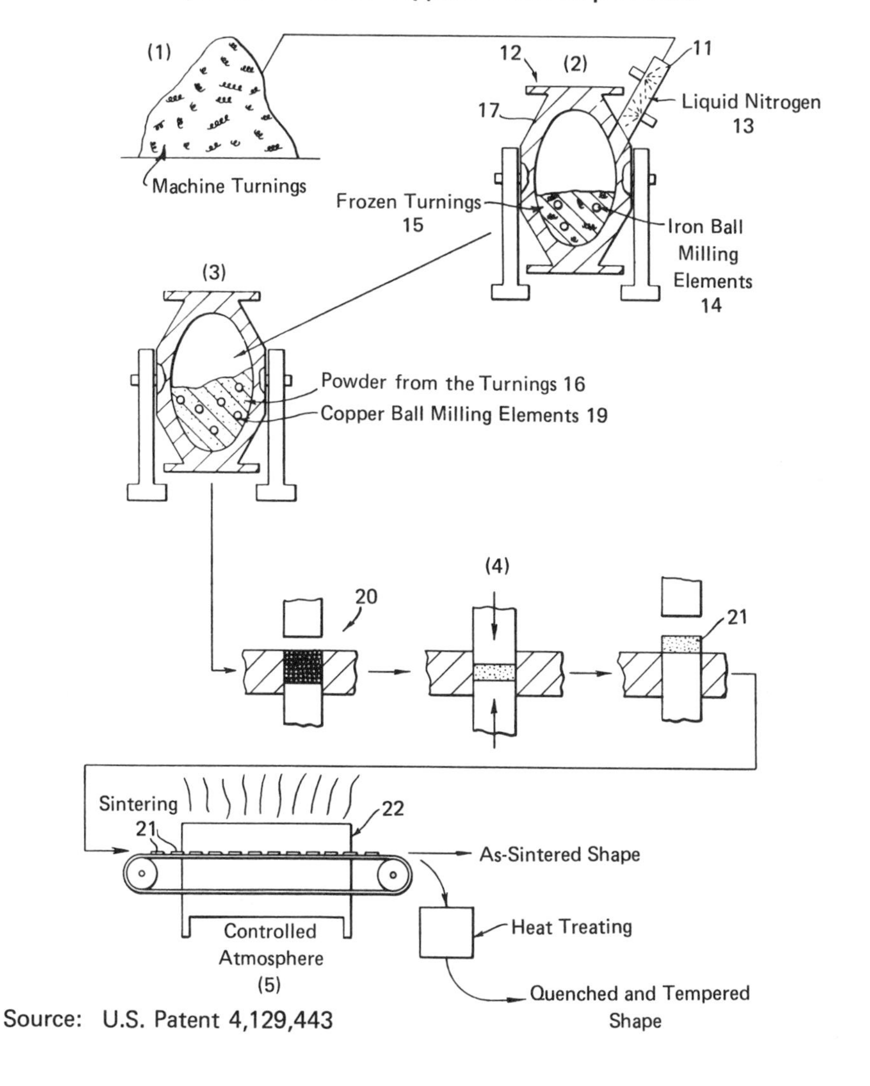

Source: U.S. Patent 4,129,443

(3) The comminuted cryogenic powder **16** is then subjected to another impacting step, but this time at ambient temperature conditions. The powder is placed preferably in another ball-milling machine, the machine having elements **19** laden with a special antioxidation coatable agent such as copper or iron. The elements are preferably in the form of solid balls of about 0.5" in diameter and consisting of iron or copper. The coatable protective agent should be characterized by (A) a hardness less than that of the coated powder to promote transfer to the particles upon impact between the elements and particles; (B) being completely soluble in the metal of the particles; (C) easy to abrade; and (D) acts as an oxidation barrier.

In trials performed herein, the interior chamber was a 3" x 6" cylinder, powder charge was 10 in^3 and the milling time was about 48 hours. Milling time and rate depend on mill volume, mill diameter size of copper or iron balls and the speed of rotation. The function of this second impacting step is two-fold: to provide an oxidation barrier on each powder particle and to cold-work each coarse particle. The balls transfer, by impact, a portion of the copper or iron ingredient carried by the ball-milling elements **19** so as to form a copper or iron shell about substantially each particle of the powder **16**. The finer powder will obtain a copper or iron or a combination coating by true abrasion or scratching with the surface of the ball-milling elements **19**. Ball-milling elements **19** should have a diameter at least 50 times the largest dimension of any of the particle shapes of the cryogenic powder **16**.

The ball-milling operation also must generate nonnatural defect sites (resulting from cold-working) in substantially all powder particles above 124 μ; the ball-milling operation herein should be carried out so that substantially each coarse particle has at least one defect site therein. This can be accomplished by rotating the housing **20** to impart a predetermined abrading force from the balls **19**.

When this step is completed, the particles will be in a condition where substantially all will have a continuous copper or iron or combination envelope (coating or shell) and be stressed sufficiently so as to have a high degree of cold work. (Defect site herein means a defect in local atomic arrangement; shell defined herein means a substantially continuous thin envelope intimately formed on the surface of the particle.) Although the shell should preferably be an impervious continuous envelope about each particle, it is not critical that it be absolutely impervious.

(4) A predetermined quantity of powder conditioned from step (3) is compacted by a conventional press **20** to a predetermined density, preferably about 6.6 g/cc. This is brought about by the application of forces in the range of 30 to 35 tsi. The presence of the copper or iron envelope about the powder particles improves compressibility. With prior uncoated powders, a density of about 6.4 g/cc is typically obtained using a compressive force of 85,000 psi; with the powder herein, densities of about 6.6 g/cc are obtained at the same force level.

The shape **21** into which such powder is compacted is designed to have an outer configuration larger than that desired for the final part. A significant and highly improved shrinkage takes place as a result of the next step (5); the shrinkage can be a predetermined known factor and allowance can be made in the compacted shape **21** of this step. Shrinkage will be in the controlled limits of 0 to 15%.

(5) The compacted shape **21** is subjected to a sintering treatment within a furnace **22** wherein it is heated to a temperature preferably in the range of 2000° to 2100°F for ferrous-based cryogenic powder. The temperature to which the compact is heated should be at least the plastic region (typically sintering temperature) for the metal constituting the powder. A controlled or protective atmosphere is maintained in the furnace, preferably consisting of inert or reducing gases.

At sintering temperatures, atomic diffusion takes place between particles of the powder particularly at solid contact points therebetween; certain atoms of one particle are supplied to fill the defect sites or absence of certain atoms in the crystal structure of the contacting particle, the defect sites being present as a result of cold-working in step (3). Diffusion is accelerated to such an extent, that an increase of more than 100 times is obtained. It is theorized that at least 60% of the improvement in physical properties of the resulting sintered shape is due to the controlled cold-working of the coarse powder particles. The increased diffusion is responsible for the increase in shrinkage.

Fe_2O_3 Dry Lubricant

A. Roy, C. Belleau and J.M. Geyman; U.S. Patent 4,110,512; August 29, 1978; assigned to Chrysler Corporation describe a process whereby various types of low friction materials and members such as turbine seals and the like may be provided which are particularly useful under high temperature, dry lubrication conditions. A wustite (FeO), magnetite (Fe_3O_4) or hematite (Fe_2O_3) porous matrix containing interstitial quantities of an interspersed trapping material comprised of an inorganic compound such as an inorganic salt or oxide comprises together a working surface which is carried by a suitable substrate.

The iron oxide matrix, during a run-in or break-in period where it is rubbed against a mating surface, undergoes abrasion causing loose particles of the iron oxide and other miscellaneous abraded debris, such as particles of the trapping material, to form between the mating surfaces. The particles embed or seat themselves in trapping material and undergo sintering, due to the rubbing and/or frictional heat generated thereby and/or elevated temperature and/or pressure exerted by the mating surfaces, to form substantially continuous Fe_2O_3 film. It has been found that the abraded particles, due to the rubbing contact, transform to Fe_2O_3 as a result of oxidizing effects and seat themselves in the trapping material in a self-aligned and highly oriented condition with the basal planes of the hexagonal cells or crystals of the Fe_2O_3 particles arranged substantially parallel to the rubbing surfaces.

The iron tends to fracture along certain crystallographic planes, i.e., the basal planes. The fractured particles align themselves so as to present the lowest resistance to sliding; the trapping material aids this self-alignment. Since the particles are of the same orientation they sinter readily at relatively low temperatures and pressures in situ to form the smooth, substantially continuous film of Fe_2O_3. Transfer of the particles and trapping material from one mating surface to the other occurs so as to form a mating continuous film.

There is thus provided at least one, but more likely two, substantially continuous Fe_2O_3 surfaces which have been found to be beneficial from the wear and friction standpoints under dry lubricated conditions, particularly at high temperatures such as those in excess of 1000°F. Moreover, once formed, the material is permanent in that it is self-renewing because as wear proceeds, it continues to form at a controlled rate.

Example: One way of obtaining the desirable ultimate Fe_2O_3 working surface is by flame- or plasma-spraying Fe_3O_4 combined with common table salt (NaCl) using four parts by weight of iron oxide powder to one part by weight of salt on a 430 stainless steel substrate; 5 to 30% by weight salt is satisfactory. Spraying may be accomplished with a Metco-Powder metallizing oxyacetylene gun, for example. Following spraying the surface is washed with an aqueous bath to remove the table salt. The resultant porous Fe_3O_4 matrix is then impregnated with fused sodium sulfide ($Na_2S \cdot 9H_2O$). The fused sulfide is applied by pouring or painting it on the matrix and allowing it to soak into the matrix interstices. Any excess is then scraped off. The structure may then be heated to about 1200°F to remove hydrated water. It is then ground substantially flat to provide a good running surface.

After run-in in a gas turbine engine of a seal member prepared according to this procedure, a substantially continuous surface of oriented Fe_2O_3 was formed over the matrix work surface. A regenerator hot-cross arm prepared in this fashion was observed to have an Fe_2O_3 film transferred to the regenerator core surface. These materials exhibit desirable wear and friction characteristics.

MAGNESIUM

Heat-Resistant, Insulating Film Containing TiO_2

In general, insulating films are formed on a grain-oriented silicon steel sheet by a method wherein a cold-rolled silicon steel strip having a desired final gauge is annealed at a temperature of 700° to 900°C for 1 to 10 minutes in wet hydrogen to remove carbon contained in the steel strip at the same time to oxidize the surface portion of the steel strip, forming a subscale containing SiO_2 on the surface of the steel strip, and then an annealing separator consisting mainly of MgO is applied on the steel strip, after which the steel strip is wound up in the form of a coil and subjected to a final annealing at a temperature of 1000° to 1200°C in a reducing or nonoxidizing atmosphere.

It is known that addition of TiO_2 to an annealing separator consisting mainly of MgO can improve properties of an insulating film. However, when TiO_2 is used, a large number of black particles are often formed and adhered to the surface of the insulating film. The black particles cannot be removed by an ordinary washing by means of a brush, which is carried out in order to remove unreacted annealing separator after the final annealing. When a silicon steel sheet having the black particles is applied with a film consisting mainly of phosphate, the appearance of the steel sheet having the film is poor due to the presence of the black particles, and further when the silicon steel sheet having such film is constructed into a transformer core, the space factor of the transformer core is decreased. Moreover, at the construction of the transformer core, the insulating film is peeled off together with the black particles due to the friction between laminated steel sheets, and the silicon steel base metal is locally exposed to decrease the interlaminate resistance.

Though the black particles can be removed by brushing the silicon steel sheet surface violently, the insulating film is peeled off together with the particles to expose the base metal, and the appearance of the film becomes considerably poor, and further, the interlaminate resistance of the steel sheet is low after the steel sheet is constructed into a transformer coil.

H. Shimanaka, T. Ichida and S. Kobayashi; U.S. Patent 4,113,530; September 12, 1978; assigned to Kawasaki Steel Corporation, Japan have found that the formation of black particles can be prevented by the use of TiO_2 having such a particle size that the content of agglomerated particles retained on a 325 mesh sieve is less than 0.5% by weight, the TiO_2 having a dispersion degree in water of at least 85%.

The 325 mesh impassable agglomerated particle content in TiO_2, the primary particle size of TiO_2 and the dispersion degree of TiO_2 in water were measured in the following manner. The 325 mesh impassable agglomerated particle content in TiO_2 was measured by the following sieve test. A predetermined amount of TiO_2 is dispersed in water and poured on a 325 mesh Tyler standard sieve, and the residual TiO_2 on the sieve is uniformly swept by means of a soft brush while pouring water on the TiO_2, sprayed with acetone and dried at 110°C. The weight percent of the residual TiO_2 based on the weight of the originally dispersed TiO_2 is measured, which is the 325 mesh impassable agglomerated particle content in TiO_2.

The dispersion degree of TiO_2 in water was measured in the following manner. 20 g of TiO_2 particles are mixed with 450 cc of distilled water or demineralized water at room temperature. After the resulting mixture is stirred for 3 minutes, the mixture is charged in a measuring cylinder of 1 ℓ capacity and distilled water or demineralized water is further added to the mixture to make up the total amount of 1 ℓ. After the resulting dispersion was left to stand for 2 hours, 250 cc of the upper layer is sampled, and the amount of TiO_2 contained therein is weighed and the dispersion degree of the TiO_2 in water is calculated by the following formula.

$$\text{Dispersion degree in water (\%)} = \frac{\text{grams TiO}_2\text{ in 250 cc of upper layer} \times 4}{\text{20 grams}} \times 100$$

The primary particle size was measured in the following manner. TiO_2 powder sample is observed by an electron microscope and the diameter of the minimum unit particles forming the primary particle is measured.

NICKEL

Electroless Bath Containing Vanadium

B. Zolla; U.S. Patent 4,167,416; September 11, 1979; assigned to Alfachimici SpA, Italy describes a composition to be used in the formation of baths for the electroless autocatalytic deposition of nickel base alloys, which includes the components of an essentially conventional formulation for the electroless deposition of Ni-P or Ni-B alloys and with the addition of ions of at least two different metals from the group consisting of Groups II-B, III-A, IV-A, V-B and VI-B of the Periodic System, in which one of these metals is vanadium. In the preferred form of the process, three or four salts of different metals from the groups indicated above are used.

The conventional composition can be any of the compositions used in electroless depositions of Ni-P or Ni-B alloys; it generally contains a nickel salt; a reducing agent, the latter being a hypophosphite or a boron reducing compound; a compounding agent for the metallic ions present in solution; and a pH stabilizing agent, which can be either acidic or basic.

The metals, from groups II-B, III-A, IV-A, V-B and VI-B of the Periodic System, from among which at least one is selected, constituting with vanadium bases of the additive components according to the process, are particularly thallium, zinc, tin and tungsten.

These metals can be added to the composition in the form of any organic or inorganic compound, preferably from among those which are water-soluble, or can be used directly in the metallic state, being introduced into the solution and being chemically soluble. The method of incorporating the various ions in the bath can thus be entirely conventional.

A bath formulated with a composition according to the process is such as to deposit a polymetallic alloy on supports such as iron, aluminum, copper, zinc and their alloys, or else on nonmetallic supports which are preventively conditioned according to the teachings of the prior art.

The total percentage of additive metals must remain between 0.4 and 9%, and preferably between 1.5 and 5%, by weight of the total deposit. More particularly, the percentage of thallium should be within the range of 0.05 to 0.4%; an excess causes increased porosity and fragility in the deposit. The percentage of vanadium should be between 0.05 and 1%; an increase beyond this quantity causes a decrease of the speed of deposition. The percentage of zinc should be between 0.1 and 2.5%; that of tin, between 0.1 and 2.1%; and that of tungsten, between 0.1 and 3%. If the indicated maximum limits are exceeded, the mechanical, and especially the hardness and ductility characteristics, will deteriorate.

Two-Layer Plating Solution

F.P. Potapov, A.K. Zorin, J.N. Sulie, A.I. Artemov, A.K. Kolchevsky; T.P. Lavrischeva, N.G. Lyandres, L.A. Toltinova and M.J. Murylev; U.S. Patent 4,150,180; April 17, 1979 describe a method for chemically nickel-plating parts having a catalytic surface, which comprises preparing a solution containing a nickel salt, a hypophosphite, a complexing agent, a buffer dopant, an accelerator and a stabilizer and heating the solution to a temperature higher than 60°C, but lower than the boiling point thereof; immersing the parts to be nickel-plated into the solution; keeping them there for as long as it takes to produce a nickel layer of desired thickness; and withdrawing the nickel-plated parts from the solution as well as adding to the solution components restoring the concentration thereof.

An alkalizing reagent is introduced into the solution, which has been cooled to below 45°C but above the freezing point of the solution. The particles of solid impurities arising in the course of nickel-plating are concentrated in the solution. In the course of nickel-plating, the upper layer of the solution is heated to the temperature indicated and the components restoring the concentration thereof are added thereto, whereas the lower layer of the solution is simultaneously cooled to the temperature indicated and the reagent restoring the acidity thereof is added.

The process permits dispensing with two solutions and using only one made up of two layers, a heated one and a cooled one; concentration-restoring reagents being supplied into the former; and a pH-restoring agent into the latter. With the quantity of the nickel-plating solution thus reduced, it is possible to cut down on the energy consumption for the heating, cooling and transporting of the solution and reduce the amount of the reagents added to the solution, thereby cutting down on their losses; and also monitor the process more effectively.

The method is desirably realized by use of an acidic aqueous solution containing a nickel salt in the form of nickel sulfate having a concentration of 25 to 30 g/ℓ, a hypophosphite in the form of sodium hypophosphite having a concentration of 15 to 20 g/ℓ, a complexing agent in the form of lactic acid having a concentration of 35 to 40 g/ℓ, a buffer and accelerating dopant in the form of boric acid having a concentration of 8 to 12 g/ℓ and a stabilizer in the form of thiourea having a concentration of 0.0005 to 0.0008 g/ℓ.

The object of the process is attained in an installation which comprises two frame-mounted communicating vessels for the solution used in the chemical nickel-plating process. One of the vessels is adapted to have the particles of solid impurities arising in the solution during the course of nickel-plating concentrated therein, an arrangement for heating and cooling the solution and a device for feeding into the solution reagents restoring the concentration and acidity thereof, wherein the vessel adapted to have the particles of solid impurities concentrated therein is provided with an arrangement for cooling the solution and communicates with a device for feeding an acidity-restoring reagent. In addition atop the vessel there is mounted the other vessel formed as a tube and provided with an arrangement for heating the solution and communicating with a device for feeding concentration-restoring reagents.

The foregoing configuration of the installation and the shape of the vessels as well as the arrangement of the means for heating and cooling the solution and correcting its concentration and acidity add up to a compact installation reliable in operation and convenient to maintain. The installation permits effecting the nickel-plating process and simultaneously correcting the solution.

The device for feeding the reagent restoring the acidity of the solution is desirably provided with a pipe inserted into the vessel which is adapted to have the particles of solid impurities concentrated therein and to cool the solution through a discharge opening formed in the bottom of the vessel, the free end of the pipe being disposed level with the junction of the vessels.

Such a position of the pipe for feeding the reagent restoring the acidity of the solution permits maintaining the acidity of the solution in the nickel-plating zone at a preset level and reduces the rate of formation of solid particles of nickel hydroxide. The position of the pipe in the median zone of the solution volume between the heated and cooled layers thereof offers an additional advantage in that it conduces to a uniform distribution of the reagent about the entire volume of the solution, with the convective flows present in the solution.

A heat-insulation layer may be provided between the vessels at the junction thereof, preventing heat transfer from the heated upper vessel to the cooled lower one and thus cutting down on undesirable heat losses.

Continuous Plating Process

M. Gulla, C.R. Shipley, Jr. and H.W. MacKay; U.S. Patent 4,152,164; May 1, 1979 have found that a metal plating solution experiencing evaporative losses of at least 1% per plating cycle is capable of infinite operation without requiring shut-down or bulk disposal of the solution provided the same is not otherwise contaminated with extraneous materials. The process comprises operation of the plating solution such that in each plating cycle volume is maintained constant, a portion of the solution is continuously or periodically withdrawn, and the solution is replenished, the process preferably being operated in the sequence of steps given though it is understood that the sequence can be changed with less efficient operation.

Operation of the solution in this manner results in withdrawal of a portion of solution by-products during each plating cycle thus preventing by-product concentration from reaching an intolerable level. Instead, by-product concentration reaches an equilibrium level, which level may be predetermined by the volume of the solution withdrawn each plating cycle.

The process also contemplates replenisher compositions which compositions differ from those of the prior art in that they are formulated to replenish solution constituents lost by reaction and drag-out, and, in addition, constituents lost by withdrawal of solution.

The total volume of liquid added to the plating solution is that amount lost by evaporation and that withdrawn less the volume added with the replenishers. The solution withdrawn may be dumped, treated to remove by-products, treated to recover all constituents or preferably used as a second stand-by or replacement plating solution. The amount of solution withdrawn can vary within broad parameters dependent upon the concentration of the components in the bath and the tolerable concentration of by-product at equilibrium conditions. Preferably, the volume of solution withdrawn is from about 1 to 60% by volume of the total volume of plating solution per plating cycle and usually varies between 5 and 25% of the solution volume.

Higher volumes of solution withdrawal assures safe operation of the plating solution, as larger quantities of by-products are withdrawn, and the solution comes to equilibrium rapidly and contains a relatively low concentration of by-products at equilibrium. However, removal of large volumes is uneconomical and hence, undesirable.

If by-products were permitted to increase in concentration without removal, their concentration would reach a level where the plating solution would no longer be suitable for use within about 3 to 10 plating cycles, dependent upon the work plated. As a guideline only, the volume of liquid withdrawn per cycle may be conveniently equated to the total volume of plating solution divided by the estimated number of cycles the solution could be used if by-products were not withdrawn. For example, using a typical electroless nickel solution to plate a mild steel substrate, dependent upon the pretreatment used, it is estimated that the solution could be used for about 7 cycles before disposal became necessary. Accordingly, while maintaining volume constant, approximately 14% of the volume of solution should be withdrawn per cycle with replenishers added to replace solution constituents removed. Following these procedures, the plating solution may be used indefinitely and plating quality will be uniform at any time during use of the solution.

To determine the amount of each component in a replenisher formulation, the concentration of such component is that amount necessary to replace that lost by reaction, drag-out and withdrawal. This can be determined by the following relationship:

$$(a) \qquad C_R = R' + xC_w + yC_o$$

where C_R is the concentration of the replenisher component in grams per cycle; R' is the amount of the component consumed by reaction in grams per cycle; x is the fraction of the total liquid withdrawn per cycle; C_w is the concentration of the component at the time of withdrawal in grams; and, if there is more than one withdrawal per cycle, the concentration at the time of each withdrawal, y is the fraction of the total concentration of the component lost by drag-out; and C_o is the total initial concentration of the component in grams per cycle. The amount of water added should be sufficient to maintain the volume of the plating solution essentially constant.

The following formulation is set forth for purposes of illustration:

Nickel sulfate hexahydrate	24 g
Sodium hypophosphite monohydrate	15 g
Sodium acetate	15 g
Lead acetate	0.02 g
Citric acid	30 g
Water	1 ℓ
pH	4.5

Replenisher 1: Replenisher 1 is to be used under the following conditions: for 1 liter of nickel-hypophosphite solution (above) with withdrawal equal to 10% of total solution per plating cycle and replenishment made when nickel is depleted by 25%.

To determine the nickel sulfate concentration from equation (a): if all of the nickel sulfate is consumed and its concentration is reduced from its orginal concentration of 24 g to 0, a cycle being the time for complete deposition of the metal being plated, R' is 24 g. The fraction of the solution withdrawn per cycle is 10% or 0.1 part of the total solution; hence, x is 0.1. The concentration of nickel sulfate at the time of each withdrawal, C_w, is 18 g as the original concentration of 24 g is reduced by 25% when replenishment occurs. Drag-out over a plating cycle comprises about 2% of the initial concentration and, hence, y is 0.02. C_o is 24 grams per cycle. From equation (a):

$$C_R = 24 + 0.1(18) + 0.02(24)$$

and the amount of nickel sulfate in the replenisher is then 26.28 grams per cycle. In comparison, the amount required for replenishment in accordance with the prior art would be 24.48 g.

The determination of sodium hypophosphite replenishment is quite similar to that for nickel sulfate. Assuming that the sodium hypophosphite is consumed at the same rate as the nickel sulfate in the reaction per cycle,

$$C_R = 15 + 0.1(11.25) + 0.02(15)$$

and the replenisher should contain 16.5 g of sodium hypophosphate monohydrate. This would compare to 15.3 g following prior art procedure.

For a supplemental component, citric acid, for example, R' of equation (a) would be 0 and the amount of acid in the replenisher would equal 3.60 g or

$$C_R = 0 + 0.1(30) + 0.02(30).$$

The total replenisher composition for this example is as set forth in the following table where Formulation A is a replenisher for a prior art operation and Formulation B is for the procedures set forth herein.

	Formulation A (g)	Formulation B (g)
Nickel sulfate hexahydrate	24.48	26.28
Sodium hypophosphite monohydrate	15.30	16.60
Sodium acetate	0.30	1.80
Lead acetate	0.0004	0.0024
Citric acid	0.60	3.60
Ammonium hydroxide, to pH	4.5-5.0	

The above Formulation B may be added in dry form, but preferably is added as a solution. For convenience, the formulations may be dissolved in an amount of water equal to the volume of solution withdrawn. In this example, for 1 liter of solution, the total volume of liquid withdrawn per cycle is 100 ml withdrawn in four equal increments of 25 ml each at each point in the cycle where the nickel solution is depleted by 25%. For replenishment, the solution would be divided into four equal portions and added following each of the withdrawals of solution.

Replenisher 2: Replenisher 2 is to be used under the following conditions: for 1 liter of nickel-hypophosphite solution (above) with withdrawal equal to 15% of solution per plating cycle and replenishment made where nickel is depleted by 33%.

	Formulation A (g)	Formulation B (g)
Nickel sulfate hexahydrate	24.48	26.88
Sodium hypophosphite monohydrate	15.30	16.80
Sodium acetate	0.30	2.55
Lead acetate	0.0004	0.0034
Citric acid	0.60	5.1
Ammonium hydroxide, to pH	4.5-5.0	

As to addition of the replenisher formulation, the same considerations apply as set forth for Replenisher 1. In this case the replenisher is subdivided into three portions.

Decomposition Products Indicator Device

Electroless nickel plating commonly is conducted in high chromium-content steel tanks (e.g., stainless steel tanks) whose interior is periodically passivated, or deactivated, by pretreatment with dilute nitric acid solution, as described by Matheny et al in U.S. Patent 2,874,083, February 9, 1959. In the course of a typical electroless nickel-plating operation, various decomposition products gradually build up on the tank. Because of further decomposition of these products and because of random chemical nickel plating, the products contain nickel which catalyzes the principal plating reaction. As a result, nickel plates upon this nickel until plating operations are terminated. If the plating operations are conducted for an extended period without removal of the decomposition products, spontaneous decomposition of the bath will occur. Thus, the tank must be cleaned periodically and sometimes be repassivated. Unfortunately, there has been no convenient and accurate way to determine when it is advisable to terminate a plating operation and treat the tank with dilute nitric acid to remove the decomposition products and, if desired, repassivate the tank.

In the absence of an accurate indication of the condition of the tank surfaces, plating operations may be terminated before conditions really require it. On the other hand, operations may be continued inadvertently to the point where spontaneous decomposition occurs; in that event, the parts being plated are spoiled and the entire plating bath must be replaced.

G.S. Petit and R.R. Wright; U.S. Patent 4,125,642; November 14, 1978; assigned to U.S. Department of Energy describe a method for generating, in the course of an electroless plating operation, a display indicative of the extent to which nickel-bearing decomposition products have accumulated on tank surfaces. This method for conducting electroless metal plating is based on the finding that if an appropriate current-conducting probe is inserted in the bath during plating, the difference in electropotential between the probe and the tank is indicative of the condition of the tank surfaces in contact with the bath. That is, the difference in potential is a reliable indicator of the extent to which nickel has plated upon decomposition products deposited on the tank.

The method can be illustrated in terms of Figure 2.2, where the numeral **1** designates a conventional nitric-acid-passivated stainless steel tank containing a conventional plating bath **3** of the nickel cation-hypophosphite anion type.

Figure 2.2: Decomposition Products Indicator Device

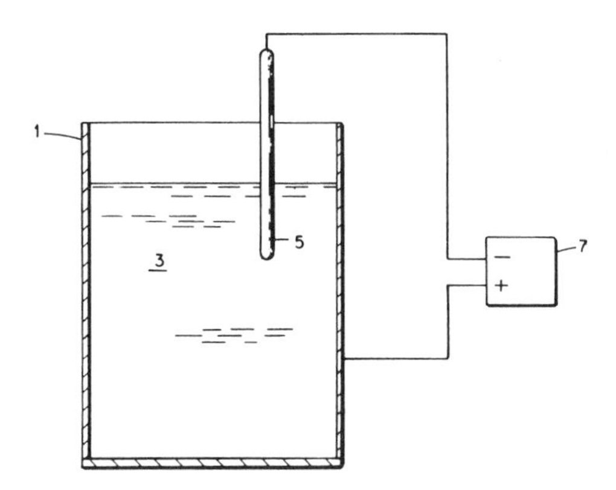

Source: U.S. Patent 4,125,642

A chrome-plated stainless steel probe **5** is mounted to contact the plating solution, but not the tank. Also, a suitable voltmeter **7**, such as a galvanometer, is connected across the probe and tank.

Using such a system, it was found that at the beginning of a typical run made with a freshly-passivated tank, the tank initially is positive relative to the probe (see the figure), the initial probe-to-tank difference in potential being relatively small. As the run progresses, the initial difference in potential decreases to zero and reverses polarity; then, without again changing polarity, it slowly increases. If the plating operation is continued indefinitely, the difference in potential continues to increase slowly until spontaneous decomposition occurs. Consequently, tests can be conducted in a given plating system to determine which voltage values correspond to various conditions in the system—especially the voltage values which prevail as spontaneous decomposition is approached.

In subsequent operations of that system, the probe-to-tank voltage can be monitored continuously, and the plating operation can be conducted until the voltmeter indicates that the buildup of the nickel on decomposition products deposits has reached a preselected level or limit. At that time, the run is terminated and the tank is retreated with dilute nitric acid to (a) remove the decomposition products or (b) remove the decomposition products and repassivate the tank.

Example: In this instance, the tank **1** was composed of stainless steel. The interior of the tank has been freshly passivated by conventional pretreatment with nitric acid. (See abovementioned U.S. Patent 2,874,083.) The plating solution was a conventional acid electroless plating bath having the following composition, the pH being 4.5±0.2 with sodium hydroxide: 0.085 mol per liter nickel sulfate, 0.25 mol/per liter sodium hypophosphite, 0.50 mol per liter lactic acid, 0.50 mol per liter sodium hydroxide, 1.5 mol per liter No. 7 Antipit (M & T Chemical Company), and 0.5 ppm Pb^{++}.

The bath was maintained at a temperature in the range of 50° to 100°C and was stirred continuously. At times, a stabilizer (Pb^{++}) was added to the bath to help prevent spontaneous decomposition, and the concentrations of the other chemicals in the bath were maintained at selected values. Throughout the run, the plating conditions were controlled in accordance with conventional electroless nickel plating practice.

The probe **5** comprised a stainless steel rod which had been treated by conventional anodic pickling and then plated in a conventional chrome-plating bath to provide porosity-free chrome plate having a thickness of 4 mils. The voltmeter **7** was a conventional high-impedance-follower amplifier driving a recorder (Speedomax G (Leeds and Northrop Corporation)]. Center-scale zero on the recorder permitted the recording of potentials of both polarities.

At the start of the plating operation, the freshly-passivated tank **1** was positive relative to the probe, and at this time the recorded probe-to-tank voltage was +0.25 V. As the run continued, the recorded value decreased to zero, reversed polarity, and gradually increased. The rate of the change of the voltage is a function of the amount of plating (mils per square foot) as well as the condition of the bath. 80 hours after the start of operations, the voltage had increased to -0.4 V. After 4 more hours of operation, spontaneous decomposition occurred, at which time the recorded voltage was -0.6 V.

Similar runs conducted with this particular system confirmed that spontaneous decomposition typically occurs at essentially -0.6 V. Thus, in this type of system it is preferred to terminate the typical run when the voltage is at a somewhat lower value in the range of approximately -0.35 to -0.5 V, as, for instance, a preselected value of -0.4 V.

Impervious Graphite Tubes

L.A. Conant, W.M. Bolton and J.E. Wilson; U.S. Patent 4,134,451; January 16, 1979 describe heat exchanger elements and other chemical processing structures or elements comprising impervious graphite materials (made impervious by filling the voids of porous graphite with one or more appropriate synthetic resins), which impervious graphite materials: (a) have been heat-treated to partially thermally degrade the polymeric resin; (b) have been externally armor-coated at an elevated temperature with a thin layer of metal; and (c) are under high compressive force, due to the differential thermal contraction of the layer of metal and the impervious graphite structure when the metal element is cooled to room temperature from an elevated temperature above about 200°C.

The heat treatment of impervious graphite tubes is undertaken in order to partially degrade the synthetic resin which is present in the pores of the graphite in order to reduce or eliminate the ordinary or natural porosity of the graphite. Ordinarily impervious graphite tubes contain from about 10 to 15 wt % of polymeric resin (to seal its pores), the remainder consisting essentially of graphite. Partial thermal degradation can be accomplished by heating the impervious graphite structures at a temperature of from about 180° to 300°C (or higher if desired) for a period of time (generally from about 3 to 30 hours, depending on the temperature) sufficient to partially thermally degrade or heat-stabilize the synthetic resin portion of the impervious graphite composition. The atmosphere in and around the article during the heat-treating step can be air or other gas such as nitrogen or carbon dioxide.

Partial degradation of the synthetic resin portion is evidenced by a loss of weight of the heat-treated article during the heat-treating step of from about 0.25 to 3 wt % or more (as compared to the weight of the element after being dried to 110°C, but prior to the heat-treating step). It is essential that the synthetic resin portion be only partially thermally degraded, in order to retain the desired imperviousness of the graphite article.

The high-temperature, metal-coating step of this process involves the application of a thin (from about 2 to 100 mils of metal or more, and preferably from about 5 to 50 mils) coating or layer of metal to the outer surface of the partially thermally-degraded, impervious graphite tube at an elevated temperature of from about 65° to 450°C (840°F), and preferably between about 175° and 300°C in accordance with the metal coating method described in detail in U.S. Patent 4,072,243. A preferred coating is a gas-plated (i.e., by chemical vapor deposition) nickel coating, having a thickness of from about 5 to 50 mils. Such a metal coating is dense and continuous over substantially the entire external surface of the tube.

Cooling to room temperature of the resulting metal armor-coated, partially thermally-degraded, impervious, graphite article is preferably accomplished in an inert atmosphere and gradually, over a period of at least a few minutes, and preferably of at least about 15 minutes.

The resulting product is stronger, tougher and much more impact-resistant than ordinary impervious, graphite tubes having the same physical measurements (wall diameters). Also the product is shatterproof and capable of being used under very high pressures, and does not rupture even if the graphite portion is notched, badly cracked or broken.

TIN

Plating of Aluminum Using Intermediate Alloy Layer

M.J. Bernstein; U.S. Patent 4,115,604; September 19, 1978; assigned to Bremat SA, Luxembourg describes the application of an intermediate alloy layer to aluminum prior to plating of the aluminum with a primary metal such as tin or a tin alloy. The material which comprises the intermediate alloy layer is characterized by good affinity for aluminum and for tin, thus assuring an efficient and durable bond of the subsequently applied metal to the aluminum. This intermediate layer alloy comprises, in percents by weight, the following materials: 83.5 to 84.5 zinc, 9 to 10 bismuth, 4 to 5 cadmium, 2 to 3 tin and 0.1 to 0.8 lead. Zinc and bismuth are used in the alloy because of their good affinity for aluminum. Cadmium is incorporated in the alloy to lower the point of fusion thereof while the tin serves to equalize and homogenize the alloy.

The process can advantageously be practiced in a plating system as shown schematically in Figure 2.3. This plating system includes tanks **6, 7, 8** and **9**, which are successively traversed by one or several objects or articles of aluminum such as the aluminum sheet indicated at **1**. The aluminum sheets or articles to be plated are delivered to the plating apparatus, in the direction of arrow A, from one or several supply reels not shown on the drawing.

Figure 2.3: Plating of Aluminum Using Intermediate Alloy Layer

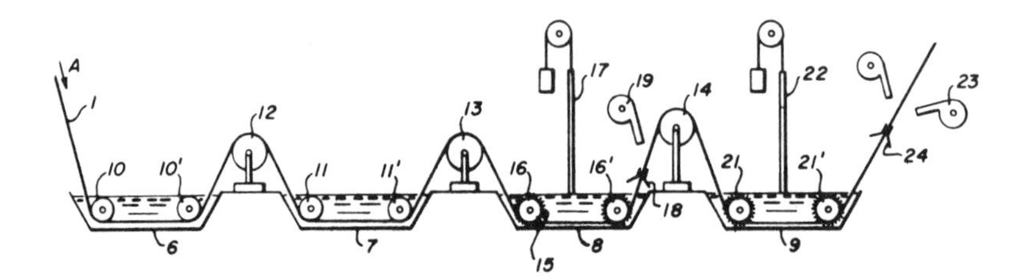

Source: U.S. Patent 4,115,604

The tank **6** will be formed of or lined with suitable plastic or an unoxidized steel. Tank **6** will contain cleansing solution as appropriate for aluminum, e.g., tank **6** will be filled with a solution of 15% caustic soda at room temperature. Two channeling pulleys **10** and **10'**, typically comprised of plastic, are immersed in tank **6** for each sheet of aluminum drawn through the tank, the channeling pulleys advancing and guiding the sheet **1** during its traversal of tank **6**. The position of channeling pulleys **10** and **10'** is adjustable in the vertical direction.

The second successive tank **7** may also be lined with plastic or an unoxidizable steel. Tank **7** contains a material in liquid form suitable for neutralizing the cleansing solution applied to the aluminum in tank **6**. In the case where the material in tank **6** is a 15% caustic soda solution, tank **7** will contain a 15% nitric acid solution which neutralizes the caustic soda. Tank **7** is also provided with a pair of channeling pulleys **11** and **11'** for each of the parallelly-disposed sheets of aluminum being drawn therethrough. The pairs of pulleys **11** and **11'** cause the sheets of aluminum to advance through the tank, while simultaneously guiding the sheets during traversal of the tank.

A further pulley or drum **12** is positioned between intermediate tanks **6** and **7**. Pulley **12**, which is preferably also comprised of an unoxidizable steel, will be mounted on bearings and will guide the sheets between tanks **6** and **7**. As in the case of the channeling pulleys in tank **6**, the pulleys **11** and **11'** in tank **7** are preferably mounted on slides so that they can be displaced in a vertical direction.

The alloy which will be deposited on the aluminum as the intermediate layer is contained in tank **8** in molten form. Tank **8** thus includes a heating mechanism, not shown, which insures a regulated temperature for the molten intermediate layer alloy. The temperature within tank **8** may attain the level of 500°C. Tank **8** is also provided with pairs of channeling pulleys **16** and **16'** for guiding the sheets of aluminum therethrough. The pulleys **16** and **16'** are advantageously provided, at their peripheries, with blades or vanes **15** of heat-resistant steel to assure continuous mixing of the alloy in the tank thereby assuring good homogeneity of the molten alloy.

In order to maintain a constant level in tank **8**, molten alloy is continuously replenished as material is deposited on the aluminum passing through the tank. Replenishment of the alloy in tank **8** is accomplished through an automatic feed system, indicated schematically in the figure, which includes a bar **17** of the intermediate alloy suspended above tank **8**. Through the use of suitable control mechanisms, bar **17** will be fed into tank **8** to maintain the desired level therein, thus achieving a continuous recharging of the tank while simultaneously avoiding a rapid drop of the temperature of the molten metal in tank **8** during the addition of material thereto.

It is possible, and in some cases may be desirable, to have means, not shown, associated with tank **8** for blowing carbonic gas over the surface of the molten alloy therein to avoid or reduce oxidation of the molten alloy.

At the exit of tank **8**, the coated sheet **1** is passed between a pair of converging plates, i.e., a drawplate mechanism which has been indicated schematically at **18**. The drawplate **18** will be formed of a very hard alloy, for example, an alloy based on tungsten carbide, and will have an exit opening corresponding to the desired thickness of the coated sheet. Accordingly, the drawplate **18** will serve to wipe and calibrate the aluminum sheet with the coating or layer of intermediate alloy deposited thereon.

After passing through drawplate **18**, the sheet **1** or other article comprised of aluminum will be cooled. The means for achieving this cooling is indicated schematically at **19** and can comprise means for generating a jet-atomized water or compressed air.

The plating with the primary metal is carried out in tank **9** which contains a bath of melted pure tin or a tin alloy. Tank **8** also contains vaned channeling pulleys **21** and **21'** which are identical to the pulleys **16** and **16'** in tank **8**. An automatic replenishment system, which preferably feeds a bar **22** of the primary plating metal into tank **9**, is also provided.

Between tanks **7** and **8**, as well as between tanks **8** and **9**, guiding pulleys respectively indicated at **13** and **14** are provided. It should be noted that the distance between tanks **8** and **9** is preferably larger than the distances separating the tanks **6** and **7** and the tanks **7** and **8**; this additional distance is provided to insure sufficient cooling of the intermediate alloy layer prior to plating with the primary metal.

The sheet of plated aluminum exiting from the plating tank **9** will traverse, successively, a drawplate mechanism **24** and a cooling apparatus **23** which are respectively similar to the drawplate **18** and the cooling device **19** associated with tank **8**. After cooling the plated aluminum article or articles exiting from tank **9** will typically be rolled onto drums or reels for storage.

Plating of Aluminum/Iron Composite

L.E. Kirman and W.A. Kruper; U.S. Patent 4,170,525; October 9, 1979; assigned to Gould Inc. describe a process for applying a thin coating or layer of tin or tin alloy on a composite structure which has one surface composed of an aluminum base metal, i.e., aluminum or an alloy of aluminum, and another surface composed of a ferrous base metal, i.e., iron or an alloy of iron. Broadly, the method requires the following minimum steps:

(1) Treating or contacting the composite-bearing structure with a mineral acid which contains either fluoride ions, fluoride-containing ions or mixtures thereof to activate the surface of the aluminum base metal; and

(2) Positioning or immersing the so-treated composite-bearing structure in an aqueous bath which contains a mineral acid, a source of either fluoride ions, fluoride-containing ions or mixtures thereof and a source of stannous ions with the stannous ions being present in an amount ranging from about 1 to 75 g/ℓ for a period of time sufficient to cause tin to be deposited on the exposed surface of the composite-bearing structure.

In the preferred process, additional processing steps are used. A typical process sequence used to apply a thin layer of tin to a bearing structure (as described in U.S. Patent 4,069,369) having a steel substrate and an aluminum base-bearing layer thereon is as follows:

(a) Vapor degrease the surface of the bearing in a chlorinated hydrocarbon solvent such as perchloroethylene;

(b) Further clean the bearing structure in an aqueous alkaline solution such as an aqueous solution of Na_3PO_4 and Na_2CO_3;

(c) Water-rinse the so-cleaned article;

(d) Soak the bearing structure in an aqueous solution of an acid, such as 10% sulfuric acid, at an elevated temperature, for example, 140°F, to remove oxides which may be present on the steel substrate;

(e) Water-rinse the so-treated article;

(f) Contact the bearing structure with an aqueous solution of a mineral acid containing either fluoride ions, fluoride-containing ions or mixtures thereof, such as a 5% hydrofluoric acid, to activate the aluminum base-bearing layer;

(g) Rinse the activated structure;

(h) Immerse the bearing structure in an aqueous plating bath containing a mineral acid, a source of either fluoride ions, fluoride-containing ions or mixtures thereof and a source of stannous ions with the stannous ions being present in an amount ranging from about 1 to 75 g/ℓ for a period of time sufficient to cause tin to be deposited on the exposed surface of the bearing structure; and

(i) Remove the article from the plating bath and rinse the same.

In certain circumstances, the tin-plated article is then immersed in an aqueous solution of $Na_2Cr_2O_7$ in order to deposit a layer of chromate on the tin to render the plated structure fingerprint-resistant.

The plating bath can contain any of the below listed ingredients within the specified ranges. The following represents the composition of a typical bath utilized in connection with the process where the desired coating is pure tin: 0 to 30 g/ℓ boric acid; 0 to 150 g/ℓ hydrofluoric acid; 0 to 150 g/ℓ sulfuric acid; 0 to 150 g/ℓ fluoboric acid; 1 to 75 g/ℓ stannous tin; 0 to 8 g/ℓ antioxidant; if used, greater than 0.1 g/ℓ nonionic surfactant; 0 to 0.5 g/ℓ grain refiner; and greater than 1.0 g/ℓ fluoride ions, or fluoride-containing ions, or mixtures thereof.

If it is desired to deposit a tin-cadmium alloy layer on the bearing structure, the above bath may contain up to 75 g/ℓ of cadmium ions. Likewise, if it is desired to deposit a tin-lead alloy, the bath may contain up to 75 g/ℓ of lead ions. However, in this latter case, the bath should not contain any sulfate. Also, if it is desired to deposit a tin-zinc alloy, the bath may contain up to 75 g/ℓ of zinc ions.

The duration of the immersion step varies with the type and thickness of metal or alloy coating to be deposited. In practice, satisfactory deposits have been obtained by using baths of the type described herein with the immersion period ranging from 3 to 4 minutes at ambient temperature.

MISCELLANEOUS PROCESSES

Electroless Plating Without Reducing Agent

S.W. Donley and P.N. Bacel; U.S. Patent 4,171,393; October 16, 1979; assigned to Eastman Kodak Company describe a method for sustained electroless plating on a metal surface that does not require the use of a chemical reducing agent in the plating solution. Any porous metal surface can be used as a substrate providing that such metal substrate has a higher oxidation potential than the plating metal, that the surface of the metal substrate is sufficiently porous to allow metal cations from the substrate to diffuse into the plating solution at a rate sufficient to sustain the deposition of plating metal on the substrate's outer surface and that the pores are small enough not to allow free flow of solution through the pores.

The method comprises immersing an article having a porous metal surface in an electroless plating solution. This solution contains cations of the plating metal, obtained by dissolving any readily available, soluble salt of the plating metal in water. In some applications of the method, these cations and product cations must be held in solution by the use of appropriate complexing agents. Such agents are well-known in the art and are chosen according to the particular plating-metal cations to be used. For example, when nickel cations are used, appropriate complexing agents may include such compounds as sodium citrate, ammonium hydroxide, ammonium chloride, ammonium sulfate, hydroxyacetic acid, and the like. One useful complexing agent, among others, for copper cations is the disodium salt of ethylenediaminetetraacetic acid (Na_2EDTA).

Additional ingredients may be necessary to keep the pH of the plating solution within an optimum range for the particular plating reaction to be effectively carried out. In one embodiment where nickel cations react with porous iron, the optimum pH range was found to be about 9 to 9.5 and ammonium hydroxide is used to maintain this pH. In another embodiment, the deposition of copper on a porous iron surface, sodium hydroxide or potassium hydroxide are present or are added stepwise during the reaction in order to adjust the pH to about 9 to 9.5 and maintain it at that level. Most well-known buffering agents can be used for this purpose and the use of such buffering agents is well-known by those skilled in the art.

The electroless plating reaction of the process is accomplished by controlling the temperature of the plating solution at a temperature effective for the particular plating reaction to take place. Each of the particular plating reactions may have an optimum temperature range at which it should be run to obtain best results. Such temperature is known or easily determined by those skilled in the art. The plating solution is heated to this temperature before immersion of the article having a porous metal surface. The nickel-iron reaction takes place best in a solution maintained at a temperature of about 90 to 95°C; the copper-iron reaction is initiated at room temperature, but takes place much faster when the plating solution reaches a temperature of about 50°C. In the copper-iron reaction, the solution temperature is allowed to rise once the reaction starts.

The particular complexing agents, buffering agents, pH control, and temperature useful for a given process are well-known in the electroless plating art and can easily be selected by those skilled in the art.

Example: The following composition may be used for plating nickel on a porous iron surface:

Water	3.5 ℓ
$NiSO_4 \cdot 5H_2O$	100 g
Sodium citrate	164 g
NH_4OH	85 ml
Hoeganaes EH iron (sponge iron)	2 kg

Except for the absence of sodium hypophosphite and the use of an iron substrate with a porous surface, the above ingredients and proportions are typical of those normally used in previously known electroless nickel-plating processes. All the ingredients except the iron were heated together in a large beaker with stirring. When the solution temperature reached 90°C, the iron was added. Nitrogen gas was blown over the solution surface to inhibit air oxidation. The temperature was maintained at 90° to 95°C, and the solution was stirred vigorously.

The reaction was complete after 20 minutes, as evidenced by the change in solution color from deep blue to light green. The plated iron was rinsed five times with water and four times with methanol and dried in air. The nickel-plated iron was analyzed and found to be 1% nickel by weight. This corresponds to a plated layer thickness of about 9.8×10^{-7} cm. No nickel was left in the solution, but iron was present in the solution in the same molar concentration as Ni originally, indicating the reaction was a direct displacement of iron by nickel.

Metal-Coated Carbon Fibers

K. Nakamura, Y. Fukube and T. Osaki; U.S. Patent 4,132,828; January 2, 1979; assigned to Toho Beslon Co., Ltd. and Nihon Shinku Gijutsu KK, both of Japan describe an assembly of a plurality of carbon fibers each coated with a matrix metal layer, the coated fibers having bonding points at the metal layers to form a two-dimensional network structure or both a two-dimensional network and a three-dimensional network structure, the bonding points being present to such an extent that individual carbon fibers do not substantially change in orientation with each other upon handling and the assembly can be handled without deviations among the individual fibers.

The metal used in this process is selected depending on the end use of the metal-coated carbon fiber assembly. For example, aluminum or magnesium can be used when the assembly is molded under heat and pressure and used as a material for aircraft bodies or shipbuilding materials. Titanium or nickel is used when the assembly is to be used as a material for turbine blades of a high-temperature, high-strength material. Copper, silver, gold, zinc, lead, tin, iron and cobalt are other useful metals.

When the matrix metal is one which tends to react with carbon fibers which are essentially carbon, or to dissolve the carbon fibers during their coating, during the production of carbon fiber-reinforced metal, or during the use of carbon fiber-reinforced metal (e.g., aluminum, titanium, nickel, iron and cobalt) an interlayer is provided between the carbon fiber and the matrix metal layer. The interlayer is a layer of titanium or silicon or a carbide or nitride of titanium or silicon. When the matrix metal is one which does not easily react with carbon fibers (e.g., magnesium, zinc, lead, tin, copper, silver and gold), the interlayer can be used to improve the wettability of the carbon fiber. If it is desired to avoid the inclusion of impurities such as titanium or silicon at some sacrifice in strength, it is, of course, possible to produce metal-coated carbon fibers without providing an interlayer. The material for forming the interlayer must be a material which does not diffuse into the matrix metal layer.

In the assembly obtained using an ion plating process or a vacuum deposition process, the individual carbon fibers are each coated with the matrix metal. Hence, the assembly has a fixed shape and a moderate flexibility. Another advantage is that a good bonding is achieved between the carbon fibers and the metal, and the metal can be coated without oxidation occurring.

Formation of the matrix metal layer by the ion-plating process is performed, for example, by using an apparatus of the type shown in Figure 2.4. In Figure 2.4, reference numeral **31** represents an air-tight chamber; **32**, a crucible or melting pot with a metal **33** therein; **34**, carbon fibers or carbon fibers having an interlayer, which are disposed on a cathode or act themselves as a cathode; and **35**, an opening for introducing an inert gas. The inside of the chamber **31** is maintained at a vacuum of about 1×10^{-4} torr or a higher vacuum, and a voltage of about -0.1 to -3.0 kV, preferably -0.5 to -1.5 kV, is applied to the cathode. An atmosphere of an inert gas, such as argon, helium, neon, krypton, or xenon with argon being preferred, is introduced into the system at a partial pressure of about 0.5×10^{-2} to 5×10^{-2} torr, and a matrix metal is evaporated therein by resistance heating, heating by irradiation with electron beams, or high-frequency induction heating, etc.

The resultant metal vapor is activated (ionized) in a plasma region therein, and is condensed as a metal layer on the surfaces of the fibers disposed on the cathode or used as the cathode. The higher the temperature of the carbon fibers, the more compact is the metal layer formed.

Figure 2.4: Apparatus for Coating Carbon Fibers with Metal

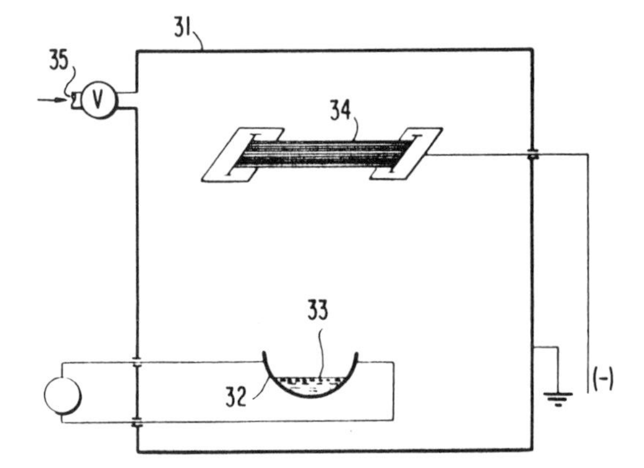

Source: U.S. Patent 4,132,828

Example 1: In a vacuum system, nontwisted tows each consisting of 6,000 carbon filaments having a diameter of 9.2 μ, a tenacity of 200 kg/mm^2 and a Young's modulus of 19.5 tons/mm^2 were aligned in a tape form in one row with the width being adjusted to 18 mm and the average number of fibers in the height direction being adjusted to 3. High purity aluminum was placed in the system, and the system was evacuated to a vacuum of less than 1×10^{-4} torr. Argon was then introduced into the system, and while maintaining the argon pressure at 2×10^{-2} torr, a voltage of -1.0 kV was applied to the carbon fibers to perform plasma-etching of the carbon fibers for 5 minutes.

After etching, aluminum was evaporated by resistance heating, and deposited by ion plating on the carbon fibers held at 100°C at a deposition rate of 1.0 μ/min.

Thus, a tape-like assembly of carbon fibers having an aluminum coating with a thickness of 4 μ was obtained. The proportion of the carbon fibers in the assembly was 30% by volume.

Twenty such assemblies were stacked in one direction and heated under pressure (900 kg/cm^2) at 560°C for 1 hour at a vacuum of 2×10^{-5} torr to produce a carbon fiber-reinforced metal having a carbon fiber content of 31% by volume. The metal partly escaped from the system at the time of heating under pressure. The resulting carbon fiber-reinforced metal had a tenacity in the direction of the fiber axis of 46 kg/mm^2 and good electric conductivity.

Example 2: In a similar manner to the procedure of Example 1, a random web which was composed of carbon fibers having a length of 1 to 4 cm, with an average length of 2 cm and having a diameter of 9.0 μ, and with a web width of 11 mm and 5 single fibers on an average in the height direction, and titanium were placed in a vacuum system. Titanium was evaporated and deposited by ion plating at a deposition rate of 0.5 μ/min. At a vacuum of 10^{-5} torr, copper was vacuum deposited onto both surfaces of the random web at a deposition rate of 3.0 μ/min.

The coated titanium layer was 0.1 μ thick and the copper layer was 8.0 μ thick. The assembly contained the carbon fibers in a proportion of 13% by volume. Ten such assemblies were stacked and heated at 1000°C under pressure (50 kg/cm^2) for 1 hour under a vacuum of 2×10^{-4} torr. A carbon fiber-reinforced metal containing 15% by volume of the carbon fibers was obtained.

Metal Coating of Wires

D.O. Gothard and G.R. Strobel; U.S. Patent 4,102,678; July 25, 1978; assigned to Huntington Alloys, Inc. describe a method for producing a metal-coated metal wire comprising the steps of drawing a metal wire while in contact with a surface-lubricated metal flake powder to form a green-coat wire, and sintering the green-coat wire to metallurgically bond the metal flake powder to the metal wire to provide the metal coated metal wire.

The metal flake powder can be a milled metal powder prepared in the presence of a lubricant. For example, the fatty acid, stearic acid, dissolved in mineral spirits was used to prepare the stainless steel, nickel, and other flake powders described in U.S. Patent 3,709,439. Other lubricants that can be used to prepare surface-lubricated metal flake powders include other fatty acids such as oleic acid, as well as camphor, paraffin, rosin and synthetic thermoplastic resins. Organic lubricants are generally used since they decompose during the sintering operation and do not lend appreciable residue.

No special preparation of the metal wire surface is required other than that it should be reasonably clean and free from surface dirt and oxides to avoid interruption of coating continuity. Conventional pickling sequences have been found to be useful for providing the desired degree of cleanliness.

The sintering temperature of the metal flake powder should be substantially below the solidus temperature of the metal wire being coated. To illustrate, for the copper coating of steel, a sintering temperature of 650°C can be used. Similarly, for the nickel coating of steel, a sintering temperature of 760°C can be used.

A suitable, dry, inert or reducing atmosphere is generally required during the sintering operation. Generally this atmosphere corresponds to that required for the sintering of the particular metal flake powder. For example, sintering of nickel flake onto the surface of a 36% nickel, balance iron alloy wire can be effected by exposure for 1 hour at 815°C in hydrogen gas having a -40°C dew point. In a continuous operation, the length of the furnace should be long enough to allow the metal flake coating to be heated to the sintering temperature for a time period sufficient to insure sintering and bonding to the metal wire.

The metal wire to be coated is generally passed through metal flake powder contained within the lubricant-holding box of a conventional draw bench and then drawn through a conventional drawing die contained within a drawing block. During passage through the die, the metal flake powder coats the wire while serving as a lubricant to provide a green-coat wire. The metallic flakes have a thin, residual layer of lubricant, e.g., stearic acid, upon their surface due to the prior manufacturing operation. In a preferred embodiment, the metal flake powder can be cleaned by washing in a suitable solvent, e.g., trichloroethane, acetone, etc. This treatment serves to substantially remove contaminants that can impede sintering and bonding without providing a negative effect on the lubricating characteristic associated with the powdered metal flake.

The elongated, flat shape of metallic flake powders aids the introduction of the flake powders to the die opening and promotes their adherence to the wire surface. The metal flake powder should be prepared from an element or alloyed powder having a U.S. standard sieve size less than about 325 mesh. Following attrition, the metal flake should have a U.S. standard sieve size less than about 100 mesh, and the thickness of the flakes should be less than about 1 μ and preferably between about 0.8 and 0.2 μ.

Following coating, the green-coat wire is passed through a conventional furnace having a dry, inert or reducing atmosphere. Exposure to this atmosphere at an elevated temperature results in densification and sintering, as well as metallurgical bonding of the coating to the wire surface.

Subsequent to sintering, the coated wire can be coiled for shipment or subjected to additional wire drawing operations or recycled, i.e., as a step toward providing a thicker or multilayered coating. In regard to multilayer coatings, it is contemplated that layers of individual, different metals can be imparted to a wire surface to promote formation of an alloyed surface layer, e.g., copper-nickel.

The coatings provided by this process are believed to be less than about 1 μ thick. Although the coating is extremely thin, it can provide enhancement in such properties as corrosion resistance, electrical conductivity, etc.

Although it is preferred to use the process for preparing coatings of nickel, nickel alloys, iron, cobalt, copper, brass, bronze, stainless steel, the platinum metals, nickel-silver alloys, aluminum, and aluminum alloys on the surface of wire substrates including steel, stainless steel, nickel-base alloys, copper, copper-base alloys, aluminum, and aluminum-base alloys, the process may be used to coat virtually any metal flake powder upon the surface of any metal wire.

Example: A 1.6 mm diameter wire, nominally containing 36% nickel, balance iron, was coated with a nickel flake powder. The -325 mesh nickel flake had been milled in a conventional attritor operating at 130 rpm for 4½ hours in a mixture of mineral spirits and stearic acid. The attritor contained 1,140 kg of 7 mm diameter balls. The liquid to powder ratio (volume) was 35:1, and the ball to powder ratio (weight) was 40:1. The nickel flake powder had an average thickness of about 0.3 μ.

The milled nickel flake was washed in trichloroethane, dried and placed in the lubricant-holding box of a conventional draw bench immediately adjacent to the 1.6 mm diameter wire drawing die. The wire was drawn through the drawing die without the use of a supplemental lubricant. The drawing operation was readily accomplished without substantial increase in wire temperature, the nickel flake serving as a lubricant. Nickel flake was firmly adhered to the surface of the wire following this treatment. The coating was sintered in place by heating to 760°C in hydrogen for 1 hour. Also, some of the green-coat wire was sintered at 980°C in hydrogen for 1 hour.

The nickel-coated wire was subsequently coated a second time in the same manner by passing a portion through nickel flake powder and drawing through a 1.5 mm diameter drawing die. Sintering treatments were imposed to bond the flake to itself as well as to the surface of the 36% nickel, balance iron wire. Sintering was accomplished during exposure of batches of wire in a hydrogen atmosphere for a period of 1 hour. Temperatures of 760° and 980°C were examined and found to provide excellent metallurgical bonding of the nickel flake to itself and to the surface of the 36% nickel, balance iron wire.

Wire coated in the manner illustrated by this example can be used as a welding filler metal. The presence of the thin layer of nickel upon the surface of the wire serves to substantially prevent oxidation during storage which, if allowed to occur, could cause improper contact in the welding machine with resultant arcing and wear of the drive rolls.

PROTECTIVE COATINGS

ALUMINUM

Coating of Carbon

R.V. Sara; U.S. Patent 4,104,417; August 1, 1978; assigned to Union Carbide Corporation describes a process for chemically bonding aluminum to a carbon substrate comprising: (a) depositing aluminum and a metal selected from the group consisting of tantalum, titanium and hafnium onto a carbon substrate wherein the metal is present in an amount greater than its solubility limit in the aluminum at the temperature of the subsequent heating step (b), and (b) heating the aluminum, metal and carbon substrate in an inert atmosphere at a temperature at which the metal is soluble in the aluminum for a time period sufficient to form an interface of metal carbide on the substrate which serves to chemically bond the aluminum to the carbon substrate. Tantalum, titanium and hafnium each have different temperatures at which the respective metal carbide interfaces will form.

Exposing an aluminum-tantalum coated carbon substrate to a heat treatment in an inert atmosphere at a temperature between about 850°C and about 1470°C for a period between about 1 second and about 1 hour will be sufficient to produce a chemically bonded aluminum coating on a carbon substrate via a tantalum carbide interfacial layer. A heat treatment of about 900°C for a period of about one-half hour is preferable.

Exposing an intermetallic phase coated, carbon substrate to a heat treatment in an inert atmosphere at a temperature between about 880°C and about 950°C for a period between about 5 minutes and about 30 minutes will be sufficient to produce a chemically bonded aluminum coating on a carbon substrate via a titanium carbide or hafnium carbide interfacial layer. A heat treatment of about 925°C for a period of about 15 minutes is preferable.

Using the above heat treatment, an aluminum layer of above about 25 μ can be obtained having a tantalum carbide, titanium carbide or hafnium carbide interfacial layer of between about 0.2 and about 50 μ. The preferred thickness for the overall coating will depend primarily on the end-use application of the coated article. For carbon electrodes, a coated layer of about 0.020 inch aluminum having an interface of about 0.0001 inch tantalum, titanium or hafnium carbide would be suitable. The interfacial layer need not be continuous to effectively bond the aluminum to the carbon with a surface of about only 25% being adequate.

Example: A slurry was prepared containing 75 g of Al_3Ta, sized 325 Tyler mesh and finer, in 100 ml of polyvinyl alcohol. The latter contained four parts water to one part polyvinyl alcohol by volume. A cylindrical graphite rod, measuring 13½" x ½" diameter was heated to 1000°C for 15 minutes in a vacuum to remove absorbed or chemically bonded volatiles. Thereafter, while the metal powder was kept in suspension by magnetic agitation, the slurry was brushed onto the graphite surface, coating only those areas which ultimately were to be covered with aluminum. The coated graphite rod was heated to 200°C and held thereat for two hours whereupon the slurry vehicle was removed.

Aluminum contained in a graphite crucible was heated in a vacuum of 10^{-4} torr to 1000°C in a 25 kW induction furnace. The coated graphite rod was slowly immersed in the aluminum and kept in a submerged condition for ½ hour, after which, upon removal, the rod was held in a cooler part of the furnace to allow solidification of the aluminum. Upon inspection, it was found that only those areas of the rod that were precoated with Al_3Ta were coated with aluminum following the dipping procedure. The aluminum coating was found to be between 0.0015 and 0.002 inch thick and the TaC interface thickness was determined metallographically to be approximately 5 μ.

Low Alloy Steel Containing Vanadium

Y.-w. Kim; U.S. Patent 4,144,378; March 13, 1979; assigned to Inland Steel Company has found that an aluminum coated low alloy low carbon killed steel article, particularly continuous steel sheets or strips, having an increased resistance to subsurface oxidation when heated at an elevated temperature in an oxidizing atmosphere can be provided economically with conventional apparatus and without using large amounts of expensive alloying elements by incorporating in a plain low alloy low carbon killed steel used to form the steel strip before aluminum coating, a small amount of vanadium sufficient to combine with or precipitate any carbon remaining in the steel base and leave an excess of uncombined vanadium in solution in the steel base.

Where the only alloying element added is vanadium, the weight percent vanadium added to the steel as the essential alloying element must be at least four times the weight percent of any uncombined carbon remaining in the steel in order to chemically combine with essentially all the carbon and nitrogen remaining in the steel and preferably provide an additional amount of vanadium sufficient to provide an excess of from about 0.1 to about 0.3% by weight uncombined vanadium in the steel.

While the vanadium content can be as much as ten times the weight percent of carbon in the steel, an amount of vanadium greater than that specified herein gives no increased benefits and merely adds unnecessarily to the cost. And, since the amount of carbon in a steel conventionally used for producing aluminum coated steel strips is small, generally less than 0.10 weight percent, the total amount of vanadium required in the process is relatively small. It is also within the scope of the process to add to the steel in addition to the vanadium, a small amount of titanium or other metallic alloying elements which will also combine with any carbon remaining in the killed steel or which will impart particular physical properties to the base steel.

When the alloying elements added are a combination of vanadium and titanium, the combined total weight percent of vanadium and titanium should preferably be about 0.4 wt % of the steel, with the vanadium and titanium alloying elements being present in any combination which will preferably provide a combined total of about 0.4 wt %, preferably providing about 0.1 wt % metallic chemically uncombined vanadium in the low carbon steel after all the carbon and nitrogen remaining in the steel are in a chemically combined form. For example, the steel in one preferred embodiment can contain 0.3 wt % vanadium and about 0.1 wt % titanium or can contain about 0.1 wt % vanadium and 0.3 wt % titanium. When less than about 0.1 wt % vanadium is used in combination with titanium, the improved wettability and reduction in segregation of titanium is not achieved.

The plain carbon or low carbon killed steel base used in the process, and to which the vanadium or vanadium and titanium are added preferably is a low carbon steel or mild steel having a carbon content of up to about 0.25 wt % maximum, usually from about 0.03 wt % to about 0.25 wt % and preferably from about 0.03 wt % to about 0.10 wt %.

Typically, a low alloy, low carbon aluminum killed steel base before the addition of vanadium or other alloying elements will consist essentially of from about 0.03 wt % to about 0.25 wt % carbon (preferably, 0.03 to 0.10 wt % carbon), from about 0.20 wt % to about 0.50 wt % manganese, 0.03 wt % sulfur, 0.02 wt % phosphorus, 0.002 wt % silicon, and 0.005 to 0.09 wt % aluminum with the balance being essentially iron with the usual amount of residuals and impurities. While the steel used is a killed steel, and preferably aluminum killed steel, a like amount of another deoxidizer, such as silicon, can be used to kill the steel.

A preferred method of aluminum coating a steel strip having the vanadium content thereof in accordance with the process is by a hot-dip coating process generally known in the art as a Sendzimir-type process, wherein a continuous steel sheet or strip which, after pickling, is free of scale and rust is fed continuously from a coil through a furnace containing an oxidizing atmosphere maintained at a temperature ranging between about 330° and 2400°F in order to burn off any oil residue on the surface of the strip and form a thin surface oxide film.

The oxide coated steel sheet then passes through a furnace containing a reducing atmosphere, such as the hydrogen-containing HNX atmosphere, having a temperature between about 1500° and 1800°F, whereby the oxide coating on the strip is reduced to form a surface layer of metal free of nonmetallic impurities to which molten aluminum readily adheres.

Following the reducing step, the strip is fed into a hot-dip aluminum coating bath through a protective hood which prevents the reduced metal surface being oxidized before entering the coating bath. The aluminum coating bath, for example, can be substantially pure aluminum or an aluminum rich alloy, such as aluminum containing up to 11% by weight silicon. After leaving the hot-dip aluminum coating bath, the coating thickness on the strip is controlled by coating rolls or preferably regulated by a pair of oppositely disposed thickness-regulating jet wipers which produce a uniform thin aluminum coating on the strip.

Complex Aluminum Halide of Alkali or Alkaline Earth Metal

R.S. Benden and R.S. Parzuchowski; U.S. Patent 4,132,816; January 2, 1979; assigned to United Technologies Corporation have found that aluminum can be caused to deposit on alloy metal substrates selected from the group of iron, chromium, nickel, and cobalt-base metals through a gas phase deposition process in which the aluminum is carried from a powder mixture to the internal surfaces of the substrate by the use of a carrier consisting of a complex aluminum halide of an alkali or alkaline earth metal. The halide portion of this complex can be any of the available halides, such as fluorine, chlorine, bromine or iodine.

Further, since under the conditions of the treating process, the complex aluminum halide vapor is an equilibrium mixture of an aluminum halide and an alkali metal or alkaline earth halide, the complex aluminum halide may be substituted by a mixture of an alkali or alkaline earth metal halide and aluminum halide. Thus, for example, instead of including in the mixture the complex Na_3AlF_6, similar results may be obtained by mixtures of sodium fluoride and aluminum fluoride. If this approach is used, the ratio of alkali metal halide to aluminum halide is preferably in the range of 1-4:3-2. The "throwing power," i.e., the ability of the activator to deposit material on internal surfaces, is substantially increased when either the mixture or the complex aluminum halide is used as opposed to using either the aluminum halide or the alkali or alkaline earth halide alone.

The coating procedure, according to this process, can be carried out at any of the conventional temperatures, preferably 1800 to 2200°F.

Leafing-Type Aluminum on Stainless Steel

A.L. Baldi; U.S. Patent 4,141,760; February 27, 1979; assigned to Alloy Surfaces Company, Inc. has found that flake forms of protective metals form particularly effective protective coatings for corrodible metals when partially diffused into the surface to be protected or when combined with special binders or added over other coatings.

As shown in the prior art, powdered aluminum has been suggested for use in applying protective layers over corrodible metals. The protection thus obtainable from layers of less than about 1 or 1.5 mg/cm^2, is greatly improved if the aluminum coating is effectively continuous over the surface being protected, a result that is obtained when leafing-type aluminum particles are applied in amounts that permit the individual aluminum flakes to partially overlap each other over the entire surface being protected. It is also helpful to subject the aluminum-coated ferrous member to a temperature that causes at least a little bit of the aluminum to diffuse into the ferrous surface.

Leafing-type aluminum particles are generally characterized by the presence of stearic acid or aluminum stearate or the like as a very thin coating on the surface of each aluminum particle, a condition which makes it extremely difficult to disperse such aluminum particles in water. A substantial amount of wetting agent will effect a suitable dispersion, although it is easier to effect such dispersions by also adding diethylene glycol or triethylene glycol or more highly polymeric ethylene glycols having a molecular weight up to about 9,000, or by adding glycerine. Very effective dispersions of leafing-type aluminum can be made from a concentrate that consists essentially of the leafing aluminum, the polymeric ethylene glycol and a wetting agent, the aluminum being present in an amount about ¼ to about 1½ parts by weight for every part of the polymeric ethylene glycol by weight, and the wetting agent concentration from about 5% to about 25% by weight of the concentrate.

The foregoing concentrate readily mixes with water in all proportions to provide an aqueous dispersion of almost any desired aluminum content. Thus a diluted dispersion containing 5% aluminum, 6% hexaethylene glycol and 0.7% p-n-octylphenyl ether of decaethylene glycol, is readily sprayed onto a stator ring of a jet engine compressor to leave a coating weighing 0.5 mg/cm^2 after drying in air to evaporate most of the water. The stator thus coated is then heated in an air oven until its temperature reaches 800°F.

The heating first causes the glycol and wetting agent to be volatilized off leaving a very adherent, continuous and shiny coating that resembles polished aluminum and significantly adds to the corrosion resistance of the stator ring even if the heating temperature goes no higher than 600°F. The increase in corrosion resistance becomes more significant when the heating carries the coating to temperatures of about 900°F, where some diffusion of the aluminum into the ferrous surface of the stator begins. The rate of diffusion and the degree of resulting corrosion resistance is further increased by confining the coated stator in an atmosphere of gaseous aluminum chloride while it is at temperatures above about 700°F.

The adhesion of 0.1 to 4 mg/cm^2 layer of a flaked metal such as aluminum is also improved by incorporating in the layer chromic acid, or a compound such as ammonium chromate or dichromate which is a salt of chromic acid with a volatile base, or magnesium chromate or dichromate, or water-soluble chromates or dichromates of other divalent metals. Dissolving in the aluminum dispersion an amount of ammonium dichromate 2% by weight of the aluminum gives a sharp increase in adhesion upon heating of a dried 4 mg/cm^2 layer of such composition on plain carbon steel to 700°F for 5 minutes. A 5% addition of the dichromate to the dispersion renders such a heat-treated coating completely resistant to wiping off.

Improvements in corrosion resistance and in appearance are also obtained when the coating is covered by a similar coating, even one that does not contain metallic aluminum.

The proportions of the ingredients in the overlying chromic acid-phosphoric acid-salt coating mixture can range as follows:

	Range	Preferred
	. . . . (mols/ℓ)	
Chromate ion	0.2-1	0.4-0.8
Phosphate ion	0.7-4	1.5-3.5
Magnesium ion	0.4-1.7	0.9-1.4
Polytetrafluoroethylene resin	2-14*	3-10*

*Grams per liter.

The magnesium ion can be replaced by aluminum, calcium or zinc, in the same concentrations.

Instead of directly applying such an overlying coating whether or not it contains metallic aluminum, it can be applied after an intervening coating of colloidal alumina or the like weighing about 0.1 to about 1 mg/cm^2.

Example: (A) Into each of four plain carbon steel retort cups 2 feet wide and 14 inches high is poured a powder pack consisting of 20% aluminum by weight and 80% alumina, both minus 325 mesh and uniformly mixed together. After the retort bottoms are covered with about one-half inch of powder, jet engine compressor blades made of AISI 410 stainless steel are laid over the powder layer, the blades being spaced about one-eighth inch apart. This layer of blades is then covered with more powder till the powder is about one-half inch above the vane tops, and another layer of blades is then laid down and the layering repeated until the entire packing is 12½ inches deep in each retort.

More pack powder is then added to each retort to assure there is about 1 inch of powder above the tops of the topmost blades following which there is sprinkled over each a very thin stratum of crystalline $AlCl_3 \cdot 6H_2O$ in an amount weighing 0.6% of the total powder weight. The retorts are then filled to their tops with additional pack powder, and they are stacked one above the other on the floor of a gas-fired bell furnace. The stacking does not seal any of the retorts shut. The top of the furnace equipped with gas inlet and outlet flush lines is lowered over the stack and sealed against the furnace floor, and a slow flow of argon gas is passed through the furnace interior to start flushing out the air within it.

After the argon purge hydrogen is substituted for the argon, and is introduced at a rate that permits it to be burned with a small flame as it emerges from the end of the outlet tube. Only a very low flow rate is necessary, about 10 to 15 standard cubic feet per hour.

The heating of the furnace is started at a rate of about 1.5°F per minute, as measured by thermocouples in each retort and connected to external meters, and when the thermocouples reach 300°F the flow of hydrogen can be reduced so that the outlet flame is very tiny. At this point the hydrogen inflow can be less than 10 standard cubic feet per hour.

As the heating-up continues, the temperatures indicated by the thermocouples increase uniformly and gradually and chemical vapors begin to appear in the burning outlet gas. By the time the thermocouple temperatures reach about 450°F the discharge of chemical vapors has subsided, the gas flow continuing till the temperatures reach 875°F where the furnace heating is set to hold.

After 16 hours at 875°F the furnace heating is terminated and the furnace permitted to cool until the thermocouple temperatures reach 300°F. The atmosphere in the furnace is then purged by switching the inflow gas to argon or nitrogen and the furnace shell then removed from the retorts, permitting the retorts to cool further in air. The contents of the retorts are then poured out, washed, dried and finally lightly blasted with fine glass particles propelled by an air stream supplied at 5 to 10 pounds per square inch, giving an aluminum pick-up of 4.0 mg/cm^2 of ferrous surface.

(B) On the lightly blasted aluminized surface there is sprayed with an air-propelled spray, a uniform very thin layer from an aqueous dispersion of 3.5% CrO_3, 2.4% MgO, 11% H_3PO_4, 5.7% leafing aluminum, 6.8% polyethylene glycol having an average molecular weight of 300 and in which the glycols range from pentamethylene glycol through heptamethylene glycol, and 0.8% p-isononylphenyl ether of dodecaethylene glycol. All percentages are given by weight.

The sprayed blades are then air-dried and baked at 700°F in an air oven for 30 minutes to give a coating weight from this spray of 0.7 mg/cm^2 of ferrous surface.

(C) The blades coated in steps (A) and (B) have their coated surfaces given a spray coating of colloidal alumina dispersed in a 20% concentration by weight in water to which a little HCl is added to bring the pH down to about 4. A very fine spray is used to leave a light coating which after drying in air weighs 0.5 mg/cm^2.

(D) The blades with the air-dried coatings are then given a top spray coating from an aqueous dispersion of 5.8% CrO_3, 4% MgO, 18.3% H_3PO_4, and 0.5% polytetrafluorethylene particles about 1 μ in size; this spray being such that upon air-drying in an oven and then baking at 700°F for 30 minutes in an air oven, the final coating weighs 0.5 mg/cm^2.

Formation of Diffusion Layer

T. Yagi and J. Yamamoto; U.S. Patent 4,150,178; April 17, 1979; assigned to Toyo Kogyo Co., Ltd., Japan describe a method of Al diffusion layer formation which comprises the steps of dipping a ferrous base alloy workpiece containing more than 8.0% by weight nickel in a molten bath of aluminum or aluminum alloy having a temperature from 650° to 750°C for 30 to 120 seconds, subjecting the workpiece thus treated to a first heat treatment at a temperature from 750° to 850°C for at least 60 minutes, and further subjecting the workpiece thus treated to a second heat treatment at a temperature from 900° to 1000°C for at least 30 minutes.

Subsequently the workpiece is subjected to a third heat treatment under predetermined conditions within the long time side region in Figure 3.1 defined by a line denoting 1050°C, a line indicating 1300°C and a line showing 10 minutes and also sectioned by a curve line connecting a point **A** (1050°C, 90 minutes), a point **B** (1100°C, 40 minutes) and a point **C** (1150°C, 10 minutes) so as to cause an Al compound layer formed on the workpiece to be removed through separation by forming only an Al diffusion layer on the surface of the ferrous base alloy workpiece.

It should be noted here that in the process, the layer formed on the surface of the workpiece by immersing or dipping the workpiece in the molten metal bath of aluminum or aluminum alloy is defined as an Al plating layer, and the layer mainly of Fe-Al compound formed on the surface of the workpiece by subjecting the resultant to heat treatment is defined as an Al compound layer, with another layer formed between the Al compound and the workpiece by diffusion of part of the Al compound layer into the workpiece defined as an Al diffusion layer.

Figure 3.1: Formation of Diffusion Layer

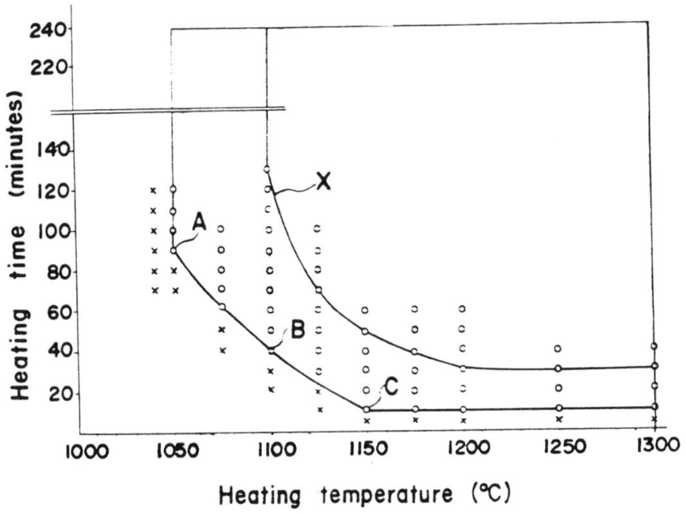

Source: U.S. Patent 4,150,178

In Figure 3.1, marks **o** denote good state of separation, while marks **x** represent poor state of separation. As is clear from Figure 3.1, it is noted that all the samples treated at temperatures over 1050°C for more than 10 minutes and lying at the high temperature and long heating side in the region sectioned by the line connecting the points **A**, **B** and **C** show favorable separating state. It is to be noted here that the line **X** represents a boundary line whereat a workpiece of 2.5 mm thickness treated according to the method is deformed beyond a permissible range, and that treating conditions should preferably be in the range below the line **X** when a steel plate is used for the workpiece.

Example: *(a) Ferrous base alloy* – austenitic stainless steel having a composition of 0.05% C, 3.5% Si, 0.3% Mn, 18.5% Cr, 13.0% Ni, 1.0% Cu and remainder of Fe was employed.

(b) Al melt plating – a steel plate prepared from the ferrous base alloy as described above (2.5 mm thick) was dipped for 60 seconds in a molten alloy bath having a composition of 10% Si and remainder of Al and maintained at a temperature of 730°C, and was then taken out therefrom.

(c) First heat treatment – the steel plate to be treated which had been subjected to the melt plating in the step (b) above was heated at a temperature of 780°C for 90 minutes.

(d) Second heat treatment – the steel plate which has been heat-treated in the step (c) above was further heated at a temperature of 950°C for 60 minutes.

(e) Third heat treatment – the steel plate which had been heat-treated in the step (d) above was further heated at a temperature of 1110°C for 60 minutes with subsequent cooling by air, and thus the Al compound layer formed on the steel plate surface was separated for removal.

The thickness of the Al diffusion layer formed on the workpiece was 150 μ. This Al diffusion layer forming method is capable of providing products superior in processability and welding performance in addition to the good resistance against oxidation, and is particularly suitable for employment in mass production as an efficient processing method, for example, of exhaust system components and parts of motor vehicles and the like.

Two-Layer Coating Process

R.D. Jones; U.S. Patent 4,150,179; April 17, 1979; assigned to University College Cardiff, and Coated Metals Limited, both of Wales describes a method of coating a substrate comprising feeding the substrate through a molten composition of a first coating material and subsequently feeding the so coated substrate into a molten composition of a second coating material whereby the coating of the second material is overlaid upon the coating of the first material. In order to prevent contact between the substrate bearing the first coating and the ambient atmosphere, the so coated substrate is preferably passed from one coating composition to another via an inert atmosphere or environment which may be liquid or gaseous. Passage through an inert atmosphere is advantageous in that surface oxidation of the first coating is inhibited, if not prevented, prior to application of the second coating.

The first and second coating compositions are preferably floatingly supported on a layer of molten material which is inert relative to the two coating compositions which are separated from one another by a partition. Where it is desired to aluminize a steel substrate, for example, a steel strip, the first and second coating compositions are preferably aluminum/silicon alloy containing between 5 and 12% silicon and aluminum respectively.

One form of apparatus for carrying out the process is shown in Figure 3.2, comprising a bath **1** containing a quantity **2** of molten lead. Floatingly supported on the molten lead are quantities of an aluminum/silicon melt **3** and an aluminum melt **4.**

Figure 3.2: Schematic Diagram of a Double Dip-Aluminizing Bath

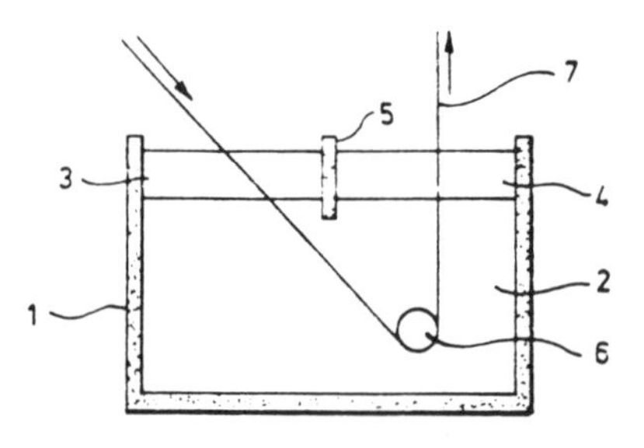

Source: U.S. Patent 4,150,179

The melts **3** and **4** are, as shown, separated from one another by a divider **5.** A steel roll **6** is mounted in the bath **1** in the position shown and steel substrate, in the form of a strip **7,** is first fed through the aluminum/silicon melt into the molten lead around the steel roll and exits from the bath through the aluminum melt **4.** The immersed steel roll serves to change the direction of travel of the strip and allows a double coating to be applied to the strip in a single operation.

An added advantage of the apparatus is that the steel roll is protected by the lead melt from the dissolution effect of molten aluminum and consequently it is expected that the roll will have an increased service life. Similarly, the molten aluminum on the exit side will be subject to less iron contamination and as such is expected to produce a more corrosion resistant coating on the steel strip.

CHROMIUM

Ferricyanide-Free Chromate Conversion Coating

Although ferricyanide activated chromate coating solutions have enjoyed appreciable commercial use, there are problems associated with their utility. For example, it has been found that the ferricyanide activated solution is subject to heat degradation at higher temperatures and also is relatively sensitive to changes in acidity. However, perhaps the most serious obstacle to the use of ferricyanide today relates to the disposal problem of the spent solutions because of the occasional occurrence of free cyanide moieties.

E.R. Reinhold; U.S. Patent 4,146,410; March 27, 1979; assigned to Amchem Products, Inc. describes a conversion coating and a method for the use of this coating on aluminum-containing surfaces. The coating solution comprises an aqueous acidic composition including the following ingredients:

(a) zinc (Zn) in an amount of from about 0.03 g/ℓ, to about 3.0 g/ℓ, preferably added as ZnO,

(b) hexavalent chromium (Cr^{+6}) in an amount of from about 0.01 g/ℓ to about 10.0 g/ℓ, preferably added as CrO_3,

(c) fluoride (F) in the amount of from about 0.01 g/ℓ to about 10.0 g/ℓ, preferably added as H_2SiF_6,

(d) molybdate (Mo) in an amount of from about 0.02 g/ℓ to about 3.0 g/ℓ, preferably added as 84% MoO_3,

(e) acid (H^+) preferably added as HNO_3, to provide a pH of 0.0 to 5.5.

A preferred solution referred to hereinafter as the bath, contemplates, in grams per liter the following:

Material	Grams/liter
CrO_3	2.00
ZnO	0.38
HNO_3 (38° Bé)	3.40 (1.9 g if 100% HNO_3)
H_2SiF_6 (23% aqueous solution)	4.56
Molybdic acid (84% MoO_3)	0.475
Water	balance

The bath is most conveniently prepared from a concentrate in which, to produce the bath, the concentrate is diluted with water. The preferred bath composition referred to above is most conveniently prepared by diluting with water a concentrate of the following composition.

Material	Grams/liter
CrO_3	40.0
ZnO	7.6
HNO_3 (38° Bé)	68.0
H_2SiF_6 (23% aqueous solution)	91.2
Molybdic acid (84% MoO_3)	9.5
Water	balance

In this process an aluminum-containing surface is contacted utilizing conventional means such as spraying or dipping with the abovedescribed composition for a time sufficient to impart thereon a chromium conversion coating.

Chromium/Plastic Emulsion

R.C. Miller; U.S. Patent 4,137,368; January 30, 1979; assigned to J.M. Eltzroth & Associates, Inc. describes a process for producing an emulsion coating composition containing chromium in the hexavalent state and at least one water insoluble particulate film-forming organic thermoplastic resin by mixing a body of an emulsion coating composition containing water in the continuous phase and in the discontinuous phase at least one water insoluble particulate film-forming organic thermoplastic resin with a gradually added dilute aqueous solution of an inorganic ionizable water soluble hexavalent chromium compound having a concentration of 1% to 10% by weight, calculated as Cr, while cooling and maintaining a pH within the range of 2 to 10.5 and maintaining the body in a nonfoaming state.

The process is preferably carried out by causing the body of emulsion to rotate and gradually adding the dilute solution to a peripheral portion of the rotating body.

In the application of the coating composition to a substrate such as a metal, those skilled in the art will recognize that cleaning and preparation of the metal is quite important and may involve the usual cleaning methods, deoxidizing of the substrate, rinsing and drying.

After the emulsion coating composition is applied to a substrate, e.g., steel, aluminum, magnesium, or a zinc surfaced substrate, it is dried or allowed to dry and cured, for example, at 400°F for from 60 seconds to three minutes or 600°F from 30 to 90 seconds. Although the overall temperature range of curing is 400° to 600°F for periods of time from 30 seconds to 3 minutes, the time of heating is longer at lower temperatures and shorter at the higher temperatures. The time-temperature relationship is preferably controlled to give a final coating having a pencil hardness of H to 2H.

The wet thickness of the applied coating is usually within the range of 0.05 mil to 10 mils (a mil equalling 0.001 inch). A preferred thickness of the wet coating is within the range of 0.1 to 0.3 mil. The greater the thickness of film the greater will be the difficulty of acquiring a uniformly "cured" film.

Example: An emulsion coating composition was prepared as follows: Four gallons of an acrylic emulsion polymer containing 46±0.5% solids and having a pH of 9 to 11 (Rhoplex MV-1) was mixed with a premix of 15 to 30 ml of tributyl phosphate and 3 to 10 ml of a defoaming agent (Nopco NXZ) and agitated thoroughly until no fish eyes or agglomerates appeared on a fineness of grind gauge. 1 to 2 gallons of water were added and to the resultant emulsion while it was being rotated with a paddle-type agitator there was added dropwise at the periphery of the emulsion body a dilute solution of sodium dichromate in water containing one gram of sodium dichromate per 10 ml of solution until a pH value of 6.6 to 6.8 was obtained. The resultant emulsion was then filtered.

A coating of the foregoing emulsion was applied using a draw bar to a thickness of 0.05 mil to 1.0 mil thickness on aluminum panels of No. 3003 alloy having a gauge thickness of 0.019 to 0.025. These panels had previously been alkali cleaned. Some were chromate conversion coated. Others were rinsed with water and then acid rinsed with chromic acid-phosphoric acid mixtures and dried prior to the application of the coating.

The coatings were cured by preheating at 120° to 160°F for 20 to 40 seconds followed by complete cure in 45 to 60 seconds at 600°F. Panels of the cured coatings were bent and the edge of the bend exposed to 60 inch-lb direct impact. There was no "pull away" using 3M-600 tape.

Diffusion Coating of Fe-Ni Base Alloy

J.J. Grisik and D.J. Wortman; U.S. Patent 4,148,936; April 10, 1979; assigned to General Electric Company describe a metallic article comprising an Fe-Ni base alloy substrate of the controlled linear thermal expansion type, characterized by the substantial absence of Cr and having a mean coefficient of linear thermal expansion of less than about 4.7 x 10^{-6} inch per inch per °F (8.5 x 10^{-6} mm/mm/°C) at the inflection temperature in the range of about 780° to 880°F (416° to 471°C).

In one form, such alloy consists essentially of, by weight, 30 to 40% Ni, 10 to 20% Co, 1 to 5% of the sum of Cb and Ta, 0.5 to 3% Ti, 0.2 to 3% Al, up to about 3% each of Hf and Zr, up to about 0.5% B, with the balance essentially Fe and incidental impurities, the substrate having diffused therein a material selected from Cr and its alloys. In a preferred form such alloy substrate consists essentially of, by weight 35 to 40% Ni, 13 to 17% Co, 2 to 4% of the sum of Cb and Ta, 1 to 2% Ti, 0.3 to 1.2% Al, up to 3% Hf, with the balance Fe and incidental impurities. It is preferred that such impurities be maintained in a range up to a maximum of 0.012% B, 0.05% Cu, 0.06% C, 1% Mn, 0.35% Si, 0.015% S, 0.015% P and 1% Cr.

According to the method, such coated metallic article is provided by diffusion chromiding the article surface in a container with a nonoxidizing, preferably reducing atmosphere such as H_2, and a powdered mixture comprising, by weight, 10 to 50% Cr powder, 0.1 to 4% of a conventional halide salt activator, particularly a chloride type such as NH_4Cl or $CrCl_3$, with the balance of the mixture being an inert powder filler such as Al_2O_3. Preferably such mixture consists essentially of, by weight, 15 to 25% Cr, 1.5 to 2.5% of a chloride salt activator, with the balance Al_2O_3. Such a method is conducted at a temperature below that which will recrystallize such an alloy, generally less than 1700°F and preferably in the range of about 1450 to 1650°F.

An article of a controlled linear thermal expansion alloy substrate into which has been diffused the element Cr at a temperature less than its recrystallization temperature is characterized by significantly improved environmental resistance without degradation of mechanical properties.

Coating of Zinc-Iron Alloys

W.C. Glassman, S.H. Melbourne, M.S. Morson and U. Soomet; U.S. Patent 4,141,758; February 27, 1979; assigned to Dominion Foundries and Steel, Ltd., Canada describe a composition for producing a chromate protective coating on zinc/iron alloy surfaces, the composition having a pH less than 1.20 and consisting of demineralized water, a source of chromium ion, the chromium ion being present in the amount of at least 20 g/ℓ and not more than 36.4 g/ℓ, a source of perchlorate ion, the perchlorate ion being present in the amount of from 0.4 to 1.2 g/ℓ, and a source of fluoride ion, the fluoride ion being present in the amount of from 0.1 to 0.35 g/ℓ, the chromium ion being the only metallic ion present therein.

The process includes applying a composition as specified above to a zinc/iron alloy surface for a period of from 1 to 10 seconds at a temperature of from 150° to 195°F (65.5° to 90.5°C).

The chromium ion is supplied to the solution most conveniently in the form of anhydrous chromium trioxide, in which the chromium is in hexavalent form. This compound can constitute a convenient medium for specifying the required concentration of the chromium in the solution, which should be present as an ion in the amount of 15.6 to 36.4 g/ℓ, preferably 20 to 24 g/ℓ. The corresponding quantities of the trioxide to be added are 30.0 to 70.0 g/ℓ, preferably 38.5 to 46.2 g/ℓ. The calculation of any other suitable material added to the solution will be apparent to those skilled in the art.

The perchlorate is added in the form of 60% perchloric acid to provide the required amount of ion of 0.4 to 1.2 g/ℓ, and for this concentration the solution should contain from 0.41 to 1.22 ml/ℓ of the 60% perchloric acid; preferably the perchlorate ion is present in the amount of 0.5 to 1.0 g/ℓ. The fluoride ion conveniently is added as 48% hydrofluoric acid to provide from 0.1 to 0.35 g/ℓ of the fluoride ion, preferably 0.12 to 0.25 g/ℓ, and this amount of ion requires the addition of about 0.19 to 0.65 ml/ℓ of the acid, preferably 0.22 to 0.46 ml/ℓ.

Carbon Steel Razor Blades

It is generally known that carbon steels have better mechanical properties for making razor blades than stainless steels. However, due to their tendency to corrode, such carbon steel blades have relatively short-use lives. It is further known that their use-lives may be increased by applying chromium coatings to the edges. Generally the most efficient way to apply such chromium coatings is to stack the blades in a magazine and apply the chromium to the edges by processes such as sputtering or vapor deposition. Although such sputtering and vapor deposition processes can be effectively used on the edges they are inefficient and/or inconsistent when used on the bodies of the blades.

It is still further known from U.S. Patent 3,490,314 that the bodies and edges of as-sharpened carbon steel blades can be made more corrosion resistant and longer lasting by coating them with nickel-phosphorous, cobalt phosphorous or nickel-cobalt-phosphorous coatings by electroless processes. Although such processes are especially effective on the bodies of the blades, there is one drawback with the edges, in that the coatings bead-up thereon causing rounding, and additional sharpening, subsequent to the coating process, is required.

S.A. Sastri, H.-Y. Chang, T.G. Decker and R. McDonald; U.S. Patent 4,139,942; February 20, 1979; assigned to The Gillette Company describe a process which comprises coating the cutting edges of the blades with chromium by any of the well known processes, such as sputtering or vapor deposition and then immersing the blades in an electroless coating bath comprising a nickel-phosphorous, nickel-copper-phosphorous, cobalt-phosphorous or cobalt-nickel-phosphorous coating material to provide a coating of the material on at least the bodies of the blades. By first coating the edges with chromium virtually none to only traces of the nickel-phosphorous, nickel-copper-phosphorous, cobalt-phosphorous or nickel-cobalt-phosphorous coatings bead-up on the edges and subsequent sharpening of the edges is not required.

It has been found especially useful after applying the metal-phosphorous coatings to apply a carboxy-functional (i.e., a carboxy bearing) silicone oil composition such as a polydimethylsiloxane oil comprising about 1 mol % of functional carboxy groups sold as Dow Corning Fluid X2-7049. Generally the amount of carboxy-functional groups present may be varied. Usually the amount present will lie between about 0.5 to 1.5 mol % and preferably about 1 mol %. In preferred embodiments the silicone oil compositions will comprise a long chain (e.g., 12 to 22 carbon atoms) aliphatic compound, wherein the compound is an acid, alcohol, amide or mixtures thereof. As examples of such aliphatic materials mention may be made of stearic acid, stearyl alcohol and octadecyl amide. The amount of aliphatic acids, alcohols or amides added usually may vary. Generally it will lie between 1 to 10% by weight of the oil composition and preferably 5%.

In applying the oil it is preferable to use a diluent or carrier such as 1,1,2-trichloro-1,2,2-trifluoroethane. A protective solution which was found especially useful for coating the blades by dipping was one comprising 1 part of the silicone oil composition containing 5% by weight of the aliphatic additive and 39 parts of the 1,1,2-trichloro-1,2,2-trifluoroethane.

Example: A magazine of commercially sharpened carbon steel blades was dc cleaned in a sputtering chamber having a 10 to 12 μ atmosphere of argon, at 1.1 amps and 1,600 volts for about 5 minutes. The cleaned blades were sputter coated with chromium in a 10 to 12 μ argon atmosphere, using 3.5 kV for 5 minutes and 20 seconds. The resulting blades having a 400 Å chromium coating on the edge were dipped into an Enplate nickel-phosphorous electroless coating composition at 95°C and a pH of 5 for 15 seconds. The coated blades were rinsed in water and in acetone and then air-dried.

The dried blades were then coated with a protective oil comprising 1 part by weight margaric acid, nineteen parts Dow Corning Fluid X2-7049 and 780 parts 1,1,2-trichloro-1,2,2-trifluoroethane. The blades were tested for corrosion by placing them in a gold plated razor, dipping them in water at 53°C for 7.5 minutes, drying them in air for 7.5 minutes and repeating the cycle sixteen times. The blades after the corrosion test were relatively rust free and were better than commercial carbon steel blades which had been lacquered. Further, in a wool-felt cutting test there was substantially little difference in the force required for blades which only had the edges coated with chromium and those which were further subjected to the nickel-phosphorous coating. Such a slight change in the force indicates that there was little beading-up of the nickel-phosphorous coating on the edge.

ZINC

Pretreatment with Salts or Metal Hydroxide

S.F. Radtke and D.C. Pearce; U.S. Patent 4,140,821; February 20, 1979; assigned to International Lead Zinc Research Organization, Inc. describe a galvanizing process whereby a molten salt or metallic hydroxide has been substituted for the fluxing techniques normally used in galvanizing. The process uses a nonfuming salt or metallic hydroxide bath as both a preheat and a surface preparation medium for the ferrous metal object prior to its entering the galvanizing bath. This process not only overcomes a number of the objectionable features of the present fluxing techniques, but offers a method by which the coating weights on silicon-containing ferrous metals may be controlled so that the galvanizer does not have to take special precautions when galvanizing a variety of ferrous metals.

Sodium hydroxide is a preferred medium for the bath, as are four particular pairs of salts, as follows: zinc chloride, from 55% (by weight) to 77 wt %, plus potassium chloride, from 45 to 23 wt %, and preferably about 65 wt % zinc chloride and about 35 wt % potassium chloride; potassium chloride, from 49 wt % to 59 wt %, plus lithium chloride, from 51 to 41 wt %, preferably about 55 wt % potassium chloride and about 45 wt % lithium chloride; lead chloride, from 75 to 95 wt %, plus potassium chloride, from 25 to 5 wt %, and preferably either about 92 wt % lead chloride and about 8 wt % potassium chloride, or about 80 wt % lead chloride and about 20 wt % potassium chloride; also, lead chloride, from 90 to 95 wt %, plus sodium chloride from 10 to 5 wt %, preferably about 91 wt % lead chloride and about 9 wt % sodium chloride.

The ferrous metal object is allowed to remain immersed in the bath in order to preheat it. The bath is maintained at a temperature above its melting point, so that immersion therein of the cooler metal object will not cause the bath to solidify. The object should be brought to at least the temperature at which the bath melts. In this way, the heat loss in the zinc bath caused by the object is lessened considerably. In the preferred embodiment, the object is heated to a temperature at least about that of the zinc bath.

The length of time required for the immersion of the object in the bath will vary, depending on the size of the object, and the size and the temperature of the preheat bath. This length of time is readily calculable by the galvanizer, who may monitor the temperature of the object while it is in the preheat bath, for example, by using a thermocouple mounted within the object. The times will be on the order of several minutes or more.

Elimination of Oxide Layer

Already known is a process for heat-treating and galvanizing steel strip including immersion in molten zinc, comprising heating the strip to a temperature higher than its recrystallization temperature, immersing the strip in an aqueous bath maintained at substantially its boiling temperature, removing the strip from the bath, heating the strip to a temperature at which it can be immersed in the molten zinc and maintaining the strip at this temperature, and then immersing the strip at this temperature in the molten zinc.

This process makes it possible to produce galvanized strip generally having good properties of ductility, drawability, and elongation; but it is necessary to prevent the sheets emerging from the hot aqueous bath from becoming covered with an oxide film, since it is practically impossible to obtain a zinc coating adhering to such a film.

In contrast *P. Paulus and M. Economopoulos; U.S. Patent 4,143,184; March 6, 1979; assigned to Centre de Recherche Metallurgique-Centrum voor Research in de Metallurgie, Belgium* do not attempt to prevent steel oxidation but rather allow or even assist the formation of an oxide layer and then eliminate the troublesome layer before hot-dip galvanizing.

Eliminating oxide from only one face of the strip makes it possible to obtain strip galvanized on one side only. Preferably, the composition of the aqueous bath is adjusted so as to ensure formation of a thin oxide layer (oxide film), e.g., less than 2 g/m^2, on the entire surface of the strip.

The oxide layer may be removed electrolytically. Electrolytic removal of oxide from only one face of the strip is possible because the electrolytic effect occurs only where the electric field is sufficient. In the case of removal of oxide by electrolysis, the strip is preferably subjected to a subsequent rinsing treatment before it is heated to the immersion temperature.

The oxide may be eliminated by reduction, which may be oxide effected by means of plasma torches producing a reducing gas. To prevent the temperature of the strip from exceeding the immersion temperature owing to heat produced by the plasma torches, a face of the strip is preferably subjected to a cooling treatment while the other is being deoxidized. This cooling treatment can be carried out either by blowing a nonreducing gas onto the face to be cooled (which also makes it possible to prevent the reducing gas of the plasma torches from coming into contact with this face), or by contacting this face with a cooled roll or cylinder.

When producing full hard galvanized strip, the predetermined temperature to which the strip is heated to give the strip desired properties is lower than the recrystallization temperature and is preferably in the range of 400° to 550°C.

In the case of the manufacture of galvanized strip for drawing or having a high limit of elasticity, the predetermined temperature to which the strip is heated to give it desired properties is higher than its recrystallization temperature. So far as galvanized strip for drawing is concerned, this predetermined temperature is in the range of 650° to 850°C, preferably 700° to 800°C. So far as galvanized strip having a high limit of elasticity is concerned, this temperature is in the range of 650° to 950°C, preferably 750° to 890°C.

Furthermore, in hot-dip galvanization, the strip should normally enter the zinc at a temperature substantially equal to or greater than the melting temperature of zinc, depending on the composition and temperature of the molten zinc. The immersion temperature is advantageously in the range of 420° to 550°C, preferably 450° to 500°C.

In the case in which cooling in the aqueous bath has brought the strip down to substantially the temperature of the bath, e.g., a temperature on the order of 100°C, the heating before immersion in molten zinc should preferably be sufficiently rapid to ensure that the strip is above 300°C for longer than 30 seconds before immersion.

The galvanization installation for treating steel strip, in particular for treating steel strip for drawing or having a high limit of elasticity, comprises a heating furnace for bringing the strip to a temperature suitable to give the strip desired properties and possibly for keeping it at that temperature for a predetermined time; a vessel containing an aqueous bath kept substantially at its boiling temperature, the strip being designed to be immersed in the bath to be rapidly cooled and possibly to be kept at the final cooling temperature for a predetermined time; means for removing or reducing a thin oxide layer formed after the strip has been immersed in the cooling aqueous bath; optionally means for bringing the strip thus cooled to the immersion temperature in molten zinc and, if necessary, for keeping the strip at that temperature until immersion actually takes place; a vessel containing molten zinc for galvanizing the strip by immersion; and means for unwinding the strip at the beginning of the treatment, and for winding the strip at the end of the treatment.

Draining of Galvanized Parts

When objects have been galvanized by hot-dipping, surplus zinc is found to adhere in certain places and holes or openings in the objects may even be covered by a film of zinc.

Such surplus zinc is not only detrimental to the appearance of component parts but will often prevent them from being assembled in the way they were designed to be, and must therefore be eliminated. Its hardness and the places where it accumulates then involve manual work with a file, and this work is not only lengthy but takes up the time of costly personnel.

In order to avoid this metal removal work, an endeavor is made to drain the parts more effectively by moving them about and even jolting them as they emerge from the baths. These operations must be carried out before the zinc solidifies, that is, within less than one minute from the time the parts leave the galvanizing baths. Not only is this method laborious, but it does not completely eliminate the need for subsequent filing.

G. Rouquié; U.S. Patent 4,129,668; December 12, 1978; assigned to Bertin & Cie, France describe a method based on the use of a range of vibration excitation frequencies capable of covering the fundamental frequencies of the parts to be drained. The basic principle consists in using narrow-band "white noise" to excite the support for the parts issuing from the galvanizing bath whereby, irrespective of their dimensions, shapes or weights, such parts are excited and made to vibrate in at least one of their natural modes. The parts then enter into resonance, whereby only a small expenditure of energy is needed to carry the method into practice.

In a first embodiment, a single exciter fed with white noise is used. In a second embodiment, a number of small vibrators are used, each of which delivers one of the frequencies in the chosen spectrum, thereby to provide a kind of white noise synthesis.

Recourse may be had to vibrators of any convenient type, such as pneumatic, hydraulic, electric or electromagnetic vibrators. Because the draining must be completed before the zinc solidifies, these vibrators must be activated very quickly, within less than a minute of the parts emerging from the galvanizing bath.

Zinc/Iron-Zinc Two-Sided Coatings

W. Batz; U.S. Patent 4,104,088; August 1, 1978; assigned to Jones & Laughlin Steel Corporation describe a method for the production of zinc coated steel strip products having a full commercial unalloyed zinc coating on one side and a fully alloyed iron-zinc coating on its other side. The iron-zinc alloy is formed through alloying of the zinc coating and the steel substrate. A product of this nature offers a highly desirable combination of properties peculiar to each type of coating for a single product.

The superior paintability and resistance spot weldability of the alloyed side of the product is attractive for applications in which the external surface would be painted and where the other, unalloyed, side would not be painted and subject to aggressive corrosion conditions during use. Typical end uses for such products would include certain parts of agricultural machinery, trucks, automobiles, and appliances.

The product may be produced by passing a steel substrate through a molten zinc bath containing from about 0.15 to 0.18% aluminum to form a zinc coating on both sides of the substrate, controlling the zinc coating to form a light coating of 0.20 oz/ft^2 maximum on one side and a heavy coating of 0.30 oz/ft^2 minimum on the other side, and then passing the differentially zinc coated substrate through a heated furnace so as to uniformly heat the product for a time and at a temperature sufficient to form a completely alloyed coating on the light side and to retain an essentially unalloyed zinc coating on the heavy side. A precise correlation or combination of aluminum content of the zinc coating bath, coating weight for each side and heat treatment is required to obtain such product.

Figure 3.3 generally illustrates combinations of aluminum content and the heat-treatment time and temperatures required to obtain complete alloying of the zinc coating. The dashed line represents the interaction of 0.13% aluminum and a coating weight of 0.18 oz/ft^2, the dotted line represents 0.18% aluminum and a coating weight of 0.18 oz/ft^2, and the solid line represents 0.19% aluminum and a 0.10 oz/ft^2 coating weight.

As may be seen, for a constant temperature, increasing aluminum contents leads to increases in alloying time. As may be further seen, an aluminum content of 0.17% is useful in combination with commercially obtainable heat-treatment times and temperatures for thicknesses within those of the process. It is preferred to maintain the aluminum content of the molten zinc bath from 0.16 to 0.17% to further ensure that complete iron-zinc alloying of the light coating side will occur at commercially feasible times and temperatures. A 0.19% aluminum content, even at a relatively light coating weight of 0.10 oz/ft^2, retards alloying to an extent that an unacceptable processing time penalty is incurred.

Figure 3.3: Temperature vs Alloying Time

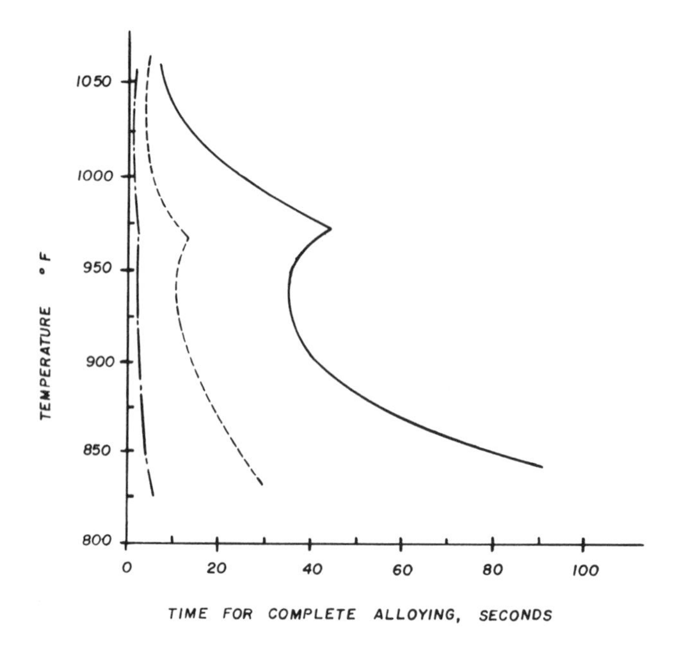

Source: U.S. Patent 4,104,088

The heat-treatment step is performed following the coating control or adjustment step that results in the differentially zinc-coated substrate. Furnace temperatures on the order of 1000°F maximum for a maximum time of about ten seconds can be utilized to produce the desired product by passing the differentially coated substrate through a continuous type furnace. This procedure subjects the moving substrate to a uniform heat treating temperature with respect to each side of the substrate. It is preferred to pass the strip through a furnace heated between about 900° and 950°F for a time on the order of ten seconds to twenty seconds because higher temperatures tend to promote a rough surface on the alloyed side characterized by sporadic nucleation of iron-zinc alloy bursts or sprays which enlarge disproportionately at the expense of the surrounding unalloyed zinc.

Following heat-treatment, the coated steel strip may be optionally temper rolled to produce a product having drawing properties equivalent to Aluminum Killed Drawing Quality steel strip. The term "Aluminum Killed Drawing Quality" is defined in a manual on carbon sheet steel published by the American Iron and Steel Institute (April 1974). Such product is highly desirable for automotive applications involving forming and one-sided painting operations.

Zinc-Aluminum Alloy

H.H. Lee; U.S. Patent 4,128,676; December 5, 1978; assigned to Inland Steel Company describes a process whereby a ferrous metal strip is continuously hot-dip coated with a zinc-aluminum alloy by immersing the metal strip in a hot-dip coating bath containing between about 0.2 and 17 wt % aluminum, between about 0.02 and 0.15 wt % lead, and between about 0.03 and about 0.15 wt % magnesium with the balance essentially zinc.

In a further embodiment the hot-dip alloy coating can also contain between 0.1 and 0.3 wt % copper. The resulting hot-dip zinc-aluminum alloy coatings when applied to a ferrous metal strip exhibit good resistance to intergranular corrosion and blistering when exposed to a high humidity atmosphere and form smooth surface coatings which have good formability both in the "as coated" state and after prolonged storage in a high humidity atmosphere.

Formation of Iron Oxide Layer

A.F. Gibson and M.B. Pierson; U.S. Patent 4,123,292; October 31, 1978; assigned to Armco Steel Corporation describe a method of preparing the surfaces of steel strip and sheet stock for fluxless hot-dip coating with molten metal, which comprises the steps of passing the stock through a first heating section under conditions which form a visible iron oxide layer on the stock surfaces within the color range of dark straw through blue, continuing the heating of the stock in a second heating section isolated from the first heating zone in an atmosphere containing less than 5% hydrogen by volume, thereby preserving the oxide layer, and cooling the stock approximately to the temperature of the molten coating metal in a cooling zone isolated from the preceding heating zones, the cooling zone containing a reducing atmosphere comprising at least 10% hydrogen by volume, in order to reduce the oxide layer completely to a metallic iron surface wettable by the coating metal.

The temperature to which the stock is heated in the successive heating sections is not critical as long as the formation of a thick oxide layer is avoided. In general the temperatures may be about 370° to 485°C in the oxidizing furnace and about 735° to 925°C in the further heating zone.

The process has particular utility in the coating of carbon steels, low carbon rimmed and aluminum killed steels and low alloy steels, with aluminum and zinc.

Sulfur/Oxygen Film

J.L. Arnold and F.C. Dunbar; U.S. Patent 4,140,552; February 20, 1979 and J.L. Arnold, F.C. Dunbar, A.F. Gibson and M.B. Pierson; U.S. Patent 4,123,291; October 31, 1978; both assigned to Armco Steel Corporation describe a method of preparing the surfaces of aluminum-killed and low alloy strip and sheet material for fluxless hot-dip metallic coating, which comprises passing the material through a furnace heated by direct combustion therein of air with gaseous fuel containing sulfur compounds ranging from about 5 to about 1,600 grains of sulfur per 100 cubic feet of fuel to produce an atmosphere of gaseous products of combustion including sulfur and from about 6% by volume free oxygen to about 7% by volume excess combustibles in the form of carbon monoxide and hydrogen, in which atmosphere the material is heated; to form a sulfur and oxygen rich film on the surfaces; passing the material into a further heating section wherein the material is brought to a maximum temperature of about 1100° to about 1700°F (593° to 927°C) in a reducing atmosphere containing at least about 10% hydrogen by volume; passing the material into a cooling section having an atmosphere containing at least 10% hydrogen by volume and the balance essentially nitrogen wherein the sulfur and oxygen rich film is reduced; and cooling the material approximately to the temperature of the molten coating metal bath.

Exemplary coating metals include zinc, zinc alloys, aluminum, aluminum alloys and terne. The coating process may be any of the conventional continuous operations currently used.

Galvanization of Selected Areas

Galvanized steel has been proposed to be used in the automotive industry to provide improved corrosion resistance for specific components. Unfortunately, the gloss of painted zinc surfaces and mild steel are difficult to match.

It has, therefore, been proposed to provide the zinc coating on the intended inner surface only of the steel strip, thereby allowing painting of the exterior steel metal surface in conventional manner.

R.F. Hunter, S.R. Koprich, R.G. Baird-Kerr and D.S. Sakai; U.S. Patent 4,101,345 July 18, 1978; assigned to The Steel Company of Canada, Limited, Canada describe a process whereby there is provided on the mild steel surface, which is not to be galvanized, a readily chemically strippable coating of chemically hydrated compounds nonwetting in molten zinc. The chemically hydrated compound results from a spontaneous reaction of the ferrous metal with inorganic acids.

By forming the coating on the mild steel surface, prior to galvanizing, zinc is prevented from being adhered to the coated surface. The coating may be removed immediately after galvanizing or may be left in contact with the mild steel surface during storage for later removal prior to finishing of the surface.

The chemical form of the coating depends on the nature of the aqueous treatment solution used. Generally, the treatment solution includes inorganic acids and treatment therewith usually is carried out under oxidizing conditions. The coating usually is composed of hydrated iron compounds, usually iron hydroxides, and hydrated iron salts of the acids used. Other salts may be present in the coating depending upon the cations present in the treating solution.

Examples of inorganic acids which may be used are solutions of phosphates, permanganates, chromates, molybdates or silicates. Owing to their ready availability and effectiveness, phosphoric acid-containing treating solutions are preferred, generally containing about 5 to about 50 g/ℓ of total phosphate.

The formation of the preferred coating layer usually is achieved by contacting the surface of the steel on which the coating is to be formed with a phosphoric acid and sodium phosphate solution having an acid pH at a temperature above about 140°F.

Oxidizing agents, such as nitrites and chlorates, may be added to the treating solution to accelerate the rate of coating deposition.

The treatment solution used preferably is a diluted form of a commercially available phosphate solution concentrate, such as Bonderite, Fosbond, Granodine or Enthox.

The contact time depends on the weight of the coating desired, the mode of contact used and in any event is very short. The coating usually has a weight of less than about 50 mg/ft^2 of surface, preferably in the range of about 20 to 40 mg/ft^2. Coating weights of this order of magnitude have been found to be satisfactory in preventing wetting of the coated surface by molten zinc in the galvanizing step and hence, in preventing adherence of zinc to the coated surface. Coating weights of this order of magnitude may be achieved by contact times of only 1 to 2 seconds at about 150°F between the treatment solution and the mild steel surface, using spray or dip application.

Where roll application is used, a more acidic solution may be required to achieve the required coating weight.

The coating layer which is provided in the preferred embodiment is a composite of a hydrated iron phosphate and iron hydroxide, such as a composite of $Fe_3(PO_4)_2 \cdot 8H_2O$ and $Fe(OH)_3$, particularly in the proportions of about 60% of the iron phosphate and 40% of the ferric hydroxide.

The coating layer may be removed from the galvanized strip after conventional adhesion of zinc to the uncoated surface and cooling to provide a clean steel surface indistinguishable from the initial mild steel surface and to which paint may be applied by conventional automobile painting techniques.

Removal of the preferred composite layer is achieved in rapid manner by mild acid treatment thereof. One preferred removal operation involves contacting the coating layer with dilute hydrochloric acid, such as about 1 to about 10% HCl, preferably about 3 to 4% HCl, at an elevated temperature in the range of about 110° to about 180°F, preferably about 150° to about 160°F. One beneficial side effect of the acid removal treatment is an apparent smoothing of the mild steel surface, thereby providing an improved surface for painting.

The ready and rapid formation and removal of the coating layer and its effectiveness in preventing adhesion of zinc to treated surfaces allows incorporation of the treatment and removal operations into a conventional continuous steel strip galvanizing operation without the necessity of expensive equipment, time-consuming operations, or significant disruption to the normal operating parameters of the operation.

One-Sided Coating Process

L.L. Franks and D.W. Gomersall; U.S. Patent 4,120,997; October 17, 1978; assigned to Inland Steel Company have found that instead of a diffusion bond forming between the surface of a ferrous metal and a thin zinc coating which is tightly adherent to the surface of the ferrous metal sheet so that the removal thereof is difficult, as would be expected, when a very thin uniform zinc film or metallic zinc coating is formed directly on the smooth surface of a ferrous metal sheet and thereafter completely converted by controlled heating into a zinc-iron intermetallic layer, the thin intermetallic layer formed is so brittle that it can be readily fractured and completely removed by applying a mild mechanical abrading force to the surface of the zinc-iron intermetallic layer, leaving the one side of the strip completely free of zinc and in condition for satisfactorily painting, enameling or welding while the protective zinc coating on the opposite side of the sheet remains unimpaired and in its as-coated condition.

In the preferred embodiment of the process the strip is differentially hot-dip coated so as to provide a protective zinc coating on one side having a thickness sufficient to impart the required degree of protection against corrosion and having on the other side the uniform thin zinc film or coating. Thereafter, and preferably before the coating solidifies, the side of the strip which has the thin zinc film thereon is exposed to a gas burner or other suitable source of heat which applies heat to only the thin zinc coated side of the strip and causes the thin zinc coating or film to be completely converted into a uniform zinc-iron intermetallic layer.

The thin zinc-iron intermetallic layer thus formed is thereafter subjected to mechanical brushing or otherwise abraded to effect the complete removal of the intermetallic layer from the surface of the strip, leaving a clean ferrous metal surface on one side of the strip and a protective zinc coating on the opposite side of the strip.

In hot-dip coating the ferrous metal strip it has been found desirable to effect removal of the excess zinc coating material to provide the ultrathin coating or film of zinc on one side by impinging gas jets onto the one side of the hot-dip coated strip as it is withdrawn from the coating bath, since it is not presently practical to obtain the desired ultrathin hot-dip zinc coating or film by means of conventional coating-weight control rolls.

Coatings of Particulate Metallic Zinc

G.D. McLeod; U.S. Patent 4,110,117; August 29, 1978; assigned to Mobil Oil Corporation describes a coating composition for ferrous material which contains: metallic zinc in particulate form, e.g., zinc dust or zinc flake; a polyol silicate; a residual polyol or polyol polymer; a coinhibitive protective (against red rust) pigment which protects in synergistic combination with the zinc, the inhibitive pigment being selected from a chromate composition (selected from an alkaline earth chromate, a comixture of zinc chromate and one or more of the finely divided silicates, carbonates, oxides, hydroxides and sparingly soluble compounds of magnesium and calcium, a comixture selected from a metal chromate compound, the metal being not lower than iron in the electrochemical series) plus a sparingly soluble metal compound filler which metals are not lower than iron in the electrochemical series.

The coating may be used advantageously without containing zinc if it is used next to a zinc coating, either under or over the zinc coating.

The polyol silicate in the coating composition is a solvent-soluble, polyol silicate ester-exchanged reaction product of: (1) an organic silicate consisting essentially of orthosilicates or siloxanes thereof having ester-exchangeable groups of 1 to 6 carbon atoms selected from the group consisting of alkyl, hydroxyalkyl, alkoxyalkyl, and hydroxyalkoxyalkyl, siloxanes thereof, and their mixtures; and (2) an aliphatic polyol which is ester-exchangeable with (1). The residual polyol or polyol polymer may be loosely reacted with the silicate in such a manner as to not cause gelation which occurs if too many OH groups on the polyol over-crosslink and polymerize the silicate; or the polyol polymer may be unreacted with the silicate.

The coating composition contains, based on the composition without the liquid organic vehicle, between about 4 and 95% by weight of the metallic zinc, between about 1 and about 80% by weight of the inhibitive or coinhibitive pigment, (the chromates being preferably lower than about 15%) and between about 1.5 and about 70 weight percent of the aliphatic polyol silicate and up to 90% of the residual polyol polymer.

The self-curing coating compositions of the process are direct substitutes for, but with improvements over, galvanized metal with the added advantage that the coating can be applied by dip, brush roll, airless or air spray and at much thinner thicknesses than have been previously possible for adequate protection by using conventional known inorganic zinc, or zinc rich coatings, and yet have highly unexpected long galvanic protection.

Example 1: A mixture of 172.4 lb of condensed ethyl silicate, containing 95% monomer and 5% dimer and having a silica (SiO_2) content of 28.4% by weight, and 87.9 lb of ethylene glycol was heated to reflux (95°C) in the presence of 8 ml of 20% aqueous sulfuric acid in a glass-lined reactor. After refluxing, the mix was cooled to 90°C and then 110 lb of toluene and 14.1 lb of methanol are added, followed by the addition of 19 lb of polyvinyl butyral while stirring. The polyvinyl butyral had a hydroxyl content expressed as polyvinyl alcohol of 9 to 13%, and an average molecular weight of 36,000, and was known as Butvar B 79. The reactants were again heated to 80°C and cooled to provide polyol silicate B.

One hundred weight parts of polyol silicate B, 4 weight parts of strontium chromate (jet milled), 50 wt parts of zinc flake (minus 325 mesh and made by ball milling zinc dust in aromatic solvent dispersion), 100 wt parts of toluene and 5 wt parts of mixed nitropropane solvent were mixed to form a coating composition.

This coating composition was stable for several months without hard settling yet gave a good hard adhesive coating which had exceptional life in the ASTM B117 salt fog test even at a 0.2 mil, dry-film thickness and provided excellent weld-thru properties. In addition, exposure (south with tipping at a 45° angle) to atmospheric conditions (the Adrian, Michigan atmosphere) for 10 months produced no rusting of any kind to a 0.1 mil dry-film on scored, cold-rolled clean steel.

Example 2: Using polyol silicate B, a series of coating compositions were prepared identical to Example 1, except that 6, 8, 10, 12, 15, 20, 40 wt parts of strontium chromate were used instead of the 4 wt parts used in Example 1. The compositions provided coating with excellent rust prevention, particularly as thin films for which the higher chromate content is preferred. In another series the zinc flake content was doubled with excellent results. Much less zinc flake is necessary than similar coatings using zinc dust. If the strontium chromate was left out of the coating composition, rusting occurred rapidly in ASTM B117 salt fog test; the coating failed in as little as 2 days when it was 0.05 to 0.1 mil dry-film thickness on clean cold-rolled steel. The composition, as a coating of such thickness, protected for more than 2 months if the strontium chromate was present.

COATING OF SUPERALLOYS

Diffusion Bonding of Superalloys

The superalloys are recognized as those alloys, usually having their basis in nickel, cobalt or iron, or some combination thereof, exhibiting good high temperature strength and oxidation resistance in environments such as gas turbine engines. Usually, these alloys also contain substantial quantities of chromium and other elements such as aluminum, titanium and the refractory metals.

It is frequently desirable to make certain gas turbine engine components by joining easily fabricable segments together into the desired configurations. However, the limited weldability of many of these superalloys has severely limited the applicability of conventional joining techniques, such as fusion welding, in the production of structural hardware. Further, many components because of their design are simply not adapted to the utilization of fusion welding. Brazing, while offering a number of advantages over fusion welding, has very limited application because of the penalties associated with the relatively low strengths and low melting points of typical brazed joints.

One key element in the transient liquid phase diffusion bonding technique is the provision between the surfaces to be joined of a thin alloy interlayer. This interlayer melts at a temperature below the melting temperature of the materials being joined, and through diffusion, solidifies at the joining temperature to form a bond. The composition of the interlayer preferably should be tailored to the alloys being joined, particularly with respect to the inclusion therein of those elements whose presence is required in the finished bond area and whose solid state diffusion rates are slow. It is also desirable to exclude from the interlayer alloy those elements which may adversely affect the bonding process or the quality of the finished joint. A melting point depressant (usually boron) is added to reduce the melting point of the interlayer to the desired point.

Since the amount of melting point depressant (boron) added to an interlayer to allow it to sufficiently melt at the bonding temperature also renders it extremely brittle; and therefore, unrollable in homogeneous form, other methods of applying such interlayers have been devised.

D.S. Duvall and D.F. Paulonis; U.S. Patent 4,122,992; October 31, 1978; assigned to United Technologies Corporation describe the production of an interlayer in lamellar form. Each of the lamellae or parts of the interlayer differ significantly in chemistry from the adjacent lamellae. Those lamellae provided in foil are free from those combinations of elements which cause brittle phases and other problems, thus the lamellar interlayer material is relatively ductile and may be fabricated by rolling. The overall composition of the interlayer (excluding the melting point depressant) is adjusted to be close to the composition of the materials being joined.

This concept may be employed in several different ways including the following:

(A) Use of two or more separate foils, produced by rolling, with adjacent foils having distinct chemical compositions. These foils may be stacked in sandwich form to produce an interlayer of the desired thickness. Even though only two distinct compositions are required to produce the desired interlayer chemistry, it may be desirable to use several sets of the two foils to reduce diffusion distances and shorten the bonding cycle, the time required to produce a homogeneous bond, especially if a thick interlayer is desired.

(B) Use of an interlayer is described in (A) above wherein the separate ductile foils have been bonded together, for example by corolling, to produce an integral lamellar interlayer material.

(C) Use of at least one rolled ductile interlayer foil in combination with at least one directly deposited layer on at least one of the surfaces to be joined. The layer would usually be metallic but might contain nonmetallic elements such as boron. An example might be one in which the reaction of chromium and carbon was a problem, in which case a layer of chromium could be directly applied to the surfaces to be bonded (plating, vapor deposition, etc.) with a wrought ductile interlayer foil containing carbon. Such a directly deposited layer need not be as ductile as the underlying foil, provided that the directly deposited layer were sufficiently adherent. That is to say, if the composite interlayer were deformed, cracking of the surface layer is not deleterious, so long as the surface layer does not spall off the interlayer surface.

(D) Use of one or more ductile rolled interlayers with a directly applied layer on at least one surface of the rolled interlayer. The layer would usually be metallic, but might contain nonmetallic elements such as boron. For example, an electroless plating scheme could be used to apply a nickel-boron alloy to the two surfaces of the interlayer, and the interlayer could be a (ductile) nickel-chromium-aluminum alloy. The same comments made above with reference to ductility of surface layers also apply to this embodiment.

In the above schemes boron would normally be used as the melting point depressant, although other depressants might be used. The boron could be applied by conventional boronizing; by vapor deposition, either pure boron or a boron metallic mixture; or by electroless plating, for example, of a nickel-boron alloy or nickel-cobalt-boron alloy. Other elements include aluminum which is desirable for precipitation strengthening and might be supplied in pure form, as part of a ductile alloy, or by various vapor deposition methods. Chromium is desirable since it lowers the melting point of some alloy interlayers, and because it increases the oxidation and corrosion-resistance of the bond region; chromium might be applied as part of a ductile alloy or by plating or vapor deposition.

Various refractory elements such as W and Mo, and carbide formers, such as Zr and Hf are desirable for high strength and creep-resistance; these materials may also be applied as part of a ductile alloy or by plating and/or vapor deposition. In general these elements (except boron) would be desirable in substantially the same concentrations as in the base alloy, and sufficient melting point depressants would be used to depress the melting point as desired.

Coating of Nickel-Base Superalloys

J.R. Rairden, III; U.S. Patent 4,101,715; July 18, 1978; assigned to General Electric Company describes a high temperature oxidation and corrosion resistant coated nickel-base superalloy article comprising (a) a nickel-base superalloy article of manufacture; and (b) a first CoCrAl(Y) coating composition consisting essentially of, on a weight basis, approximately 26 to 32% chromium, 3 to 9% aluminum, and 0 to 1% yttrium, the rare earth elements, platinum or rhodium, and the balance nickel.

The CoCrAl(Y) coatings can be applied to the nickel-base superalloys by means such as physical or chemical vapor deposition, or any other means well-known to those skilled in the art for the application of CoCrAl(Y).

In general, the CoCrAl(Y) coated nickel-base superalloys can have any coating thickness sufficient to give a desired oxidation and corrosion resistance. Generally economic and effective coating thicknesses are 1 to 20 mils for most commercial applications. In preferred embodiments, where electron-beam techniques are employed, the coating thicknesses range from 1 to 5 mils and where plasma flame spray techniques are employed the coating thicknesses range from 3 to 10 mils.

In another preferred embodiment where an aluminide overcoating is employed, the aluminide overcoating, including any duplex heat treatment where the aluminide overcoating is heated for periods of time of from 30 or 60 to 120 minutes at elevated temperatures of 850° to 1200°F in air, argon, etc., for the purpose of diffusing aluminum into the CoCrAl(Y) coating—the aluminide process is carried out in a manner which limits the aluminum penetration into the CoCrAl(Y) coating to a distance no nearer than a 0.5 mil measured from the interface of the nickel-base superalloy and the CoCrAl(Y) coating.

This aluminide diffusion penetration limitation is essential to the integrity of the CoCrAl(Y) nickel-base superalloy interface since an increase in the aluminum content of the CoCrAl(Y) coating to levels of 10% or more deleteriously affects the integrity of the coating composition.

Coating for Carbide-Reinforced Superalloys

In a similar process *M.R. Jackson and J.R. Rairden, III; U.S. Patent 4,117,179; September 26, 1978; assigned to General Electric Company* describe an article of manufacture having improved high temperature oxidation and corrosion resistance comprising: (a) a superalloy substrate containing a carbide reinforcing phase; and (b) a coating consisting of chromium, aluminum, carbon, at least one element selected from iron, cobalt or nickel and optionally an element selected from yttrium or rare earth elements. Another embodiment comprises an aluminized overcoating of the coated superalloy.

Broadly, any of the superalloy compositions included within the "Compilation of Chemical Compositions and Rupture Strengths of Superalloys" described in the ASTM data series publication DS9E, which include carbon within the alloy and rely on carbides for at least a portion of their reinforcing strengths, e.g., (1) carbide reinforcement of grain boundaries in (a) monocarbide form, commonly referred to as MC, and (b) chromium carbide forms, commonly referred to as $M_{23}C_6$ and M_7C_3, (2) refractory metal carbides, etc., in platelet or fiber form strengthening grain interiors, aligned or nonaligned in accordance with the method of casting using conventional or directional solidification casting techniques, are included within the scope of the process.

The coating compositions can be described by the formulas:

MCaAlC or MCrAlCY,

in which M is base metal element, e.g., iron, cobalt or nickel. Any amount of base metal element, chromium, aluminum, and optionally yttrium or a rare earth element can be employed in accordance with the amounts well-known to those skilled in the art with regard to oxidation and corrosion resistant coatings containing the aforesaid elements subject to the proviso that the coatings contain an amount of carbon (1) sufficient to saturate the solid state phases of the coating composition; (2) sufficient to essentially equilibrate the chemical potential of carbon in the coating with that in the substrate with minimum interaction; and (3) insufficient to form substantial quantities of carbides in the coating composition. The function of the carbon in the coating is to avoid denudation of the carbide reinforcement in the substrate which has been found to occur very rapidly at service temperatures equal to or greater than 1100°C, during periods of time in the order of magnitude of 1 to 3 hours. Denudation will occur at lower temperatures over longer time exposures.

Preferred carbon stabilized MCrAlY coatings are of the compositions in wt % set out below. The preferred aluminum content depends strongly on whether a duplex aluminizing treatment is to be given to the coated superalloy substrate.

Ingredients	General	Preferred	More Preferred	Most Preferred
Chromium	10-50	10-30	15-25	19-21
Aluminum	0-20	2-15	4-11	4-11
Carbon	0.01-0.5	0.01-0.2	0.05-0.15	0.05-0.15
Yttrium	0-1.5	0-1.5	0-1.5	0.05-0.25
Iron, Cobalt, Nickel	..................Balance..................			

In general, the carbon saturated MCrAlY coatings can be applied by any means such as (1) Physical Vapor Deposition (PVD) (subject to the proviso that the carbon be deposited from a separate carbon source since carbon, which has a very low vapor pressure, if contained in the MCrAlY melt source would not be transferred to the superalloy substrate); (2) Chemical Vapor Deposition (CVD) wherein organometallic compounds are employed wherein during decomposition of the organometallic compounds, the carbon residue incorporated into the coating is present in amounts sufficient to saturate all phases of the coating; and (3) carburization, wherein the MCrAlY coating is saturated with carbon by pack carburizing or gas carburizing the PVD coating in an atmosphere containing carbon such as an atmosphere of carbon monoxide or carbon dioxide, etc.

A preferred method of preparing the coated superalloy substrates employs a flame spraying procedure wherein an alloy wire or powder of a carbon saturated MCrAlY composition is deposited on a superalloy surface. Flame spraying or arc plasma spray deposition involves projecting liquid droplets onto a superalloy substrate by means of a high velocity gas stream. To minimize the oxygen content of the coating, deposition is often done in an inert atmosphere, such as argon, or in a vacuum.

Example: Pins of NiTaC-13 were electrodischarge machined from directionally solidified NiTaC-13 ingots which had been melted with a radio frequency graphite susceptor system and solidified at 0.635 cm per hr. Prior to deposition of the coating the pin specimens were centerless ground and lightly abraded with alumina powder. The NiTaC-13 pin samples were 4.4 cm long and 0.25 cm in diameter. The TaC fiber direction was along the axis of the pin specimens.

Ingots of carbon-containing and noncarbon-containing MCrAlY coating source alloys were prepared by induction melting high-purity metals in a low-pressure, nonoxidizing environment with subsequent casting of the alloys in an argon atmosphere. The alloys containing carbon were hot swaged to 0.33 cm diameter wire for flame-spraying purposes. For electron beam deposition of carbon-free coatings, two 0.25 cm diameter pin specimens were mounted approximately 10 cm from the deposition source and were rotated at approximately 10 rpm during deposition of coatings. Specimens coated using flame-spraying techniques were mounted approximately 15 cm from the carbon-bearing wire spray source and were rotated at approximately 200 rpm during deposition.

The coating composition for the electron beam coating employed a nickel-20 chromium-10 aluminum-1 yttrium source which deposited a composition of nickel-20 chromium-10 aluminum approximately 0.1 yttrium coating on the superalloy substrate. The flame-spraying source alloy contained nickel-20 chromium-5 aluminum-0.1 yttrium-0.1 carbon and was used for MCrAlCY coating of the superalloy substrate. The MCrAlCY coated pins were subsequently aluminized by duplex coating techniques employing pack-aluminization in a 1% aluminum pack at 1060°C for 3 hours in dry argon. Sufficient aluminum-aluminum oxide (Al_2O_3) mixed powder was used to produce approximately 6 mg/cm^2 of aluminum deposition during the pack cementation process.

Flame-Sprayed, High Energy Milled Powder

In another similar process, *H.H. Hirsch and J.R. Rairden, III; U.S. Patent 4,101,713; July 18, 1978; assigned to General Electric Company* describe a flame-sprayed high energy milled coated article of manufacture having improved high temperature oxidation and corrosion resistance comprising: (a) superalloy substrate; and (b) a coating consisting of chromium and at least one element selected from iron, cobalt or nickel. Optionally, the coating can contain other elements, e.g., aluminum, carbon, yttrium, or, the other rare earth elements, etc.

Especially useful substrates are superalloys which include carbon within the alloy and rely on carbides for at least a portion of their reinforcing strengths.

The coating compositions can be generically described by the formulas:

MCr, MCrAl, MCrAlY or MCrAlCY,

in which M is the base metal element, e.g., iron, cobalt or nickel; Cr represents chromium; Al represents aluminum; C represents carbon; Y represents yttrium and the other rare earth elements.

Some preferred coating compositions include the following, set out on a weight percentage basis.

	Coating Compositions									
Ingredients	1	2	3	4	5	6*	7*	8*	9	10
Cr	10-50	10-30	15-25	19-21	10-30	10-50	10-30	10-30	20-50	–
Al	1-20	2-15	4-11	4-11	0-20	0-20	0-20	0-20	–	5-20
Ta	–	–	–	–	1-10	–	–	–	–	–
Pt	0-10	–	–	–	–	–	–	–	–	–
Hf	0-10	–	–	–	–	–	–	–	–	–
Y	0-1.5	0-1.5	0-1.5	0-1.5	–	–	–	–	–	–

(continued)

Ingredients	Coating Compositions 1	2	3	4	5	6*	7*	8*	9	10
C	0–0.5	0–0.2	0–0.15	0–0.15	–	–	–	–	–	–
ThO_2	–	–	–	–	–	–	–	0.1–5.0	–	–
Al_2O_3	–	–	–	–	–	–	0.1–5.0	–	–	–
Y_2O_3	–	–	–	–	–	0.1–5.0	–	–	–	–
Fe/Co or Ni	Balance									

*Dispersion strengthened, oxidation resistant alloys.

The uniqueness of the process is related to the fact that:

(a) high energy milled powder, although not completely alloyed on a submicroscopic scale and which essentially contains all of the alloy constituents of a coating, is flame-sprayed on a superalloy substrate;

(b) flame-spraying of the powder releases exothermic heat of reaction as intermetallic alloys form during flame-spraying;

(c) flame-spraying of the powder releases via exothermic heat at least a portion of the work mechanically introduced by compressive forces during attriting of the powders; and

(d) flame-spraying of the powder substantially homogeneously disperses a hard phase or dispersoid in the coatings, when the coating powder contains elements that form dispersions under high energy milling conditions.

Ingot Containing Reactive Element

For adequate protection of some alloys exposed to heat and corrosive gases, such as some of the high temperature alloys in gas turbines, a protective coating for the alloys has been found to provide significantly longer life for the turbine parts. This is particularly true of turbine blades and vanes. Such a coating is regularly applied by vapor deposition of the coating alloy, the vapor produced as by electron beam vaporization, in a vacuum, of an ingot made of the coating alloy as described, for example, in U.S. Patents 3,620,815 and 3,667,421.

One problem has been the preparation of the ingot to be evaporated, especially when one of the elements of the alloy is reactive such as yttrium, hafnium or other elements of this type. The coating, to be most effective, must have such reactive element or elements present in a precise percentage and thus the ingot must have the proper amount of this reactive element.

To obtain the desired composition, the ingots are produced by investment casting the desired alloy and then machining the casting to ingot size. Such techniques are much more expensive than casting large billets and subsequent extrusion to size. With certain alloys, however, the presence of second phases associated with the reactive element of the alloy makes this latter technique impractical.

Further, some of these alloys with the reactive element cast in are very difficult to extrude into ingot size and shape for use in vapor deposition. Also in making the casting, the reactive element promotes decomposition of the crucible material by reacting with it, frequently resulting in impurities in the casting and loss of the desired quantity of the element in the finished casting. Additional quantities of the reactive element in the crucible added during melting to compensate for the lost quantity of the element merely increase the detrimental effects. The reactive element, particularly yttrium also tends to segregate in the casting and can produce compositional variations in the vapor during the coating process.

The presence of impurities in the ingot, as from the decomposition of the crucible during casting, can cause eruptions in the pool of molten alloy during vapor deposition, thereby introducing globules of molten pool alloy on the coated part. These particles can cause rejection of the finished part. Although the coating on the rejected part may be stripped and the part recoated, it is an expensive process and greatly increases the cost of the finished part.

R.C. Elam and N.E. Ulion; U.S. Patent 4,110,893; September 5, 1978; assigned to United Technologies Corporation describe a process to avoid these problems by forming a large billet as by casting the alloy minus the reactive element, placing the reactive element in a form of a wire or rod in an axially extending hole in the casting, and then extruding the billet to the desired dimension for the ingot. The diameter of the rod or wire of the reactive element provides a cross-sectional area that has the same relation to the area of the billet as the proportion of the reactive element in the alloy. The resulting ingot thus has the composition required in the coating alloy. Although the reactive element is desirably located centrally of the ingot it is not necessarily so located so long as it extends axially of the ingot and with the proper cross-sectional area.

The billet may have a cored hole to receive a rod of the reactive element. Alternatively, certain alloys which are machinable without the reactive element present may have the center hole drilled or the billet may be split, a groove machined in the matching faces, and the halves of the billet then put back together around the central rod.

The ingot to be used in vapor deposition may be made by other techniques as by forming powdered alloy into ingot shape by hot isostatic pressing. In this event, the powdered alloy, minus the reactive element, is encapsulated in a container of a material not reactive to the alloy and capable of withstanding the temperatures and pressures of the heating and pressing treatment necessary in forming the ingot. In this technique the rod of reactive metal is positioned centrally of the powdered alloy in the capsule and is thus centrally located during the forming of the ingot into shape.

The result is a simplified procedure for producing a usable ingot for vapor deposition in which the reactive element is separate from the rest of the metals of the alloy during the forming of the ingot to finished shape thereby avoiding both loss of this element and undesirable second phases in the alloy and precise control of the percentage of the reactive element in the ingot. With the desired percentage of the reactive element in the ingot, the same percentage will be in the coating vapor and in the coating on the coated article. The ingot of the process also assures a greater homogeneity in the coating and the proper quantity of the reactive element throughout the length of the ingot.

Aluminum Bearing Overlay Alloy Coatings

H.A. Beale, T.E. Strangmen, and E.W. Taylor; U.S. Patent 4,109,061; Aug. 22, 1978; assigned to United Technologies Corporation have found that the composition and structure of aluminum bearing overlay alloy coatings, such as

MCrAlY-type coatings, can be altered during deposition from a metallic vapor, such as during vacuum vapor deposition, sputtering and the like, by biasing the substrate at a small negative potential relative to ground while at least a portion of the metallic vapor is ionized. More specifically, it has been found that the aluminum content of the coating can be considerably reduced and that leader defects can be substantially eliminated when deposition occurs under such coating conditions.

The coating layer thus deposited has been found to possess a highly useful combination of properties. Namely, because of its substantial freedom from leader defects, the layer can serve as a preferred secondary substrate to reduce leader defect formation in an oxidation resistant MCrAlY layer subsequently deposited thereon under conventional coating conditions and also because of its reduced aluminum content, the layer exhibits improved ductility and can serve as a ductile barrier layer against thermal crack propagation during high temperature service. It is of significance that these compositional and structural alterations can be achieved even though a single ingot source of homogeneous chemistry is utilized to generate the vapor for coating.

In one embodiment of the process, the abovedescribed coating conditions will be established during the initial portion of an otherwise conventional coating process cycle to deposit first on the substrate a reduced aluminum, ductile and leader-free MCrAlY layer and the biasing then removed to allow deposition of an MCrAlY layer of normal composition in conventional fashion atop the first layer, the relative thicknesses of the layers being varied as desired. Not only does the first compositionally and structurally modified MCrAlY layer subsequently serve as a preferred secondary substrate during the conventional portion of the coating cycle to reduce the number and severity of leader defects in the MCrAlY layer of normal composition but it also serves as a ductile barrier layer during the high temperature service to minimize thermal crack propagation.

In another embodiment, the abovedescribed coating conditions will be established periodically for short time intervals during an otherwise conventional coating process cycle to deposit an overlay coating having a lamellar structure of alternating ductile and oxidation resistant MCrAlY layers and an overall composition corresponding essentially to the MCrAlY composition, the coating being substantially free from leader defects throughout a majority of the overall coating thickness. In this embodiment, it is preferred that the first layer be bias-deposited and that the last layer be deposited under conventional coating conditions.

In still another embodiment, the abovedescribed coating conditions are maintained throughout the entire coating cycle to produce an oxidation resistant MCrAlY layer which is essentially free from leader defects. In this embodiment, the ingot source preferably contains an increased aluminum content to counteract the reduction in aluminum caused by substrate biasing.

Chromized Nickel Aluminide Coating

N.S. Bornstein, and M.A. DeCrescente; U.S. Patent 4,142,023; February 27, 1979; assigned to United Technologies Corporation describe a method for producing a chromized aluminide coating on a nickel-based superalloy substrate in which the chromium is totally dissolved in the nickel aluminide coating and, therefore, does not exist as a separate phase or layer.

The process comprises forming a chromized nickel aluminide coating on a nickel-base superalloy substrate by a first step of enriching the surface of the superalloy substrate with chromium and a second step of growing an aluminide coating by the outward diffusion of nickel from the substrate into the surrounding pack rather than by the inward diffusion of aluminum from the pack, as has heretofore been accomplished. Certain critical limitations must be observed in the implementation of the process.

In the first step, it is essential that the nickel-base superalloy surface be enriched with chromium without the formation of a coating of either chromium or a chromium alloy with nickel. The art of chromizing superalloy substrates is well developed and, in general the pack diffusion techniques utilized by the prior art can be applied, provided that certain restraints are observed. Since it is desired to merely enrich the surface of the superalloy substrate with chromium rather than to form a chromium or a chromium alloy coating on the surface, it is necessary to limit the rate at which the chromium is diffused into the substrate.

Since the conventional chromizing techniques are based upon a diffusion of chromium into the substrate from a pack containing a source of chromium, an inert diluent and a carrier, the rate at which the chromium diffuses into the surface can be controlled by appropriate selection of the composition of the chromium-containing material. Since it is the intention of this process to prevent the formation in the substrate of the body-centered cubic crystal structure, which is characteristic of chromium-based alloys (i.e., nickel in chromium) and retain the face-centered cubic structure of nickel (i.e., chromium in nickel) the desired results can be obtained if the source of chromium in the pack is formed from a prealloyed nickel-chromium powder containing not more than the maximum solid solubility of chromium in nickel.

Thus, the upper limit on the chromium concentration in the chromium source in the pack is 40 wt % and it is preferred that the process be operated with a composition as near to this as possible in order to reduce the time involved in the chromizing step. Any of the inert diluents, such as Al_2O_3, Cr_2O_3, ZrO_2, HfO_2 and TiO_2 can be used as well as any of the known carriers such as NH_4Cl, NH_4F, NaCl, LiCl and $AlCl_3$. The time and temperature for the chromizing step can also be varied over wide limits and temperatures of 1400° to 2200°F, at times ranging from 5 to 49 hours have been found to produce the desired results. In a preferred embodiment alumina is the preferred inert diluent and ammonium chloride is the preferred carrier at 2000°F for 16 hours.

In the second step it is essential to form the aluminide coating by the outward diffusion of nickel from the substrate into the surrounding pack. Since aluminizing is based on a pack diffusion concept similar to that described with the chromizing process, it again becomes necessary to carefully control the concentration of the aluminum in the pack in order to get diffusion of nickel outwardly from the substrate into the pack, rather than the diffusion of aluminum inwardly from the pack into the substrate.

In theory this means that the activity of the aluminum in the pack must be less than that associated with the delta phase aluminide, Ni_2Al_3. In translating the theory to practice, this means that if the source of aluminum, whether it be pure aluminum or a prealloy aluminum powder such as 60% chromium, 40% aluminum, is reacted with nickel and the delta phase forms, the powder is not acceptable.

Thus, in the preferred embodiments of the process, the source of aluminum for the formation of the coating described herein is either the stoichiometric intermetallic compound, NiAl, or a powder of nickel and aluminum having the overall activity of aluminum associated with NiAl either as a uniform mixture of Ni and Al powders or a core of pure aluminum powder surrounded by a jacket of pure nickel powder.

Any of the known inert diluents and accelerators for the pack diffusion aluminizing process, which are similar to those described above with respect to the chromizing process, can be employed and temperatures and time of treatment ranging from 1 to 48 hours and temperatures of 1400° to 2200°F have produced satisfactory results. A preferred embodiment is operated at 24 hours at 2000°F.

Example: Two 3" long, 0.5" diameter erosion bars of NX1888 (74% nickel, 18% Mo, 8% Al) were treated according to the process by being packed into canisters containing 50 wt % of a prealloyed NiCr powder having a nickel-chromium weight ratio of 60/40, 48% alumina and 2% ammonium chloride and were heat-treated for 16 hours at 2000°F in an inert environment to produce a chromium-enriched surface layer. The erosion bars were then removed from this pack mixture and packed into another canister containing 50 wt % of powdered NiAl, 48% alumina and 2% ammonium chloride and heat treated at 2000°F for 16 hours to produce a 4 mil coating and 50 hours to produce a 6 mil coating. Microscopic examination of the coatings so produced showed that the aluminized coating existed as a single-phase of NiAl with no observable chromium precipitates.

CLADDING PROCESSES

Angled Strip System

J.J. Barger; U.S. Patent 4,149,060; April 10, 1979; assigned to Combustion Engineering Inc. describes a method and apparatus for strip cladding into a corner while magnetically agitating the weld deposit. The clad strip is angled away from the corner in the plane of the electrode strip, and means are provided for using the corner-forming obstruction as a pole piece for the electromagnet. The welding head can thereby be positioned to clad into the corner without interference from an obstructing pole piece or from the welding head itself.

Figure 3.4 shows an embodiment of the process that eliminates some of the problems encountered in the prior art. A work piece **64** is shown with an obstruction, flange **44**, that would ordinarily prevent cladding into the corner formed by work piece **64** and its flange **44**. The electrode-strip feeding apparatus **54**, performs function of feeding strip electrode **56** to the work piece. The ordinary application of this apparatus would involve the deposition of a flux material **58** in the weld area that would become molten flux **60** and protect the pool **62** of molten metal. An electromagnet **68** with coil **70** is provided. The molten pool **62** is formed adjacent a previously laid cladding strip **66** during the cladding operation, and a relatively smooth tie-in between strips **62** and **66** is provided by a pulsating magnetic field.

Unlike the prior art, electrode strip **56** is inclined away from obstruction **44** at an angle in the plane of electrode strip **56**. This allows the apparatus to fit into the corner, but the incline alone is not sufficient to allow for provision of a magnetic pole piece in the vicinity of the weld. According to the process, a means **52** is provided for transmitting magnetic flux into the typically ferromagnetic obstruction **44**.

Figure 3.4: Angled Strip System

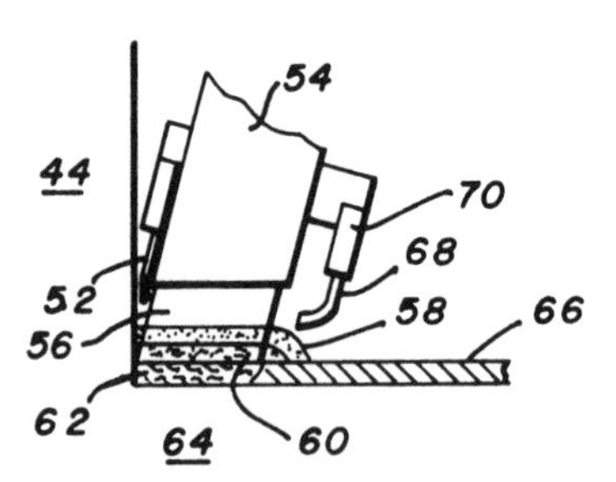

Source: U.S. Patent 4,149,060

It has been found that the transmission of flux into the obstruction in the area shown will result in the performance of obstruction **44** as a magnetic pole complimenting pole piece **68**. The pole thus formed is positioned on the side of electrode strip **56** opposite magnetic pole piece **68**, and this is the proper position for the operation of the apparatus.

Cladding of a Dished Surface

D. Mort and B.B. Hood; U.S. Patent 4,149,061; April 10, 1979; assigned to Westinghouse Electric Corp. describe an apparatus for continually cladding a dished surface which comprises a pedestal, a turntable rotatably and pivotally mounted on the pedestal, a device for rotating the turntable at varying angular velocities and a device for tilting the turntable generally through at least 90°. The apparatus also comprises a welding manipulator pivotally mounted on the turntable base and connected thereto by a parallel bar 4-bar linkage. The welding manipulator has a base portion, a boom pivotally mounted to the base and an upright member slidably connected to the boom and depending therefrom. A welding head is cooperatively associated with the upright member.

The parallel bar 4-bar linkage cooperates with the pedestal, the turntable and the base of the manipulator to keep the base of the manipulator generally horizontal as the turntable tilts. The welding manipulator further comprises a drive for moving the upright generally vertically, a drive for swinging the boom about its pivotal axis, and drives for moving the welding head horizontally in two directions at right angles to each other. The apparatus further comprises sensors responsive to the tilt angle of the turntable to position the weld head generally above a bottom dead center position of the dish and sensors responsive to the tilt of the turntable to vary the angular speed of the turntable to provide a generally constant linear speed of the dished surface relative to the weld head and a control which is capable of responding to each revolution of the turntable to change the tilt angle thereof a predetermined amount depending on the shape of the dished surface and the previous angle at which the turntable was tilted, whereby cladding is disposed on the dished surface by the weld head in successively disposed connected angular rings to continuously clad the dished surface except for a relatively small portion adjacent the center thereof.

Referring to Figure 3.5, apparatus **1** to continuously apply cladding to the dished surface **2** of the head **3a**, comprises a pedestal **4** fixed to a foundation (not shown).

Figure 3.5: Cladding of a Dished Surface

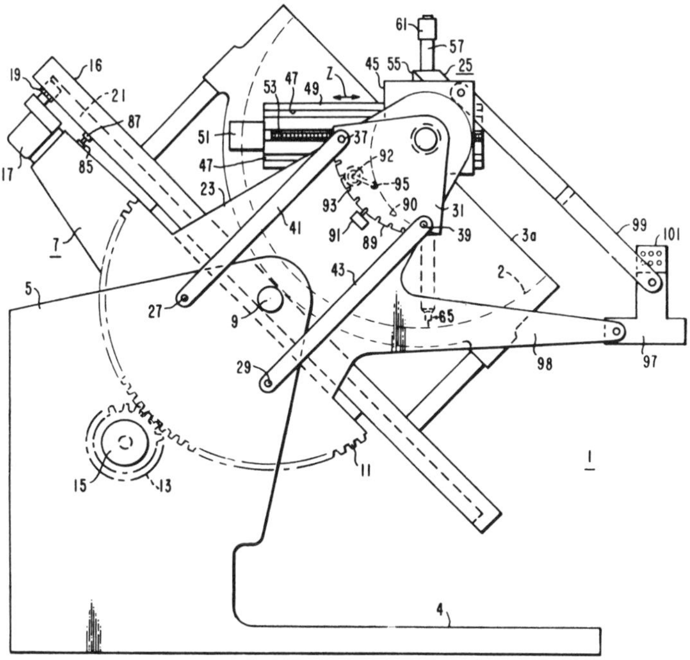

Source: U.S. Patent 4,149,061

The pedestal **4** is generally U-shaped and has two legs **5** which extend generally vertically and upwardly. A turntable assembly **7** is pivotally mounted on the pedestal **4** by trunnions **9**, which are rotatably disposed in bearing surfaces within the legs **5**. Affixed to the turntable assembly **7** is a segment of a gear **11** disposed generally parallel to and adjacent one of the legs **5**. The axis of the gear **11** is disposed coaxially with the trunnions **9**. A pinion **13** is disposed to engage the gear segment **11** and is directly connected to a reversible motor **15** or other drive means, which rotate the pinion **13** and gear segment **11** to tilt the turntable assembly **7** to any desirable position from a vertical to a generally horizontal position. The turntable base **7** shown tilts approximately 180°; however, for this process it is required that the turntable assembly **7** tilt approximately 90°.

A turntable **16** is rotatably disposed about a central axis on the turntable assembly **7**. A gear reduced variable speed motor **17** driving a pinion **19** and ring gear **21** is disposed on the turntable base **7** to rotate the turntable **16** about its central axis in this embodiment, however, other drive means well known in the art may be utilized. Being a variable speed motor, it is capable of varying the angular velocity of the turntable **16** to produce generally constant linear speed of the dished surface relative to a generally fixed position for welding.

An arm **23** extends upwardly from the turntable assembly **7** and a welding manipulator **25** is pivotally mounted on the arm **23**. The welding manipulator **25** is also connected to the pedestal **4** by a parallel bar 4-bar linkage or other means such as a chain or gear arrangement for maintaining the welding head in a generally vertical or fixed orientation as the turntable assembly **7** tilts. The parallel bar 4-bar linkage consists of two sets of parallel bars, each set consisting of two bars of identical length. The first set of bars consists of a portion of the pedestal **4** disposed between two pivot points **27** and **29** and constitutes a fixed bar of the linkage.

The other bar in the first set consists of a portion of a cam **31** disposed between the points **37** and **39**. The cam **31** is affixed to the manipulator **25** and pivotally mounted on the arm **23**. A line between the points **37** and **39** will always remain parallel to a line between the points **27** and **29**, which are fixed. The second set of bars consists of bars **41** and **43** which are pivotally connected adjacent their ends to the points **27, 37, 29** and **39**, respectively.

The welding manipulator **25** comprises a base portion **45** affixed to the cam **31** and pivotally disposed with respect to the arm **23**. Because of the cooperation of the bars of the parallel bar 4-bar linkage and the pivotal connection of the base **45** to the arm **23**, horizontal and vertical surfaces on the base **45** remain generally horizontal and vertical as the turntable assembly **7** tilts. Slidably connected to the base **45** by dovetail grooves **47** or other means for producing rectilinear motion therebetween, is a horizontal slide portion **49**. A reversible motor **51** and screw **53** are cooperatively associated with the horizontal slide **49** to move the horizontal slide portion **49** horizontally with respect to the base **45**. A boom **55** is pivotally mounted on a vertical axis on the horizontal slide portion **49**.

A reversible motor or other drive means swings the boom about its vertical axis. An upright member **57** is slidably connected to the distal end of the boom **55**. A horizontal drive motor and slide or other drive means moves the upright horizontal with respect to the boom **55** and a vertical drive motor and slide **61** or other drive means moves the upright member **57** vertically with respect to the boom **55**. The motor **51** and screw **53** move the horizontal slide **49** to provide movement of the upright **57**. A welding head **65** is disposed on the lower end of the upright **57**.

The various members forming the manipulator **25** are representative of the cooperative association of members necessary to provide motion relative to the welding head **65**, to allow the welding head **65** and boom **55** to swing out of the way for loading and unloading. The motion in any direction may be provided by a plurality of drives, particularly where it is desirable to provide a long stroke at high speed and a short accurate adjustment in the same direction. Such drives are well known in the art and it is understood that they may be utilized to provide such movements in any desired combination.

The head **3a** having a hemispherical surface is placed on the turntable assembly **7** so that it is centrally located on the turntable **16** and the flange face of the dished surface is generally parallel to the turntable **16**. The center of the major radius of the head **3a** is disposed on the central axis of the turntable **16** adjacent a horizontal line extending from the axis about which the manipulator **25** is pivotally disposed.

MISCELLANEOUS PROCESSES

Orbiting and Rotating Substrate

G.R. Scheuermann; U.S. Patent 4,122,221; October 24, 1978; assigned to Airco, Inc. describes a rotatable device for holding at least two substrates, such as turbine engine parts, whereby each substrate may also be rotated about its individual axis during a high temperature vacuum coating process.

The holder illustrated in Figure 3.6 enables each substrate to be rotated at the individual axis desired, as the holder **20** has a hollow housing **24** which is conveniently formed from two parts, walls **24a** and **24b**, held together by bolts **60**. Wall **24b** is welded to one end of a hollow shaft **26**. A quick disconnect means (not shown) is provided for attaching the other end of hollow shaft **26** to the end of main shaft **25**. An interior shaft **38** is concentrically mounted for rotation inside hollow shaft **26**. One end of interior shaft **38** extends into housing **24** through an opening in wall **24b**. The other end of interior shaft **38** has a tongue **58** which engages a shaft (not shown) mounted for rotation inside main shaft **25**. Conventional means are used for rotating the unshown shaft independently of main shaft **25**.

A stub shaft **28** is provided for mounting each substrate. In the preferred embodiment, four stubs are mounted symmetrically about the main axis of hollow shaft **26**. One end of each stub shaft **28** is provided with means to receive and to hold a substrate. Preferably, stub shaft **28** is provided with a hole **32** for receiving a pin which may be inserted to secure a substrate or a substrate adapter to the stub shaft **28**. A thrust washer **36** is placed around stub shaft **28** and between the wall **24a** and the substrate or adapter in order to assist in bearing the load of a substrate.

Each stub shaft extends into housing **24** and passes through a recess **31** in bearing **30** mounted in wall **24a**. If desired, an additional bearing may be mounted in wall **24b**. A gear **34** is mounted on each stub shaft, and a central gear **40** is fixed to interior shaft **38** so as to engage each of the gears **34**. When each stub shaft is mounted at angle, typically 10° to 45°, to the axis of the interior shaft, it is preferred that each of gears **34** and **40** be conical. The engagement of the gears is adjustable by means of a lock nut **48** and a set screw **46** which extends through the wall **24a**. The set screw positions an end bearing **42** which is mounted in recess **44** in the wall **24a** and which restrains the end of interior shaft **38**.

Figure 3.6: Orbiting and Rotating Substrate

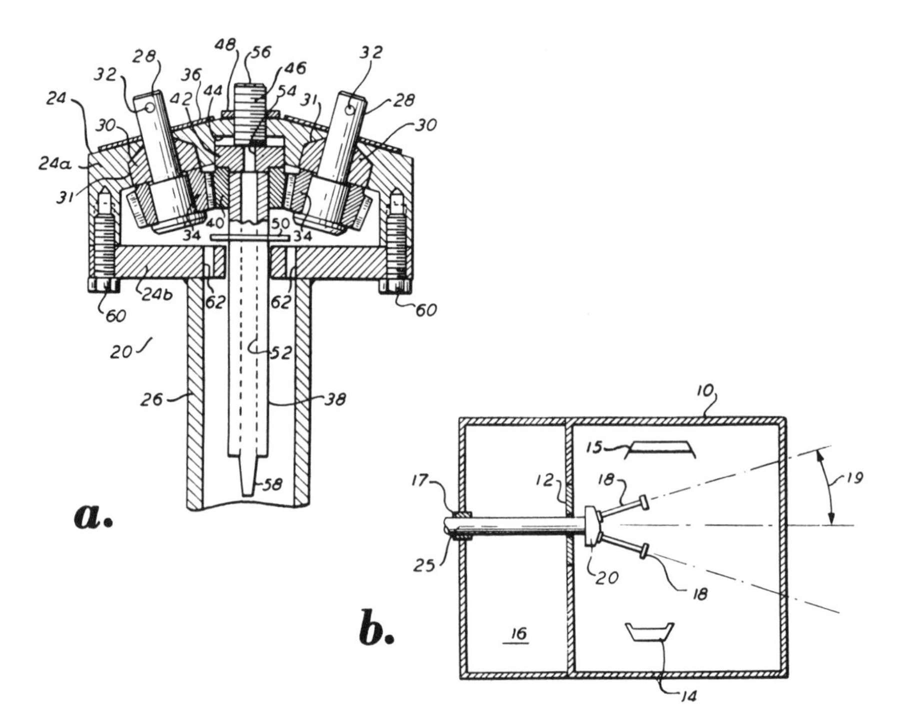

Source: U.S. Patent 4,122,221

A washer or plate **50** is preferably affixed to the interior shaft **38** inside of the housing **24** to limit the possible travel of the interior shaft **38** along its axis. If desired, interior shaft **38**, bearing **42** and set screw **46** may have a channel or hole **52**, **54**, and **56** respectively, for the insertion of a thermocouple or other sensing device along the main axis.

Preferably, housing wall **24b** is provided with a number of openings **62** which provide communication between the interior of housing **24** and the interior of hollow shaft **26**. Holes **62** insure that any gases which may be inside of head **24** may be evacuated through hollow shaft **26** rather than leading past the bearings **30** and **42** in the direction of the coating process.

Housing **24** and hollow shaft **26** may be 304 stainless steel although a material such as Hastelloy B alloy that will withstand the high-temperature environment of the coating process better, is preferred. The gears **34** and **40**, the stub shaft **28**, and the interior shaft **38** must be of a material such as Hastelloy B alloy, which has high strength at the operating temperature. Further, the gear teeth must be relatively large in order not to bend under load; a dimetral pitch of 16 is preferred. Gears of 316 stainless steel and a dimetral pitch of 24 were bent during approximately 100 hours of operation in which the housing temperatures approached 1000°C.

In order to prevent binding at both room and operating temperatures, the bearings **30** are preferably molybdenum, graphite, or other material which has a thermal expansion coefficient significantly less than that of the material of stub shaft **28**. The various parts are dimensioned such that at room temperature the shaft turns inside bearing **30**, which is stationary in the recess **31**, while at operating temperature stub shaft **28** and bearing **30** rotate together inside recess **31**. Such construction is useful because the loading forces are then distributed over a much larger bearing surface at operating temperature where the materials are more prone to deform.

The required dimensions and clearances of the parts can be readily calculated from their relative thermal expansion coefficients.

End bearing **42** is preferably molybdenum so that relative thermal expansion of the parts will tend to free rather than restrict the bearing.

In order to prevent seizing of shaft, bearing, and gear surfaces, all the parts are preferably coated with a low vapor pressure, antiseizing compound, such as molybdenum disulfide, tungsten disulfide, or boron nitride, before assembly. Molybdenum disulfide is particularly preferred because it can be applied with an aerosol spray.

The use of materials which will maintain their strength at the operating temperatures is preferred because the need for special cooling of the holder is eliminated. Water cooling of parts, especially rotating parts, is particularly disadvantageous in vacuum because of problems with leaks.

The number of stub shafts and the angles of displacement of the stub shafts **28** will depend upon the nature of the coating source and the requirements of the particular coating desired. As illustrated, the central axis of holder **20** is coaxial with the main axis of shaft **25**. However, in some cases holder **20** may be mounted at an angle provided that appropriate means are included so that hollow shaft **26** and interior shaft **38** are rotatable as desired.

In a typical application, holder **20** is placed in load lock **16** and attached to main shaft **25**. With load lock **16** open, and chamber door **12** closed, substrates **18** are mounted on the ends of stub shafts **28** and locked in place by means of pins inserted in holes **32**. The load lock **16** is closed, door **12** is opened, and the substrates and holder are moved into the vacuum chamber **10**. In one method of use, both the hollow shaft and the interior shaft **38** are rotated together thus rotating the housing and all of the substrates about the main axis.

After a period of coating, the interior shaft is rotated with respect to hollow shaft **26** so that each substrate is turned 180° about its individual axis. The coating process is continued with the hollow shaft **26** and the interior shaft rotating together as before. Thus, each substrate receives a uniform coating of material around its circumference. It should be apparent that both the hollow shaft and the interior shaft may be turned continuously but at different rates or in different directions throughout the entire process.

After the substrates are coated as desired and the holder and substrates are withdrawn into load lock **16**, the door **12** is closed, the lock is vented to atmosphere and opened so that the substrates may be removed from the stub shafts. Several load locks and substrate holders may be provided for a single coating chamber so that one set of substrates is being coated at all times while other sets of substrates are being loaded or unloaded.

Refractory Metal Chlorides Vapor

F.A. Glaski; U.S. Patent 4,138,512; February 6, 1979; assigned to the U.S. Secretary of the Army describes a process whereby a coating of refractory metals alloy, e.g., an alloy of tantalum and tungsten, is deposited on a substrate, such as the bore of a gun barrel, by reacting an alloy of the refractory metals with chlorine to form a mixture of the vapors of the refractory metal chlorides and reacting the resulting mixture of vapors with hydrogen to reduce the metal chlorides to the free metals. The alloy coating thus deposited possesses a finer grain structure and is more homogeneous and ductile than one obtained by chlorinating the tantalum and tungsten metals separately.

As shown in Figure 3.7, the apparatus consisted of a fire brick insulated, split half resistance furnace **10** standing on a support table **12** and provided with two carbon steel supports **14, 16**. The furnace contained four separately controlled Nichrome resistance heaters of the clamshell type **18, 20, 22, 24** (to permit the rapid simultaneous removal of all four heaters from the retort tube at the conclusion of the deposition run), which surrounded the entire length of an Inconel 600 retort tube **26**, 1.5" (3.8 cm) i.d. and 28" (71.1 cm) long.

The retort tube was flanged at both ends with top plate flanges of carbon steel **28, 30** and bottom plate flanges of 304 stainless steel **32, 34** and contained a retort cooling copper water channel **36** located below top flange **30**. The gun barrel or rod specimen **38** was held in the retort tube by means of a specimen holder **40.**

A carbon steel tube **42** for effecting the chlorination and reduction reactions was mounted in the top flange **28** of the retort tube. The upper end of the tube **42** contained a chlorine inlet tube **44** and a rubber seal **46**, below which copper cooling coils **48** were mounted. A lower section of the tube contained a bed of refractory metal alloy chips **50**, which was supported by an annular rim **53** and surrounded by a resistance heater **52** with a carbon steel support **54**. A hydrogen inlet tube of copper **56** was located below the bed of metal alloy chips.

Depositions on the Exterior Surface of Molybdenum Rods: Ta-W alloys were deposited on the exterior surfaces of solid molybdenum rods in the foregoing apparatus as follows:

A stream of chlorine gas was passed through a heated bed of Ta-10W alloy chips **50** consisting of 90% tantalum and 10% tungsten. The resulting gaseous mixture of Ta and W chlorides was mixed with a stream of hydrogen gas from inlet tube **56** and the total gas mixture was passed through the retort tube **26** through positive containment measures, wherein deposition of the Ta-W alloy was effected on the exterior surface of the heated molybdenum rod. The residual gases flowed from the exhaust end **58** of the retort tube **26** into the vacuum system below the tube (not shown).

Depositions on Tube Bores: The foregoing procedure using Ta-10W alloy chips was employed to deposit Ta-W alloy coating on the bore surface of chromium plated tubes of 4130 steel 22" (55.9 cm) long and 0.3" (7.62 mm) i.d. The specimen holder **40** conducted the gaseous mixture through the tube bore which was chromium plated.

Figure 3.7: Vapor Deposition Apparatus

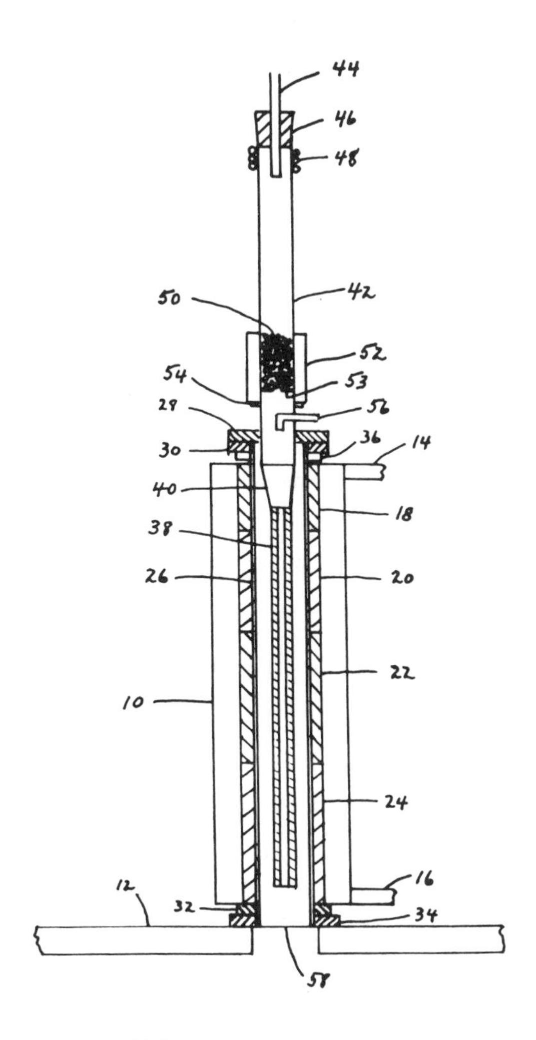

Source: U.S. Patent 4,138,512

Antifouling Coating for Aluminum

I.R. Kramer; U.S. Patent 4,130,466; December 19, 1978; assigned to the U.S. Secretary of the Navy describes a method for inhibiting the fouling of aluminum and aluminum-base alloy structures.

In applying the process to an aluminum alloy object, the following sequence of steps can advantageously be employed. The object to be treated is first cleaned and degreased, then anodized, for example, in a sulfuric acid bath to provide an oxide layer on the surface thereof greater than about 7.5 microns in thickness. Next, the object is withdrawn from the anodizing bath, rinsed in water to remove any excess acid, and then dried. This is followed by immersion of the anodized object into a molten bath containing the antifouling compound at a concentration greater than about 20% by weight. Other methods may also be effectively used to impregnate the pores of the anodized layer. For example, the impregnant may be applied by brushing, rolling, or spraying. Finally, the object is cleaned to remove any excess impregnant.

The anodizing step is advantageously carried out in an aqueous solution containing about 15 to 50% by weight of sulfuric acid at a temperature of about 20° to about 40°C. These temperature and acid concentration ranges are applicable to any aluminum or aluminum alloy, although a particular alloy may react best to a specific combination of parameters within these ranges. It has been found that, when using the sulfuric acid method, the object must be anodized for at least about 10 minutes to obtain an oxide layer greater than about 7.5 microns in thickness on the surface of the aluminum, and the subsequent impregnation step is particularly effective when the thickness of the anodic oxide layer is between about 7.5 and 25 microns.

The anodizing process produces an oxide layer having pores therein on the order of about several hundred Angstroms in diameter. In some cases, depending upon the particular aluminum alloy, the pore size may be too large to properly retain an impregnant. For these alloys it is necessary to partially seal the pores in the anodized layer by means of water, thereby reducing the pore diameter to a size that will effectively contain and retain the impregnant. Sealing by means of water includes such methods as sealing at the boiling point of pure or slightly acidic (pH 6) water and sealing in steam at temperatures above the boiling point of water.

The partial sealing step should be applied for between about 10 to 30 minutes in order to effectively reduce the diameter of the oxide layer pores. Examples of aluminum alloys which require partial sealing prior to impregnation are the 5000 to 6000 series alloys, as designated by the American Standards Association (ASA). ASA 2000 and ASA 7000 series alloys, on the other hand, may be effectively impregnated without partial sealing.

Following the partial sealing process, or following the anodizing step for alloys which do not require partial sealing, the object is dried and then immersed in a molten bath containing the antifouling compound. The temperature should be chosen so that the impregnant is very fluid but not so high as to cause decomposition of the constituents in the bath. To enable adequate impregnation of the antifouling compound into the oxide layer pores, the object should remain in the molten bath for at least about 1 minute.

Effective antifouling compounds that may be used include: tributyltin, tripropyltin esters of vinyl, vinyl maleic acid copolymer resins, polymers of tributyltin, tripropyltin esters of acrylic acids, tripropyltin esters of methacrylic acids, and copolymers with other acrylic and vinyl copolymers. Alternatively, monomers of the abovementioned resins can be impregnated into the anodized pores and polymerized in situ.

The impregnation can advantageously be carried out by incorporating one of the aforementioned antifouling compounds at a concentration of at least about 20% by weight into a molten bath of long-chain fatty acids, alcohols, or amines. Although the alcohols and amines are effective, the fatty acids are preferred because of their low cost. The long-chain compounds that are effective are those with at least about 10 carbon atoms up to about 24 carbon atoms in the chain.

In general, the life of the antifouling protection is dependent upon the rate at which the impregnant dissolves in the water. The dissolution rate, however, decreases with increasing chain length. Compounds such as sodium salts of the long-chain acids as well as dibasic acids are also effective but are more costly to use than the simple straight carbon compounds. It is contemplated that the molten bath may also contain such additives as antioxidants, chelating agents, and the like which are normally employed in small but effective amounts. Fatigue and corrosion resistance may also be imparted to the anodized aluminum object by including aliphatic compounds such as stearic acid in the molten bath along with the antifouling compound.

The antifouling coating process is completed, following impregnation, by rinsing or otherwise removing any excess impregnant from the treated object.

Magnesium Coating for Saline Environments

Jet and gas turbine engine compressor components, for example, discs and blades, are subject to corrosion in highly saline atmosphere at the air intake end of the engine and also to direct impact of abrasive particulate matter, such as coral dust. Additionally, compressor discs and blades among other components are subjected to tremendous mechanical stresses from centrifugal forces, thermal shock, vibration and other sources of stresses. Thus, corrosion can accelerate catastrophic failure, since pits and other corrosion defects can act as stress raisers.

M.F. Dean and R.L. Blize; U.S. Patent 4,125,646; November 14, 1978; assigned to Chromalloy American Corporation describe a method of protecting a metal substrate against the corrosive effects of saline, marine and other corrosive environments. The metal substrate of interest is first coated with a magnesium-reacting matrix metal selected from the group consisting of silver, copper, nickel, cobalt, cerium, silicon, tin and zinc which is capable of forming an intermetallic compound with magnesium.

Following the application of the metal coating, magnesium is then thermally diffused into the metal coating to form a sacrificial coating anodic to the metal substrate comprising at least one magnesium-containing intermetallic compound bonded to the metal substrate.

A nonmetallic layer may then be applied to the sacrificial coating as an overcoat comprising a solution of soluble silicate salt selected from the group consisting of sodium silicate, potassium silicate, lithium silicate and ethyl silicate which is dried and then cured at a temperature of about 150° to 430°C. In a preferred embodiment, a conversion coating is applied to the cured silicate layer using a solution containing phosphoric acid, chromic acid and at least one chromate- and phosphate-forming metal, such as aluminum and/or magnesium which is thermally cured (about 150° to 500°C) to provide in effect a duplex coating, that is to say, a sacrificial coating of a magnesium-containing alloy and a glassy nonmetallic overcoat.

Preferably, the sacrificial coating is produced by a magnesium pack diffusion process, by means of which magnesium is thermally diffused into the selected coating matrix. The coating is sacrificial to all steels and also to some aluminum alloys. It is corrosion resistant, oxidation resistant, abrasion resistant, substantially uniformly applicable over complex geometries and also can be deposited over a thickness range of about 0.0001" to 0.005" (i.e., from 0.1 to 5 mils).

A particularly preferred coating is the system magnesium-nickel. First, a nickel coating is applied by any suitable method, such as by electroplating, electroless plating, and the like. Electroless plating is preferred since this method enables the consistent production of a uniform nickel layer on the surface of a complex shape.

In the case of certain other elements such as silicon, cerium, etc., these can be plated using gas plating techniques, for example, by transfer to the metal substrate from a halide vapor of the metal, this method being a very well known method. One method in particular is referred to in the art as siliconizing. A still further method is a vacuum plating method from the vapor of the coating metal of interest.

Example: An AMS 6304 low alloy steel part or substrate, such as a compressor disc, is generally degreased by chemical cleaning, if necessary, and then mechanically cleaned by grit blasting with 220 mesh silicon carbide powder at a pressure of 40 psig and a distance of about 6" to 12" from the steel workpiece prior to nickel plating.

Prior to applying the nickel coating, the clean part is subjected to an activation step comprising immersing the part in a 50% by volume hydrochloric acid solution for about two minutes to activate the surface. The part is then rinsed to remove any adhering HCl residue and placed immediately into a dimethylamine borane electroless nickel plating bath of the following composition:

20 g/ℓ sulfate ($NiSO_4 \cdot 6H_2O$)
10 g/ℓ citric acid monohydrate
25 ml/ℓ conc HCl
NH_4OH, add to raise pH to 7
2.5 to 3 g/ℓ dimethylamine borane (DMAB)
0.5 to 2 mg/ℓ 2-mercaptobenzothiazole (MBT)
15 mg/ℓ sodium lauryl sulfate
Temperature, 100°F

The citric acid is employed as a complexing agent, DMAB as a reducer, the MBT as a stabilizer and the sodium lauryl sulfate as an antipitting agent.

The part is maintained in the electroless plating bath for one hour to provide a thickness of about 0.0005". After removal from the plating bath, the part is raised and oven dried at 400°F (205°C) for about 30 minutes.

In preparing the nickel-plated part for pack diffusion, the part is grit blasted with 320 mesh Al_2O_3 powder at 20 psig at a distance of from about 6" to 12" to remove the sheen from the outer surface of the nickel plate and obtain a matte finish.

Thereafter, the part together with other parts similarly prepared is packed in a steel retort containing approximately a 50–50 mix by weight of -20 to +40 mesh magnesium powder and 28 to 40 mesh Al_2O_3 (U.S. Standard Screen) that has been previously energized with approximately 3% by weight of NH_4Cl.

The retort is closed substantially airtight, except for allowing gases to escape therefrom during pack diffusion, and then placed in an oven and the temperature of the retort brought up to about 875°F (470°C) for a time of about 48 hours to produce the sacrificial coating. The temperature employed is below the melting point of magnesium.

The coated parts are removed from the retort and then subjected to oxide removal treatment by dipping in an approximate 30% by weight chromic acid solution for a time ranging up to three minutes, the parts being thereafter water rinsed and oven dried at 400°F (205°C) for 30 minutes.

The cleaned parts are then provided with a potassium silicate sealing coat by spraying the outside surface of the diffusion coated parts at ambient temperature with a 25% by volume potassium silicate solution (formed from a 29.8 Bé solution) followed by drying and oven curing at a temperature of about 400°F (205°C) for 30 minutes. The thus-treated part is then subjected to a subsequent dip in a 10% by volume potassium silicate solution (also from 29.8 Bé solution) consisting of three dips, with an air blow off after each dip to remove excess solution. This is then followed by a second spray application of the 25% potassium silicate solution.

After the latter treatment, the parts are oven cured at 400°F (205°C) for 30 minutes following which the temperature is raised slowly to 750°F (400°C) for five minutes. The 29.8 Bé potassium silicate solution has a weight ratio of SiO_2/K_2O of 2.5:1 and contains 8.3% K_2O and 20.8% SiO_2.

Following the curing of the silicate coating, a conversion coating is optionally applied by spraying three applications of an aluminum phosphate-chromate solution, with each application receiving a subsequent oven cure at 750°F (400°C) for 30 minutes. This provides a hard glass overcoat on the parts.

Refractory Lined Cylinders

A.J. Pignocco and R.H. Kachik; U.S. Patent 4,117,868; October 3, 1978; assigned to United States Steel Corporation describe a method of producing a refractory-lined cylinder by an exothermic reduction reaction, such as aluminothermic reduction.

Tubes, pipes, cylinders, and tanks lined with acid-resistant, corrosion-resistant or abrasion-resistant materials intimately bonded to the metal shell are required in many industrial applications. The shells of some tanks used in the oil industry, for example, are protected by a cured mixture of furnace cement and sand containing short asbestos fibers. The interiors of steel pipes and tanks exposed to corrosive water, salt solutions, or oils containing sulfur compounds are also often coated with cement to inhibit the attack on the steel. In cases where tanks are not subjected to high abrasive wear, but still require protection from an agressive environment, expensive stainless steels are often used in place of carbon steel.

The process consists of filling a metal tube with an exothermic reduction reaction mixture such as an aluminothermic reduction (ATR) mixture, rapidly rotating the filled tube about its longitudinal axis, initiating the ATR reaction, and continuing to rotate the tube until the reaction products have solidified. The centrifugal forces developed effect a separation of a metal phase from a slag phase, propelling the heavier metal phase toward the tube wall where it bonds metallurgically to the metal. The lighter molten slag layer, being displaced toward the center of the pipe by the metal phase, subsequently solidifies to form a continuous layer of refractory. Upon cooling, the metal pipe walls will contract to a greater extent than the ceramic liner, thereby locking the liner into the pipe. The chemical reaction involved in this method includes the very energetic reduction of oxides by such metals as aluminum, magnesium, silicon, and calcium-silicon alloys and mixtures thereof and can be generally represented by:

$$2Al + 3MeO \longrightarrow Al_2O_3 + 3Me + \Delta H$$

where ΔH represents the evolution of a large quantity of heat per mol of reductant (Al). The metal oxide (MeO) should be low in cost, readily available in a dry form and easily reduced by the reductant (Al, Mg, Si, or SiCa) to generate a substantial heat of reaction. Common iron ores, containing magnetite (Fe_3O_4) or preferably hematite (Fe_2O_3) are well suited for the process, and a stoichiometric mixture of iron ore fines and aluminum is preferred for most applications. In some applications, particularly where the pipe wall is rather thin and in danger of being burned through by the ATR reaction, it is preferable to include a small amount of alumina, typically up to about 20%, to retard the ATR reaction, and thus minimize the risk of burn-through.

The metal oxide preferably has a size at least as fine as -35 mesh, and advantageously is no finer than +200 mesh. The fuel powder should have a size at least as fine as -100 mesh, but no finer than +325 mesh.

Example 1: A 1¼" o.d. by 4" long steel pipe was packed with a stoichiometric mixture of powdered aluminum and hematite iron ore (Fe_2O_3) without cleaning or preparing the pipe in any manner. The ends of the pipe were partially closed with pipe reducer sections, thus providing a ¼" high lip to contain the molten reaction products. The pipe was then rotated at 1,500 rpm. Upon igniting the charge, the ATR reaction rapidly propagated through the charge. When the reaction was complete, and the products had solidified, rotation was stopped and the assembly was allowed to cool to ambient temperature. The pipe was then sectioned to show a fairly uniform and continuous slag layer firmly attached to the inside of the pipe.

Example 2: 115.7 g of a mixture consisting of 88.9 g of nickel oxide sinter, 21.3 g of 20 mesh aluminum and 5.5 g of powdered alumina was charged into a length of 1" steel pipe. The pipe was not cleaned or prepared in any manner. The pipe was then rotated at a speed of 2,250 rpm in a horizontal position, and the ATR reaction initiated. Rotation was continued until the reaction was complete and the reaction products solidified. The resulting ceramic alumina layer was firmly secured inside the pipe.

Example 3: 101.7 g of a mixture of 71.0 g of cobalt oxide (Co_3O_4), 21.4 g of 20 mesh aluminum and 9.3 g of powdered alumina was identically processed in a 1" steel pipe as described above for the nickel oxide. Again a secure ceramic alumina coating resulted within the pipe.

Prevention of Stress Corrosion Cracking

K. Yamaguchi and K. Kawakami; U.S. Patent 4,135,013; January 16, 1979; assigned to Nippon Mining Co., Ltd., Japan describe a method for the prevention of stress corrosion cracking of machinery comprising coating the metal part of the machinery which is in contact with corrosive materials with a metal more base than that of the machinery metal, or with an alloy essentially consisting of the base metal in a corrosive environment with extremely small amounts of water.

Examples of the coating metals are: aluminum, magnesium, zinc, or an alloy containing one or more thereof. As used herein, a coating metal is more base than that of the foundation metal if it is more electrochemically positive, i.e., in a local cell, the more base metal will give up positive ions to the electrolyte.

According to the process, a corrosion resistant base metal need only be applied to the part of machinery which actually requires the sacrifice anode so that the stress corrosion cracking which might lead to serious trouble may easily and economically be prevented.

High-Temperature, Abrasion-Resistant Ferroboron Coating

R.H. Kachik and A.J. Pignocco; U.S. Patent 4,131,473; December 26, 1978; assigned to United States Steel Corporation describe a method of facing crash decks and other objects subject to extreme wear from high temperature, high abrasion uses, by metallurgically bonding a unique ferroboron hard facing to a ferrous metal substrate by a rapid and relatively inexpensive aluminothermic reduction (ATR) deposition method. It was found that when from about 2 to about 8 wt % of boron is present in the final outer surface of the composite that surface is hard and wear resistant. Unexpectedly, the wear resistance remained good at elevated temperatures, i.e., 1000° to 1600°F.

As shown in Figure 3.8, a steel substrate **10** is placed on a bed of sand **12**, and the level of the sand is brought up to the level of the upper surface of the substrate.

Figure 3.8: Sectional View of Refractory Lined Perimeter

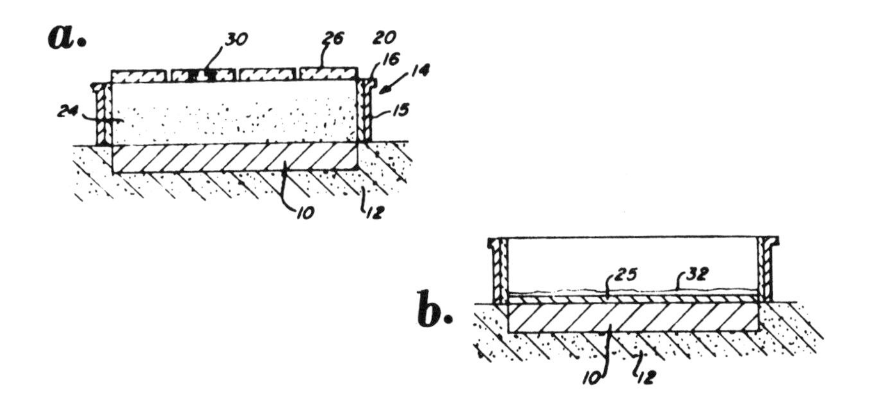

Source: U.S. Patent 4,131,473

A refractory-lined perimeter **14** comprising a steel exterior with flanges **16** for crane hooks and a refractory lining **20**, which is in this case graphite, has an interior dimension identical to the exterior dimensions of the steel substrate. The perimeter is positioned on the sand base **12** surrounding the substrate. An aluminothermic reduction charge or mixture **24** is placed within the perimeter on the substrate to a generally uniform depth of at least about ½" and up to a depth of about 12". The resulting hard facing layer **25** will have a thickness about ⅙ the average depth of the original powdered charge. The charge consists of about three parts powdered iron oxide, which is preferably Fe_2O_3 but can be Fe_3O_4, preferably having a size at least as fine as -35 mesh, one part aluminum powder preferably having a size between about -20 mesh and +325 mesh, and sufficient ferroboron to provide from 2 to 8% boron in the final hard facing composition.

The ferroboron is added in the form of crushed pellets, preferably having a size of -20 mesh. The composition range of the basic ATR charge is 65 to 85 parts iron oxide and 15 to 35 parts aluminum. Stoichiometric amounts are ordinarily used, but up to 5% excess of either component has been employed with good results.

Optionally, refractory plates, such as graphite plates **26** shown in Figure 3.8 are placed atop the refractory perimeter. One of the plates is provided with a hole **30** through which the charge is ignited. These refractory plates prevent splashing, contain the heat of reaction of the charge and force such heat into the substrate to enhance the adherence of the hard facing. The charge is ignited by a convenient means, such as a welding torch inserted in the hole which ignites the fuel powder, in this case aluminum. Other fuels that might be used instead of aluminum are magnesium, calcium, silicon and calcium silicon alloy. These fuels may replace only a portion of the aluminum powder, if desired.

The reaction is very exothermic which produces products having a high degree of superheat from which the dense metal phase separates and metallurgically bonds to the substrate. The less dense slag layer **32** collects on top of the metal phase. After the reaction is complete, the graphite plates are removed from atop the refractory perimeter and the product is insulated. Insulation (not shown) is provided by placing a blanket of Kaowool or pouring sand on top of the slag crust. This causes the metal to solidify from the bottom and promote a sound, pore-free hard facing. The product is allowed to cool until the hard facing has solidified at which time the insulation and the perimeter can be removed. The slag is removed merely by breaking it to leave a ferrous metal substrate **10** having a boron-containing abrasion-resistant surface **25**. 4 to 7% boron is preferred with the optimum boron level being 5.5 to 7%.

A ferroboron hard facing is formed which consists of an iron-base matrix containing from about 20 to 90 volume percent of Fe_2B, preferably 45 to 80 volume percent Fe_2B, with the optimum range of 60 to 80 volume percent of Fe_2B. This hard facing has a minimum thickness of about 0.1" and preferably is not less than 0.25". Hard facings 2" thick or more can be formed.

While in the preferred embodiment an iron-base matrix containing Fe_2B is formed on a ferrous metal substrate, hard facing layers of any matrix-forming metal or alloy can be formed on most metallic substrates. Copper, tin, nickel, chromium, cobalt and molybdenum as well as brass, bronze, ferrous and nonferrous alloys and stainless steels are all suitable substrates.

The substrate should be preheated prior to placing it on the sand base. While the substrate can be preheated to any temperature below its fusion point, it is preferred to preheat to within the range of 1400° to 2000°F, with an optimum temperature of 1800°F.

Boride Coating

The boriding of iron materials and nonferrous metals has been known for a long time as a process for producing wear-preventing coatings. Of the processes described in the literature only the powder boriding process has been practiced to a significant extent. In this process the production part to be treated is packed into a mixture of different materials and subjected to a temperature treatment. As the boriding agent for the most part there is used a mixture which consists of boron carbide as the boron yielding substance, silicon carbide or another filler for regulating the activity and potassium borofluoride as the activator.

This mixture furthermore contains in part amorphous carbon and other additives which should increase the activity. It is used as a powder or granulate. The temperature treatment is carried out nearly exclusively in oven furnaces, muffle furnaces or pot furnaces.

Although trouble-free boride coatings are produced by these procedures, they have several severe disadvantages. The packing of the production pieces in the boriding agent and the unpacking are only possible by hand. The area of use of the process from the beginning is limited to the treatment of individual pieces or small series. However, the process is only used reluctantly in practice with large or complicated shaped individual pieces since the consumption of boriding agent is very high in those cases. Finally, the partial boriding, i.e., the treatment of separate parts of the production pieces is either impossible or is only possible with considerable difficulty.

H. Kunst and C. Scondo; U.S. Patent 4,126,488; November 21, 1978; assigned to Deutsche Gold- und Silber-Scheideanstalt vormals Roessler, Germany describe a process using a paste of a boron-yielding substance (i.e., a boron donor), a filler, an activator and water as a binder, wherein the paste additionally contains 2 to 8 wt % of pyrogenic silica, i.e., silica produced by flame hydrolysis.

As the boron-yielding substance there can be employed boron or boron carbide. Also, there can be used ferroboron, boric anhydride or borax. As fillers which simultaneously act to regulate the activity of the paste, which only form monophase coatings of Fe_2B, there can be mentioned, for example, aluminum oxide, magnesium oxide or similar inert materials, e.g., graphite.

Finally, as an activator there can be used in known manner potassium borofluoride. Other conventional activators such as ammonium, alkali metal and alkaline earth metal chlorides, bromides and fluorides can be used, e.g., ammonium chloride, potassium chloride, sodium chloride, calcium chloride, barium chloride, potassium fluoride, barium fluoride, magnesium fluoride, sodium bromide, sodium fluoride, calcium bromide and sodium borofluoride.

The proportions of the materials other than the silica are not critical and can be those conventionally employed in the art. Thus, the boron-yielding substance can be 5 to 45%, the filler 10 to 60%, the activator 2 to 15% and the water 15 to 50% by weight.

The portion of pyrogenic silica can be varied within the given limits, according to the operational requirements. For example, if the paste is applied by dipping, a thicker consistency is selected, i.e., the portion of pyrogenic silica is selected to be relatively high. On the contrary, if the paste is applied to the production piece by spraying, a small portion of pyrogenic silica is used. It has proven especially advantageous to use 2 to 5 wt % of pyrogenic silica.

The pastes described herein have a series of substantial advantages over the state of the art. They are stable and are not inclined to settle. Besides, they are not combustible. Their consistency is variable inside wide limits. Upon cooling from the boriding temperature at the end of the treatment, the paste nearly completely falls off or scales off of the production pieces. If with complicatedly shaped production pieces there remain residues, they can be removed without trouble by using warm water; in treating a large series of pieces, in a washing machine. When using the pastes, well-formed, homogeneous boride coatings are formed.

Example: A paste of the following composition was prepared:

	Weight Percent
Boron carbide	20
Silicon carbide	40
Potassium borofluoride	6.7
Water	30
Pyrogenic silica	3.3

Into this paste were dipped the faces of small pieces measuring 50 x 30 x 20 mm^3 of an unalloyed steel Ck 15. These pieces had strong abrasive wear on one face.

The production of the paste was taken up while the powdery components, boron carbide, silicone carbide and potassium borofluoride were first intensively mixed and then stirred into the aqueous suspension of the silica. After the dipping, the parts, without drying, were placed on the conveyor of an automatic conveyor furnace, namely, to the surface which is opposite to the one coated with the paste. The furnace was operated with nitrogen as a protective gas. The speed of the conveyor through the furnace was so regulated that the parts after the preheating were exposed for 3 hours to a temperature of 900°C and were cooled to about 400°C until the end of the furnace was reached. From the conveyor end the parts were discharged into a crucible where they were allowed to grow cold. No paste residues adhered to the smooth parts present in the crucible.

The boriding fully answered all requirements. On the treated face there was formed a well-formed, homogeneous boride coating having a thickness of 80 to 90 μ. Yet it is worth mentioning that in the described process (boriding only the functional surface with paste, continuous furnace under protective gas) 3.3 g of boriding paste were consumed per part. In comparison in the conventional powder boriding (embedding the entire part in powder) there are required 130 g of boriding agent per part.

Three-Component Flame Spray Powder

M.S. Patel; U.S. Patent 4,118,527; October 3, 1978; assigned to Eutectic Corporation describes a metalliferous flame spray powder comprising a mixture of ingredients in the form of free-flowing agglomerates, the agglomerates having an average composition in which the ingredients range by weight from about 3 to 15% particulate aluminum, about 2 to 15% particulate refractory metal silicide and the balance essentially nickel-base, cobalt-base, iron-base and copper-base metals. The material when flame sprayed to provide a bonded coating on a metal substrate, e.g., a ferrous metal substrate, is characterized by improved machinability, improved bonding strength, and improved wear resistance.

Preferably the refractory metal silicides are selected from the group consisting of disilicides of Ti, Zr, Hf, V, Nb, Ta, Cr, Mo and W.

In producing the powdered agglomerates, it is preferred that the average particle size of aluminum powder and the refractory metal silicide powder not exceed about one-half, and more preferably, not exceed about one-fourth of the average particle size of the metal powder selected from the group consisting of nickel-base, cobalt-base, iron-base and copper-base metals, these metals being hereinafter referred to as the core metals. These core metals include nickel, cobalt, iron and copper per se as well as nickel-base, cobalt-base, iron-base and copper-base alloys. The agglomerates generally comprise the aluminum and refractory metal silicide adhesively bonded to the core metal.

The preferred core alloys may comprise alloys containing self-fluxing alloying agents, such as silicon and/or boron. The self-fluxing agents may be present in the core metals in an amount ranging by weight from about 0.5 to 6% silicon and/or about 0.5 to 5% boron.

It is preferred that the aluminum content of the spray material range by weight from about 3 to 10% and the refractory metal silicide from about 2 to 10%, with the balance essentially the core metal.

The average size of the aluminum and silicide powder is preferably less than about 30 microns, the average size of the aluminum powder generally ranging from about 0.1 to 15, e.g., about 2 to 10 microns, and the silicide powder from about 0.1 to 25, e.g., 0.1 to 10, microns.

The core powder may have an average size less than 140 mesh (U.S. Standard), for example, at least about 80% ranging from about -200 to +325 mesh, and when formed as an agglomerate with the other ingredients, the average size of the agglomerate preferably ranges from about -100 to +325 mesh and, more preferably, from about -140 mesh to +325 mesh.

The powders may be sprayed using various types of metal spray torches well known in the art. As regards such torches, the powder formulation is injected into the stream of burning gas and emitted from the torch and applied to the metal substrate.

A preferred torch is that disclosed in U.S. Patent 3,620,454 which is adapted for gravity feed of the powder externally to the flame issuing from a nozzle.

Coated Building Components

M. Krejci and P. Kresse; U.S. Patent 4,117,197; September 26, 1978; assigned to Bayer AG, Germany describe a process for coating a preformed component based on an inorganic binder containing standard aggregates with a glaze-like silicate-containing and/or phosphate-containing coating. In the process a component is molded from a workable mass comprising an inorganic binder, water and standard aggregates and an aqueous paste containing water glass and/or a phosphate and a metal oxide is applied to the preformed component in a quantity of from 190 to 400 g/m^2 of surface to be coated. This is done in the presence of at least one soluble inorganic salt, the salt being present in a quantity of at least about 0.5% by weight based on the inorganic binder or, in the case of lime-sand bricks, based on the binder plus aggregate. The inorganic salt converts the aqueous paste into a gel-like nonflowing form. The preformed component and coating are subsequently hardened.

The preformed components to be coated may be, for example, concrete roof tiles, asbestos cement slabs or lime-sand bricks. The coating process is preferably applied to concrete roof tiles and lime-sand bricks.

The composition of the paste with which the preformed components (either alone or provided with a preliminary layer) are coated in accordance with the process may vary within relatively wide limits. Preferred coating pastes contain alkali metal silicate, for example sodium silicate, in aqueous solution (water glass), metal oxides, for example ZnO, MgO, PbO, CaO, B_2O_3, Al_2O_3, either individually or in any combination, the SiO_2-content amounting to between about 42 and 63 mol %, the Na_2O-content to between about 11 and 27 mol % and the total metal oxide content to between about 19 and 42 mol % (based on the total weight of these components).

It is also possible to use oxide-containing compounds for example, carbonates or phosphates for the necessary metal oxide content in the paste. The paste may also contain pigments, for example TiO_2 red, yellow or black iron oxides and/or iron oxide hydroxides, chromium oxide pigments, conventional fillers, for example, kaolin and calcium carbonate. Pigments, fillers and water are added in such quantities that a readily processible, sprayable and spreadable paste is obtained. In addition, the pigment and/or filler content in the paste should not exceed about 25% by weight and preferably amounts to between about 10 and 15% by weight.

The various components of a water glass paste are processed, for example, in a dissolver or in ball mills, to form a homogeneous paste and in this form are sprayed, injected or spread in a thin layer (approximately 40 to 80 microns) onto the preformed components. Approximately 190 to 400 g and preferably about 250 to 300 g of water glass paste are used per square meter of the surface of the preformed component to be coated.

After the water glass paste has been applied to the preformed component, the water glass paste is initially left to solidify into a gel-like nonflowing form. In the case of concrete roof tiles, asbestos cement slabs and lime-sand bricks, this solidification takes about 0.5 to 3 hours, the necessary solidification time being at the lower end of the abovementioned time range in cases where inorganic salts are added in relatively large quantities to the mass from which the components are preformed. After solidification, the preformed component and solidified coating can be hardened.

Hardening is preferably carried out in an indirectly electrically heated autoclave at temperatures in the range of about 150° to 210°C, preferably at temperatures in the range of about 170° to 180°C, under pressures of about 4 to 19 bars and preferably under pressures of about 7 to 10 bars.

In the case of concrete roof tiles, the hardening time at the abovementioned temperature amounts to between about 4 and 8 hours, in the case of asbestos cement slabs to between about 8 and 12 hours and in the case of lime-sand bricks to between about 4 and 8 hours.

Basically, however, it is not absolutely essential to carry out hardening in an autoclave in order to obtain a hard weatherproof glaze-like coating according to the process. Thus, it is possible in the hardening of concrete roof tiles or asbestos cement slabs to apply the water glass paste coating to the fresh nonset concrete roof tile or asbestos cement slab, to harden it in the usual way at ambient temperature in air (maximum relative air humidity during solidification of the silicate coating preferably below about 60%) for about 14 to 28 days and then to complete hardening by a heat treatment at temperatures in the range of about 200° to 400°C.

Example: *Production of Coated Concrete Roof Tiles* – Portland cement and Rhine sand (particle size up to 3 mm) were intensively mixed with the following additives in a mixing ratio (ratio by weight) of 1:3 in the presence of water [water-binder value (cement) 0.37]: (a) 1% by weight of calcium chloride, based on the cement; and (b) 2% by weight of calcium formate, based on the cement.

The mixture was then processed in a Type 270065 test concrete roof tile machine (built in 1970 by Ing. Kurt Schade) to form 20 x 30 cm concrete roof tiles.

A water glass paste with the following composition was then applied to the concrete roof tiles thus produced: sodium water glass 37° to 40° Be', 70 parts by weight; ZnO, 15 parts by weight; kaolin, 5 parts by weight; pigment (iron oxide), 5 parts by weight; and water, 5 parts by weight.

The water glass paste was sprayed on in a thin layer in a quantity of 300 g/m^2 of concrete roof tile surface.

After standing in air for approximately 90 minutes, the paste on the surface of the concrete roof tiles was no longer free-flowing and had solidified to such an extent that the concrete roof tile and the coating could be hardened. Hardening was carried out in an indirectly heated autoclave at temperatures of from 170° to 180°C under a pressure of from 8 to 9 bars. After a hardening time of 4 hours, hardening was complete. The hardened concrete roof tiles had a uniform impervious coating.

PLASTIC SUBSTRATES

ABS RESINS

$PdCl_2$-$SnCl_2$ Catalytic Activator

M.N. Jameson, J.D. Klicker, G.A. Krulik and J.F. York; U.S. Patent 4,120,822; October 17, 1978; assigned to McGean Chemical Co. describe catalytically active compositions for rendering the surface of a nonconductive substrate such as ABS receptive to an electroless plating solution to form a uniformly adherent layer of metal.

In the process, catalytically active compositions are prepared by reacting a palladium salt dissolved in an aqueous halide solution with a molten tin salt, or a solution thereof, in an aqueous nonacid solution. A principal advantage is that no acid is used with either the palladium salt or the tin salt solution. While the reactants may be considered acids, the compositions are free from extrinsic sources of acid, such as hydrochloric or sulfuric acids, which the prior art indicates are absolutely necessary in the preparation of palladium-tin catalyst systems.

Halide ions, particularly the chloride and bromide ions, from any compatible water-soluble salt, are used to prepare the palladium salt solution, most commonly in the form of the chloride. The tin solution may also contain a compatible halide and any amount of water up to that which causes precipitation of the tin salt. Typical solutions of the tin component include pure molten $SnCl_2 \cdot 2H_2O$; mixtures of anhydrous stannous chloride and molten stannous chloride dihydrate; mixtures of either containing a compatible halide salt; and water, if desired, under the limitations mentioned above. No acid is needed in this process.

The resulting catalytically active product may be either a liquid or a solid depending on the process conditions used during the manufacture thereof. However, for reasons of stability and ease of handling, it is preferred that a substantially solid product be produced. These catalysts are effective initiators of electroless nickel, copper, and other conventional electrode plating solutions. They may be used on any suitable nonconductive substrate requiring sensitization, such as acrylonitrile-butadiene-styrene graft polymer (ABS), polypropylene, polyphenylene oxide based resins, epoxies, etc.

Example: In this example standard test plaques were sequenced through a preplate cycle which included the following steps: (1) preliminary etching of the plaque in a chromic-sulfuric acid etch bath; (2) rinsing in water; (3) neutralizing any remaining acid upon the surface; (4) sensitizing in the catalytic solutions as described above; (5) acceleration of the sensitizer; and (6) immersion in an electroless nickel bath which contained a source of nickel cations, a hypophosphite reducer, and various stabilizing and buffering compositions.

A mixture of 25.2 g of stannous chloride dihydrate ($SnCl_2 \cdot 2H_2O$) and 2.51 g of potassium chloride (KCl) was melted and maintained at approximately 85°C, which is above the melting point of the salt mixture. A solution containing 3.36 g of KCl and 2.0 g palladium chloride ($PdCl_2$) in 17.79 g of water was added to the molten salt mixture. The resulting mixture was maintained at 85°C for 1 hour with constant stirring. At this point 106.19 g of $SnCl_2$ (anhydrous) was added and the solution heated at 85°C for an additional hour. The dark brown solution was allowed to cool to room temperature yielding a friable, dry product having a brownish-black appearance.

Upon completion of the first step described above, an excess of water was present. If the solution were allowed to cool to room temperature, the product would be a liquid and the components would tend to crystallize. Consequently, anhydrous stannous chloride is added in the second stage to react with the excess water to yield stannous chloride dihydrate which is a solid at room temperature. An excess of stannous chloride above that which is needed to react with excess water is actually added in order to get an even drier product. The solid component has an actual water deficit of about 10%, being a mixture of about 90% $SnCl_2 \cdot 2H_2O$ and 10% anhydrous $SnCl_2$ (along with the other components).

To 1 liter of a 3 N solution of HCl was added 18 g of the solid catalyst described above. The solution was stirred until all the catalyst dissolved and the working bath became a dark brownish-red. An etched and neutralized standard ABS plaque (Borg-Warner EPB-3570) was immersed in the catalyst for 3 minutes. The plaque was then accelerated with dilute HCl and placed in a room temperature, electroless nickel bath (Borg-Warner N-35) for 6 minutes. The ABS plaque had 100% nickel coverage, showing that the catalyst had excellent activity.

Hydrous Oxide Colloids of Nonprecious Metals

N. Feldstein; U.S. Patent 4,180,600; December 25, 1979; U.S. Patent 4,131,699; December 26, 1978; U.S. Patent 4,132,832; January 2, 1979; and U.S. Patent 4,136,216; January 23, 1979; the last one assigned to Surface Technology, Inc. describes a process comprising the coating of a dielectric substrate, such as ABS, which has preferably first been etched by conventional means to improve adherence, with a hydrous oxide colloid of nonprecious metal ions preferably selected from the group consisting of copper, nickel and cobalt followed by reduction to a reduced or elemental state with a suitable reducing agent to activate the surface for subsequent electroless coating.

More specifically, the process is comprised of the following sequence of steps followed by electroless plating.

(1) Immersing a dielectric substrate in a solution comprised of a hydrous oxide colloid of a nonprecious metal, preferably selected from the group consisting of copper, nickel, cobalt and mixtures thereof;

(2) Rinsing the substrate with water to remove excess colloid;

(3) Immersing the substrate after rinsing in a solution comprised of a reducing agent capable of reducing the metallic ionic portion of the adsorbed colloid to a lower or zero oxidation state; and

(4) Optionally rinsing the substrate with water prior to electroless plating.

In addition to the water employed in steps (2) and (4) of the above process, the objects of the process are achieved using the following system of solutions:

(1) A hydrous oxide colloid of a nonprecious metal preferably selected from the group consisting of cobalt, nickel and copper and mixtures thereof; and

(2) An aqueous solution of a reducing agent which will reduce nonprecious metal ions in the colloid following adsorption of the colloid onto the substrate to a lower valence or metallic state.

Example 1: An ABS substrate was etched in a solution comprised of 400 g/ℓ chromium oxide and 350 g/ℓ concentrated sulfuric acid for approximately 4 min at a temperature of 70°C. Thereafter, the etched substrate was immersed in a primer solution for 5 minutes, the primer solution being prepared by dissolving 1 g of $NiCl_2 \cdot 6H_2O$ in 100 ml of deionized water and raising the pH to 6.5 with the slow addition of 1 M NaOH. The primed substrate was then rinsed and immersed in a developer solution comprised of 1 g/ℓ of KBH_4 for 2 minutes. The substrate was then rinsed and immersed in an electroless copper bath at a temperature of 40°C having the following composition to effect plating:

$CuSO_4 \cdot 5H_2O$	15 g/ℓ
EDTA (40%)	68 cc/ℓ
NaOH	9 g/ℓ
NaCN	3 ppm
Tergitol TMN	0.2 (wt %)
H_2CO (37%)	22 cc/ℓ

Example 2: The electroless plating procedure of Example 1 was followed with the exception that the primer solution, i.e., the hydrous oxide colloid, was prepared by adding 2 g of sucrose and 15 ml of 1 M ammonium hydroxide to 4 ml of 1 M cupric chloride. Good plating resulted at 25°C.

Example 3: A hydrous oxide colloid was prepared by adding 5 ml of 0.05 M ammonium hydroxide to 20 ml of 0.25 M copper acetate with good mixing. This colloid, when substituted for the colloid employed in Example 1, produced good electroless plating.

Tin-Copper Priming Solution

N. Feldstein; U.S. Patent 4,150,171; April 17, 1979; assigned to Surface Tech-

nology, Inc. describes a similar process for the coating of dielectric substrates. In general, the process comprises the following steps:

(A) Priming a dielectric substrate, which has preferably first been cleaned and etched by conventional procedures, by coating the surface of the substrate with an aqueous solution containing stannous and cuprous ions; and

(B) Developing the substrate primed by step (A) by reducing the valence state of the cuprous ions present on the surface, preferably by treating the primed substrate with an aqueous solution containing a reducing agent capable of reducing the valence state of the cuprous ions.

Alternatively, step (A) may be divided into two steps, i.e., the substrate may be treated with an aqueous solution of stannous ions followed by contacting of the surface with an aqueous solution containing copper ions.

Following priming and developing of the substrate in the aforesaid manner, the substrate may be plated with a desired metal such as copper, nickel, gold or cobalt and combination alloys by immersion of the developed substrate in an electroless plating bath of the desired metal.

Example: An etched ABS substrate is immersed for about 1 minute in a primer solution comprising $SnCl_2 \cdot 2H_2O$, 72.4 g; CuCl, 12.5 g; HCl (conc), 75.0 cc; and distilled water, 50 cc. The substrate is then rinsed with water and immersed in a developer solution comprising 5 g/ℓ dimethylamine borane (DMAB) and NaOH and having a pH of 12. The primed and developed substrate is then rinsed with water and immersed in an electroless copper bath at 40°C. A continuous conductive electroless copper film was readily deposited which could subsequently act as the base for electroplating.

Two-Layer Nickel Coatings

H. Narcus; U.S. Patent 4,160,049; July 3, 1979 describes a process to provide an effective and economical process for electrolessly applying attractive, bright metallic coatings of nickel to nonconductive surfaces, particularly a wide variety of synthetic resins, for example, acrylonitrile-butadiene-styrene (ABS), without the necessity of subsequent electroplating or electrodeposition.

Other nonconductive substrates described in the prior art, including the broad range of thermoplastic and thermosetting resins, glass, ceramics, etc., may be suitably electrolessly coated with nickel in accordance with the process. In practice, these substrates are surface conditioned or etched in the manner known to the artisan prior to the electroless process in order to improve the adherence of the metallic nickel.

The etched substrate is then sensitized and activated in a one-stage or two-stage process (preferably one-stage) utilizing a noble or nonnoble activator or catalyst. The pretreated substrate is then coated with electroless nickel, termed in this process, the Primary Electroless Nickel, applied from a special composition to produce a dense, leveled electroless nickel deposited in proper thickness to act as a suitable foundation metal for receiving a Secondary Electroless Nickel possessing attractive brightness. This combination of a primary and secondary electroless nickel gives a resulting bright, attractive uniform deposit to the plastic substrate. No subsequent electroplating such as with electroplated copper, nickel, or chromium, as necessary in the prior art, is further required in the process.

Parts of the primary electroless nickel coating may be mechanically removed prior to applying the secondary nickel coating by immersion of the substrate in the secondary nickel bath. The secondary nickel coat (overlayer) is produced only in those substrate surface areas having primary nickel underlayer. The secondary nickel coat can be supplemented by other metal coatings.

The primary nickel bath is maintained basic (8.0 to 10.0 pH, preferably 8.5 to 9.0) and the secondary nickel bath is maintained acidic (4.0 to 6.0 pH, preferably 4.5 to 5.5).

In both the primary and secondary nickel coating baths, solutions of nickel salts in polar solvents (preferably aqueous) are employed together with a source of phosphorus, preferably alkali metal or ammonium hypophosphite producing a nickel-phosphorus alloy in the final product throughout both primary underlayer coat and secondary underlayer coat.

Example:

(1) Etch the part in a conventional chromic acid-sulfuric acid solution of the following composition:

Chromic acid–184 g/ℓ
Sulfuric acid (66° Bé)–368 g/ℓ
Fluorad FC-95–1 g/ℓ
Fluorocarbon surfactant

Treatment–5 minutes at 65° to 70°C;

(2) Water rinse;

(3) Neutralize: Sodium bisulfite, 150 to 200 g/ℓ
Treatment–1 minute at 25° to 30°C;

(4) Water rinse;

(5) Activate (nonnoble catalyst): acidified cuprous chloride solution
Treatment–15 minutes at 40°C;

(6) Hot water rinse for 5 minutes;

(7) Acceleration:

Sodium borohydride–1 g/ℓ
Sodium hydroxide–1 g/ℓ
Fluorad FC-95 surfactant (1% solution)–1 ml/ℓ

Treatment–10 minutes at 30°C;

(8) Water rinse;

(9) Primary electroless nickel bath immersion

Nickel sulfate–45 g/ℓ
Ammonium chloride–30 g/ℓ
Sodium citrate–5 g/ℓ
Glacial acetic acid–10 g/ℓ
Sodium hypophosphite–40 g/ℓ
Ammonium hydroxide (26° Bé)–45 g/ℓ

Operating conditions:

Temperature, 25° to 30°C
pH, 8.5 to 9.0
Time, 10 minutes;

(10) Secondary electroless nickel bath immersion

Nickel sulfate–20 g/ℓ
Lactic acid (80%)–30 g/ℓ
Propionic acid–2 g/ℓ
Lead acetate–0.001 g/ℓ
Sodium hypophosphite–25 g/ℓ

Operating conditions:

Temperature, 70° to 90°C
pH, 4.5 to 5.5
Time, 20 to 30 minutes.

OTHER SPECIFIC SUBSTRATES

Pretreatment of Epoxy Resin for Nickel Coating

R.A. Henry and E.M. Summers; U.S. Patent 4,113,899; September 12, 1978; assigned to Wear-Cote International, Inc. have found that a cured epoxy resin article filled with finely divided metal particles can be prepared for electroless plating in a commercially practical manner without a stannous chloride/palladium chloride activation if (a) resin encapsulation of the metal particles at the surface of the article is removed by impingement with a suitable fine abradant, and (b) metal oxide coatings on the exposed surfaces of the resin particles are removed with an extremely brief pickling treatment. The thus-treated article can then be plated by electroless techniques in plating baths maintained at temperatures above 160°F.

The metal particles distributed uniformly through the article should be small (in U.S. Standard mesh, -60 mesh and preferably -100 mesh), and should preferably comprise 35 to 85% by weight of the article to be plated. The impingement step involves the use of a more finely divided grit or abradant than normally used in the initial sandblasting of metal articles; instead of the usual -30+50 or -40+60 U.S. mesh sand or grit, a -60 or even -100 U.S. mesh grit is preferred. Glass beads or silica sand of this degree of fineness are of adequate hardness, and very soft abradants (e.g., those with a Moh's scale hardness less than 4.5) are ordinarily impractical.

The pickling treatment should not have any significant oxidative effects, either upon the resin or plastic matrix or the metal filler of the article to be plated. Indeed, the purpose of the pickling is to remove oxide films or coatings on the exposed surfaces of the metal particles. For this purpose, hydrochloric acid (especially for iron fillers) or 7 to 40 wt % nitric acid (especially for aluminum fillers) can be used, provided the pickling treatment is sufficiently brief; best results are obtained with pickling times less than 30 seconds, although up to a minute can be used with hydrochloric acid acting on an iron filler. After pickling, the electroless nickel plating procedure described in Canadian Patent 962,899, can be followed, so that a continuous plate with a thickness of at least 0.00025" (more than 5 μ) is obtained. Thicknesses of 25 to 75 μ are preferred.

Aluminum Orthophosphate Coating on Polyolefin

R.A. Dratz; U.S. Patent 4,140,822; February 20, 1979; assigned to Thilmany Pulp & Paper Company describes a method which is suitable for providing polyolefin films with a paintable, printable, glueable and flexible coating comprising aluminum orthophosphate consistent with high-speed operation.

In accordance with the process, the polyolefin film is uniformly coated with an aqueous dispersion comprising aluminum chlorhydroxide, phosphoric acid and high molecular weight polyvinyl alcohol.

The aqueous dispersion is applied to the surface of the polyolefin film at a level sufficient to provide a coating level on a dry solid basis, which is in the range of from about 0.03 to about 0.75 pound per ream (3,000 sq ft) and preferably will be in the range of from about 0.05 to about 0.5 pound per ream of the polyolefin material per coated surface of the film.

The aqueous dispersion is prepared by dissolving the phosphoric acid and aluminum chlorhydroxide in separate portions of the water to be used in preparing the dispersion. The high molecular weight polyvinyl alcohol is dissolved in the water prior to adding the phosphoric acid or the aluminum chlorhydroxide thereto. By aluminum chlorhydroxide is meant the compound $Al_2(OH)_5Cl$ or mixtures of aluminum hydroxide and hydrochloric acid wherein the atom ratio of chlorine to aluminum is between about 0.2 to 1 and 0.7 to 1. The aluminum chlorhydroxide and phosphoric acid can be added in quantities which result in atom ratios of aluminum to phosphorus within a wide range, preferably from about 0.3 to about 1.5.

By high molecular weight polyvinyl alcohol is meant a resin made by the hydrolysis of polyvinyl acetate and having a molecular weight of at least about 100,000 preferably from about 115,000 to about 125,000. The polyvinyl alcohol useful in the process has an acetate content of less than about 25% and preferably is fully hydrolyzed.

The use of the high molecular weight polyvinyl alcohol at a level of from about 0.25 to about 0.5% by weight of the dispersion provides a dispersion with a viscosity in the range of from about 5 to about 10 cp. Such viscosity is suitable for use with transfer coating and Mayer rod application equipment.

The dispersion can also contain additives such as resins that improve the wettability and adhesion of the coatings to the polyolefin film to which they are applied. Additives which have been found to be useful include melamine formaldehyde resins, urea formaldehyde resins, and amino acids such as glycine and alanine alone or in combination. In general, the additives can be used at a level of up to about 50% by weight of the phosphate present in the dispersion. The additive is preferably added to the dispersion after the addition of the phosphoric acid and before the addition of the aluminum ion.

After application of the dispersion, the coated polyolefin film is dried at elevated temperatures to remove excess water from the dispersion. Drying time and temperatures can be varied over a wide range depending, for example, on the composition of the polyolefin film, the chemical composition of the coating, the concentration of the coating, the coating thickness, and the airflow in the dryer.

The coated polyolefin film is conveniently dried by passing through a hot air tunnel with countercurrent airflow.

Example: A solution of polyvinyl alcohol in water was prepared. ¼ lb of Vinol 165 (high molecular weight polyvinyl alcohol) was added to 100 lb of water and the mixture was heated to a temperature of 195°F with stirring. The polyvinyl alcohol solution was divided into two equal parts, referred to as part A and part B. 3.1 lb of 50% aluminum chlorhydroxide (Chlorohydrol) was added to part A with stirring. 3.6 lb of 85% phosphoric acid and 0.7 lb of Accobond 3524 resin was added to part B with stirring. Part A was then combined with part B to form an aluminum orthophosphate colloidal dispersion of aluminum orthophosphate in situ.

The dispersion had an aluminum/phosphorus ratio of 0.45. The dispersion was applied to a polyethylene film having a thickness of ¾ mil by means of a transfer roll. The dispersion was applied at a level sufficient to provide 0.03 lb of aluminum orthophosphate, dry basis, per ream of the polyethylene film. The coated film was dried in a hot air tunnel with countercurrent airflow having an initial temperature of 375°F and supplied at a rate of 5 ft^3/ft^2 of film.

After drying and cooling, it was found that the coating was well anchored to the polyethylene surface and showed excellent resistance to removal by the conventional Scotch tape adhesion test. The resulting light weight coating permitted application and excellent adhesion of a variety of conventional water base and solvent base inks. The inks also resisted removal by the Scotch tape test. Latex, solvent base and enamel paints also adhere well and resist removal by boiling water for more than 30 minutes.

Elastomeric Polyurethane Laminate

R.E. Dunning and V.H. Rampelberg; U.S. Patent 4,101,698; July 18, 1978; assigned to Avery International Corp. describe a transfer laminate having a flexible transparent, or translucent elastomeric layer, and a layer of metal bonded to the elastomeric layer in separate microscopically discontinuous planar quantities of high reflectivity. The metal layer is deposited so it forms an apparent visually continuous reflective surface. The metallized elastomeric layer is attached to a substrate, either by an adhesive, or by being integrally bonded to the substrate surface, to provide a reflective metal finish which is capable of being deformed. The laminate can either be used as an impact-absorbing surface which distorts and returns to its original position, without destroying the reflectivity of the metal surface; or it can be distorted sufficiently to provide a reflective metal surface finish for three-dimensional contoured shapes, without disrupting the continuous reflectivity of the metallized surface.

Referring to Figure 4.1, an elastomeric laminate **10** according to this process includes a carrier sheet **12** having a release coat **14** overlying the carrier sheet. A layer **16** of synthetic resinous elastomeric material, such as polyurethane, is coated over the release coating **14** to form a relatively thin, continuous, planar flexible and foldable elastomeric film or skin coat after the coating sets. A layer of metal **18**, to be described in detail below, is applied to the surface of the elastomeric film **16**.

Preferably, the carrier sheet **12** may be any carrier sheet or web. For example, it may be a polyester film or a web of other plastic sheeting such as polyvinyl chloride, ABS, cellophane, or cellulose acetate.

Figure 4.1: Cross Section of Laminate

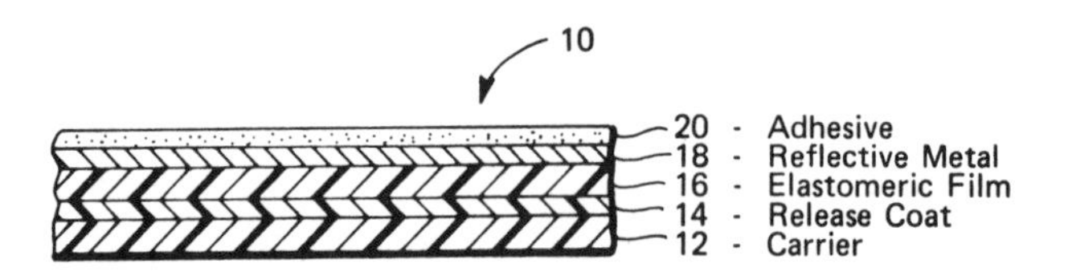

Source: U.S. Patent 4,101,698

The release layer **14** may be any conventional release coating, such as those having a wax, paraffin or silicone base, for enabling the carrier **12** to be stripped from the elastomeric film **16**.

The elastomeric film **16** preferably is a relatively thin, flexible and foldable sheet which is capable of being stretched or deformed to a desired shape under pressure, but will return to substantially its original shape after the stretching or deformation force is released. The elasticity of the film **16** may be of the type which has a memory, i.e., is capable upon removal of a deforming load, of returning to its set configuration with time and relaxation. The flexible or stretchable materials to be avoided are those which stretch readily, with slight deformation, beyond their elastic limit.

Preferably, the elastomeric film is transparent, or at least translucent, relatively durable and not relatively abradable. The elastomeric film also has a sufficient refractive index when transparent to allow the metal layer **18** to appear as though it is a surface layer and a smooth skin, so as not to disrupt reflection from the metal layer.

The preferred elastomer is a thermoplastic or thermosetting polyurethane film. Other plastic films such as polyvinyl butyral, polyvinyl acetal, transparent vinyls such as polyvinyl chloride, flexible polypropylenes or polyacrylates are suitable. Mylar or Melinex are not suitable. These films are extensible for the most part, but they do not have the elastomeric qualities necessary for the process. Natural or synthetic rubber such as styrene butadiene thermoplastic rubber or polypropene rubber also are not desirable because they are not sufficiently transparent.

The layer **18** is made from a highly reflective corrosion and abrasion-resistant metal, such as chromium. The metal is preferably applied by vacuum deposition techniques which bond the metal to the elastomeric film. Preferably, the reflective metal layer is deposited in a layer having a thickness of 0.01 mil or less, the metal layer being vacuum-deposited in discontinuous quantities, or separate planar reflective segments such as dots, which are discontinuous but which are deposited so close together that they give the optical visual effect of a continuous, highly reflective metallized surface.

A layer **20** of adhesive overlays the reflective metal layer **18** to provide means for adhering the laminate **10** to a substrate. The adhesive **20** is one which not only bonds to the metal layer **18**, but also is a flexible permanently thermoplastic adhesive which will not disrupt the ability of the laminate to elastically deform and return to its original shape. In certain instances the layer of adhesive need not be provided when adherence to a substrate may be provided by other means.

To maximize the desired continuous bright reflective effect of the metallized layer **18**, a variety of means can be used to reinforce the opacity of the metallized layer. For example, the adhesive layer **20** may include a black pigment, white pigment, or aluminum flakes to provide the necessary opaque undersurface for the reflective metal layer.

This process provides an improvement over the common chrome-plated metal trim parts for automobiles, or other articles having three-dimensional surface configurations. For example, contoured trim parts which fit around the headlamps or taillights of automobiles can be made from injection-molded plastic rather than metal castings. The metallized elastomeric laminate of this process can be stretched to conform to the three-dimensional contoured surfaces of these plastic trim parts without disrupting the reflectivity of the metal layer.

Pretreatment of PVC

C.A. Deckert; U.S. Patent 4,131,698; December 26, 1978; assigned to RCA Corporation has found that adherent, nongrainy conformal metal layers can be applied to PVC plastic substrates by pretreating the substrate surface before electroless deposition.

The method involves the immersion of PVC plastic such as a vinyl record, in an aqueous solution comprising at least about 10 g/ℓ to about 400 g/ℓ of an alkali metal hydroxide, such as LiOH, NaOH, and KOH; about 70% to about 95% by volume of water; and about 5% to about 30% by volume of a mono-, di-, or polyhydric water-soluble alcohol for a sufficient time to pretreat the surface of the plastic to enable subsequent electrolessly deposited metals to better adhere to the plastic and form nongrainy conformal coatings.

The mono-, di-, or polyhydric alcohol useful herein may be any alcohol known in the art which is soluble in water. Suitable mono-, di-, or polyhydric alcohols include, but are not limited to, methanol, ethanol, isopropanol, 1,4-butanediol, ethylene glycol, propylene glycol and glycerin. Due to their cost and availability, methanol and ethanol are preferred alcohols.

The alcohol is present in a concentration from about 5% and preferably about 10% to about 30% by volume of the total solution. If the alcohol concentration in the solution is less than 5%, then the solution's ability to wet the plastic surface is decreased; alcohol concentrations of about 5% to about 10% cause the solution to marginally wet the plastic surface. The surface wetting helps the alkali metal hydroxide to react more readily with the plastic surface. When relatively higher concentrations of the alkali metal hydroxide are present, an alcohol concentration greater than about 25% by volume may cause the formation of a two phase solution. The alcohol concentration should be adjusted to avoid a two phase solution. Thus, 25% to 30% by volume alcohol is the highest acceptable concentration. The water concentration varies from 70% to about 95% by volume of the total solution.

Heating the alcoholic hydroxide solution to a temperature from about 35° to 40°C accelerates the pretreatment of the PVC plastic substrate. Temperatures higher than about 40°C may, however, tend to soften and degrade the plastic substrate.

Ultrasonic agitation of the solution in which a polyvinyl chloride substrate is immersed will also shorten the treatment time. The combination of heating and ultrasonic agitation reduces treatment time from overnight or several hours to several minutes.

Treated PVC plastic substrates can be processed by electroless deposition methods known in the art. After electroless deposition, the metal layer is sufficiently adherent to pass the Scotch tape test. Cellophane tape is cut long enough to curl. The curled section is attached to the metallized substrates and then ripped off. Alternatively, a section of tape is cut and pressed onto the metallized plastic substrate and pulled off. Coatings which have not adhered well to the plastic surface will be pulled off by the cellophane tape.

Example: A wedge of plastic was cut from a standard PVC plastic record containing about 13% to about 15% by weight polyvinyl acetate and immersed in an alcoholic hydroxide solution comprising 200 g of potassium hydroxide, 250 ml of ethanol, and 750 ml of deionized water overnight.

After pretreatment, the plastic wedge was rinsed in water for about 5 minutes and immersed and ultrasonically agitated in Shipley 1160, a mild cleaner, diluted 1:19 with water, for about 3 minutes. The wedge was rinsed in water for about 2 minutes and immersed in a solution of about 8 parts water and about 1 part Shipley 9F, a palladium-tin catalyst, for about 2 minutes. Following the Shipley 9F solution, the wedge was water rinsed for about 2 minutes and immersed in a solution of about 3 to about 4 parts water and about 1 part Shipley Accelerator 19 for about 1 minute and water rinsed for about 1 minute.

Thereafter, the wedge was immersed in an electroless nickel bath comprising about 1 part solution A, wherein solution A comprises 50 g $NiSO_4 \cdot 6H_2O$, 100 grams $Na_4P_2O_7 \cdot 10H_2O$, and 15 ml concentrated NH_4OH; about 1 part solution B, wherein solution B comprises 3 g/ℓ dimethylamine borane, $(CH_3)_2NH \cdot BH_3$; and about 6 parts H_2O. Finally, the wedge is water rinsed for about 1 minute then rinsed with isopropanol for about 30 seconds and spun dry. After drying, the metal layer was sufficiently adhered to the polyvinyl plastic to pass the Scotch tape test.

Plating of Polyvinylidine Fluoride

L.M. Schiavone; U.S. Patent 4,180,602; December 25, 1979; assigned to Bell Telephone Laboratories, Incorporated has found that continuous, adherent metal layers on polyvinylidene fluoride (PVDF) are achievable if a very particular series of steps prior to the actual electroless plating are performed. First the plastic must be treated with a solvent for PVDF. Exemplary of suitable solvents are N,N-dimethylformamide, cyclohexanone, dimethyl sulfoxide or propylene carbonate. It is expedient before processing to first clean the plastic substrate to remove any grease or dirt which might interfere with the solvent treatment or other subsequent steps. Conventional methods are used for cleaning. For example, the plastic is immersed in a series of suitable solutions for removal of oil or other surface contamination such as trichloroethylene and acetone. After cleaning, the treatment solvent such as DMF is then generally used for a period between 2 and 10 minutes, preferably between 3 and 5 minutes. After treatment with the solvent, the plastic substrate is rinsed and dried in preparation for subsequent steps.

The next essential step to the process after the solvent treatment is sensitization with elemental silver. This is conveniently done by first subjecting the plastic to a reducing agent and then to a silver salt. The reducing agent absorbed on the plastic subsequently reduces the silver from the salt solution onto the substrate.

The particular salt used is not critical, however, for convenience it is desirable that it is soluble in a common solvent. Exemplary of suitable salts is silver nitrate. This salt is easily soluble in water and is applied to a plastic by simple immersion. Concentrations of between 1 M and 10 M, preferably between 1.5 and 2.5 M are acceptable. The plastic is left immersed, typically between 5 and 15 minutes until the entire surface of the plastic has been treated with silver. Treatment temperatures between 36° and 70°C, preferably between 60° and 70°C are adequate. Lower and higher temperatures than those specified, although usable, are less efficacious and generally produce less adherent or less uniform coatings.

The silver salt is preceded by treatment with a reducing agent for silver. Compounds such as $SnCl_2$ have been found adequate for this purpose. Again the reducing agent should be soluble in a common solvent to facilitate treatment. For stannous salts, solutions in the concentration range of 1.5 M to 2.5 M are adequate when the solution is maintained at a temperature in the range of 36° to 70°C, preferably in the range of 60° to 70°C. For the conditions disclosed, adequate treatment times generally are from 5 to 15 minutes.

The electroless plating by the desired technique is then performed. Most electroless plating solutions require that the substrate to be plated first be treated with a catalyst. For gold, it is known that a palladium catalyst in conjunction with a gold plating solution containing $KAu(CN)_2$, KOH, KCN, and KBH_4 is needed. The form of palladium used to activate the pretreated plastic is again conveniently a soluble salt. For example, $PdCl_2$ is easily soluble in a dilute acid solution, e.g., a water solution of HCl and glacial acetic acid. The pretreated plastic is immersed into this solution at temperatures typically between 25° and 50°C, preferably in the range of 40° to 50°C. Concentrations of palladium salts between 0.001 M and 0.002 M for these temperature ranges are adequate.

It should be stressed that the process resides in the two critical steps of pretreatment with a solvent for PVDF followed by sensitization with silver. Known electroless plating solutions and their associated catalysts are then used following these necessary steps. The following example exemplifies the conditions and steps previously discussed.

Example: A thin strip of PVDF measuring 0.75" by 0.75" and having a thickness of 0.005" was first cleaned by immersion in trichloroethylene for 4 minutes. The plastic was then immersed in acetone for an additional 3 minutes. After this cleaning, etching was performed by immersion in DMF for 3 minutes. The plastic was then rinsed thoroughly with distilled water.

Sensitization was then performed by first immersing the plastic in a 2.5 M $SnCl_2$ aqueous solution for 10 minutes. The sample was then rinsed thoroughly with distilled water and immersed for an additional 10 minutes in a 2.5 M $AgNO_3$ aqueous solution. Both solutions during these steps were kept at 70°C. After the $SnCl_2$ bath, the sample was again thoroughly rinsed in distilled water followed by drying with nitrogen gas and radiant energy from a xenon lamp.

An activation solution was prepared by first dissolving 0.3 g of $PdCl_2$ in 9 ml of conc HCl. This solution was diluted with 9 ml of distilled water and added to 864 ml of glacial acetic acid. This combined solution was then added to 18 ml of conc HCl and thoroughly mixed. The activating solution was heated to approximately 50°C and the treated plastic was immersed in it for approximately 30 minutes. The sample was thoroughly rinsed in distilled water to insure that all of the solution was removed.

A gold plating solution was prepared by first dissolving 56 g of KOH in 50 ml of water. Approximately 54 g of KBH_4 was then added and the mixture agitated until all the solids were dissolved. A second solution was prepared by dissolving 32.5 g of KCN in 200 ml of water followed by addition of 4.3 g of $KAu(CN)_2$. The mixture was agitated until all the solids dissolved. The first solution was then added to this second solution in a volumetric flask and distilled water was added to produce 1 liter of solution. This combined solution was then filtered through a Whatman No. 41 filter paper. (If the solution is to be stored before use, the storage container should be an inert plastic, such as Teflon or polypropylene. Additionally, the cap to the storage container should not be tightened because hydrogen is liberated during storage.)

The gold plating solution was heated to 85°C and the plastic was immersed in it for approximately 30 minutes. No agitation was used since this produced bubbles in the solution which interfered with the uniformity of the plating. The plated plastic was then thoroughly rinsed in distilled water followed by two rinsings in warm methanol and then dried with nitrogen gas.

The resulting product has approximately a 1 μ thick layer of gold which had a resistivity of approximately 2.5×10^{-6} ohm-cm. The plating appeared uniform and passed the adhesive tape test. That is, a piece of adhesive tape was placed onto the gold plating and then removed. None of the gold adhered to the tape.

Metal Coated Vinyl Disc

J.L. Vossen, Jr., F.R. Nyman and G.F. Nichols; U.S. Patent 4,101,402; July 18, 1978; assigned to RCA Corporation have found that copper improves the adhesion of nickel and chromium alloy thin films to polymeric substrates and coatings. The copper can be applied as a very thin film between the polymeric material and the metal to be deposited, or can be first admixed with the metal to be deposited as an alloy constituent. The greater the amount of copper added, the greater the adhesion provided, but a film as little as about 25 Å thick at the polymer-metal interface or the addition of as little as about 10 atomic percent of copper to an alloy composition, improves metal-polymer adhesion by several orders of magnitude.

According to one method of preparing the films, the polymeric substrate to be coated with a conductive adherent metal film is placed in a vacuum chamber and connected to a positive source of current, such as a planar magnetron source. The vacuum chamber is also fitted with a negative electrode of copper and another electrode of the nickel-chromium alloy to be deposited. The chamber is then evacuated to a pressure of about 5 x 10^{-6} to 3 x 10^{-5} torr and a small amount of an inert gas, such as argon, is fed into the chamber to a pressure of up to about 15 millitorr. The pressure is not critical however, and can vary from about 2 to about 100 millitorr.

When a planar magnetron is employed in the chamber as a source of current, the voltage can be varied from about 300 to 1,000 V and the current can be up to 10 A, depending upon the rate of deposition desired and the size of the electrode.

The copper electrode is activated first to initiate sputtering on the substrate and is continued until a thin layer of about 25 to 50 Å of copper is deposited. The current to the copper electrode is then discontinued and the nickel-chromium alloy electrode is activated so as to sputter a layer of alloy of the desired thickness, generally from about 200 to about 400 Å thick. In the event that a polymeric coating is to be deposited over the nickel-chromium alloy, a third layer, also of copper, can also be deposited over the nickel-chromium alloy to provide a thin film of copper between the metal-polymeric coating interface.

Copper can also be cosputtered with the nickel-chromium alloy in a similar vacuum chamber, except providing an electrode of a nickel-chromium alloy in which pure copper has been inserted into spaces cut for that purpose. The size and location of the copper in the electrode is chosen to deposit the amount of copper desired onto the substrate to be coated. The current to the single alloy-copper electrode is then turned on and sputtering is continued until a layer generally about 200 to 400 Å thick, has been deposited onto the substrate.

Example 1: A vacuum chamber was fitted with two planar magnetron sputtering cathodes, one made of copper and the other of Inconel-600, an alloy containing 76.8±3% of nickel, 13.8±3% of chromium, and 8.5±2% of iron (plus minor amounts of impurities). Both cathodes were 8.25" x 3.56" in size. A vinyl video disc 12" in diameter was suspended about 2" above the electrodes and rotated at 40 rpm. The chamber was evacuated to a pressure of 3 x 10^{-5} torr and backfilled through a valve to a pressure of about 15 millitorr with argon.

The copper electrode was activated with 360 V, 0.3 A of current. The average deposition rate on the record under these conditions was about 80 to 100 Å/min. Copper deposition was continued for about 30 seconds or until a layer of about 50 Å thick of copper had been deposited, when this electrode was inactivated.

The Inconel-600 electrode was then activated with 650 V, 1.5 A of current, resulting in a deposition rate of about 330 to 400 Å/min. Deposition was continued for about 30 seconds or until a layer of about 200 Å thick had been deposited when the electrode was inactivated. The copper electrode was then reactivated to apply another layer of copper about 50 Å thick over the Inconel-600 layer.

The metal film was tested for adhesion by storing for 120 hours at 90°F and 90% RH in air and applying Scotch tape to the surface. No film was removed when the Scotch tape was pulled off.

Stress measurements were made in known manner by depositing films as above on very thin aluminum oxide discs and noting the bending of the discs microscopically. Whereas a film of Inconel-600 about 225 Å thick had a compressive stress of 30 x 10^9 dynes/cm^2, a trilayer coated disc prepared as above had a compressive strength of only 6 x 10^9 dynes/cm^2.

Example 2: The procedure of Example 1 was followed except that a sputtering electrode was prepared of Inconel-600 in which two slots were machined 0.252" x 6" in size, the first one 1.225" from and parallel to one of the long edges of the electrode, and the other 1.245" from and parallel to the other long edge of the electrode. A ¼" wide copper bar the length of the slot was fitted into the slot so that the edge was flush with the electrode surface. These dimensions were chosen so that the center line of one of the copper bars is in the center of one eroded or sputtered track and the other bar is on the inside edge of the second track in the electrode.

The chamber was evacuated to a pressure of about 3 x 10^{-6} torr and backfilled with argon to a total pressure of 1.5 x 10^{-2} torr. The electrode was then activated with 650 V and 1.5 A of current, resulting in a deposition rate of about 330 to 400 Å/min. Deposition was continued for about 30 seconds or until a layer about 200 Å thick had been deposited.

The resultant metal film was tested for adhesion by the Scotch tape test as in Example 1. No film was removed when the Scotch tape was pulled off. The compressive stress of this metal film, measured according to the procedure of Example 1, was only 5 x 10^9 dynes/cm^2.

The metal coated vinyl disc as prepared above was coated with a polymer of styrene as follows: a vacuum chamber fitted as above was evacuated to a pressure of about 3 x 10^{-3} torr and backfilled with nitrogen to a pressure of about 8 to 10 x 10^{-3} torr. Styrene monomer was then added to a pressure of 13 to 15 x 10^{-3} torr. The metal coated disc was suspended about 2" above a planar magnetron source having an electrode 3.5" x 7" in size at a power supply frequency of about 10 kHz and a voltage of 680 V. The power was turned on for 30 seconds and the disc was lowered to face the electrode and rotated at about 40 rpm for 2 minutes so as to deposit a styrene polymer film about 350 Å thick. Compressive stress for the resultant film was only 3 x 10^9 dynes/cm^2.

Metallizing of Resin-Coated Paper

R.W. Steeves; U.S. Patent 4,177,310; December 4, 1979; assigned to King Seeley Thermos Company describes a metallized paper product made by coating a thin metal layer onto a resin-coated paper surface.

Paper substrates suitable for use in the process will be flexible and can be provided in roll form so as to be particularly adapted for a continuous process. Typically, the paper will be about 2 to about 20 mils thick and be a 15 to 80 pounds (per ream) paper and will have a surface pretreated to provide a smoother surface upon which the resin film will be coated. Suitable pretreatments include paper sizing, calendering techniques, or machine glazing or polishing techniques. For example, the paper substrate can be passed over a heated drum rotating at a speed different than the paper substrate thereby polishing, i.e., smoothing the surface of the paper substrate. Speed differentials on the order of about 1:10 or 10:1 are suitable for this purpose.

In order that the final product will be flexible and to economize on materials, it is necessary that the resin film on the paper substrate be very thin. In general the cross section thickness of the resin film should be from about 0.05 to about 3.0 mils, preferably from about 0.1 to about 0.3 mil.

Suitable resin precursors have viscosities of from about 1,000 to about 5,000 cp at the temperature at which they are applied to the paper substrate. Precursors having viscosities of from about 2,000 to about 3,000 cp are preferred while those having viscosities of from about 2,600 to about 2,700 are most preferred. Resin precursors of these viscosities can be applied to the paper substrate by means of a finely etched reverse gravure roll of 100 to 300 quad, depending on desired film thickness although any method capable of coating a thin film of the resin precursor onto the paper substrate can be used.

It has been found that the desired smooth film of precursor which requires a relatively low viscosity and the desired limited penetration which requires a relatively high viscosity can be obtained by initiating the curing of the precursor immediately after coating the paper substrate. This immediate cure can be accomplished by employing radiation curable resin precursors and curing by means of electron beam radiation. The precursor must consist essentially of reactive monomers or oligomers which will substantially completely polymerize. By this is meant that less than 10% and preferably less than 1% of monomeric precursor material will remain after the precursor has been cured. Resin precursors comprising a nonpolymerizing solvent are not satisfactory as the solvent contributes to outgassing and is detrimental to achieving the desired smooth film surface.

There are many suitable resin precursors specifically designed for electron beam curing and which are commercially available including polyester, urethane, acrylic, epoxy and vinyl-based resin precursors and mixtures thereof. Acrylate substituted urethane resin precursors are preferred.

After the resin has been substantially cured, a thin layer of metal is deposited thereon in a metallizing step. The metal layer will be on the order of 1 to 2 microinches thick and is less than 1/100 the thickness of the resin film. Suitable metals are those well known in the art and include aluminum, copper, gold, silver, etc. While the curing step must be carried out immediately after application of the resin precursor to the paper substrate, the metallizing step can be carried out at any time after the resin precursor is substantially cured.

Preferably metallizing is carried out by vacuum metallizing which is conventional in the art. Alternative metallizing methods include other metal depositing techniques such as thermal or catalytic decomposition, electrolytic and electrophoretic deposition, sputtering and ion deposition techniques. The coated substrates of this process can be metallized at high rates normally associated with the processing of plastic films.

The modified paper end products made through this process can be used for decorative furnishings (e.g., drapes, wallpaper), wrapping purposes, such as Christmas wrapping paper, in graphic and printing arts, and in technical applications such as reflective optics, thermal insulation, electrical circuit and component production, food and chemical wrapping or conveying with controlled moisture or other fluid blockage advantages and in clothing exterior layers or liners. Adhesives can be applied over the metallized surface for lamination to other layers (of paper, plastic or metal) or to objects such as boxes, crates, walls. Protective coatings can be applied over the metallization in a manner well known in the metallizing art.

Example: A roll of a paper substrate (28 pounds per ream paper coated on one surface with a sizing material) is unrolled and passed through an offset gravure printing station having a 200 quad impression roller and which coats a thin film of a radiation curable acrylic resin precursor onto the sized surface of the paper substrate. The precursor has a viscosity of about 2,650 cp and is applied to the paper in an amount of about 2 pounds per ream.

Less than one second after application to the paper, the paper is passed through an electron beam apparatus where the resin precursor (Mobil 76X414B) is contacted with electron beam radiation until the resin is cured as is evidenced by a lack of tack or sticky feel when touched. The paper is then wound onto a roller and transported to a conventional vacuum metallizing chamber maintained at about 5×10^{-4} torr and in which the paper is passed over a source of aluminum heated to about 1350°C with the coated side of the paper facing the source of aluminum. The speed of the paper is about 500 ft/min and the uncoated side of the paper is maintained, as much as reasonably possible, in contact with chilled rollers to minimize outgassing from this uncoated side of the paper.

A 1 microinch layer of aluminum is deposited onto the resin film and the paper is wound onto a take-up roller and removed from the vacuum chamber. The resulting metallized paper product has good flexibility and has a decorative and shiny metallic layer on one surface. It is found that either or both surfaces of the paper are printable with conventional paper printing techniques.

COATING VARIOUS SUBSTRATES

Pre-Etch Conditioning

A considerable demand exists for metal plated nonconductive articles, particularly plastic articles. In the finished product the desirable characteristics of the plastic and metal are combined to offer thereby the technical and aesthetic advantages of each. For instance, superior mechanical properties of polysulfone resins may be aesthetically enhanced by a metal coating.

Although the polysulfones, like most polymers, are electrically nonconductive, they should be readily platable using an electroless plating operation. This operation is typically accomplished by conditioning the surface for plating by inorganic etching with a strong oxidizing acid or base, seeding the surface by contact with a noble metal salt solution, e.g., a palladium chloride solution, then immersing the seeded surface in an autocatalytic electroless solution where an initial coating of a conductive metal, e.g., copper or nickel, is established by chemical deposition. The metal coating formed acts as a bus which allows a thicker metal coating to be built up electrolytically. Adhesion between the metal plate and a polymeric substrate is, however, dependent on the strength of the resin-metal bond.

L.P. Donovan, E. Maguire and L.A. Kadison; U.S. Patent 4,125,649; November 14, 1978; assigned to Crown City Plating have found that the adhesion of electrolessly deposited metal coatings for polymeric substrates, especially polysulfone, can be induced or markedly improved by preceding the inorganic etch with a strong base or oxidizing acid with contact of the substrate with an aqueous pre-etch solution comprising at least one water compatible halogenated compound containing a functional

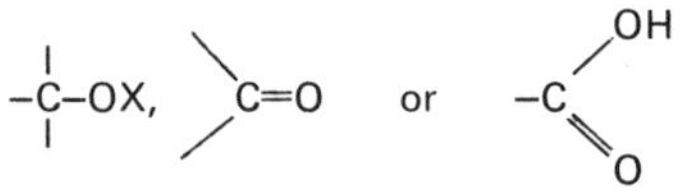

$$-\overset{|}{\underset{|}{C}}-OX, \quad >C{=}O \quad \text{or} \quad -C\begin{matrix} \diagup OH \\ \diagdown\!\!\diagdown O \end{matrix}$$

group wherein X is hydrogen or an alkali metal and in which a halogen is separated from the group by no more than 3 carbon atoms. Aliphatic halogenated compounds in which the halogen is separated from the functional group by no more than 2 carbon atoms are preferred. The preferred aliphatic halogenated compounds are polyhalogenated containing at least 2 halogen atoms and from 1 to about 6 carbon atoms.

Chlorinated compounds are preferred. The particularly preferred aliphatic chlorinated compounds are dichloropropanol, dichloroacetone, dichloroacetic acid and trichloroacetic acid. The alkali metal salts of halogenated phenols may also be used. Dichloropropanol is especially preferred for polysulfones, particularly filled polysulfones.

Residence time in the pre-etch is sufficient to modify the surface to enable etching in a strong base or oxidizing acid and surface absorption of noble metals to permit electroless plating. Contact times for this purpose will, depending on the nature of the resin and temperature, range from 0.5 to about 10 minutes, preferably from about 1 to 3 minutes. Temperatures generally range from 80° to about 160°F. For polysulfone, the optimum temperature range is from about 130° to 150°F, preferably 135° to 145°F. Temperature, in any instance, should not exceed the melting point of the polymer.

The water compatible halogenated compounds are water-soluble or water-miscible. Concentrations will vary depending on the compound selected and will range from about 5 to about 70% by weight.

While especially adapted for surface treatment of polysulfones, polycarbonates and polyesters, the pre-etch compositions of this process may also be used to modify the surface of other polymers and resins such as polystyrene, styrene-acrylonitrile, acrylonitrile-styrene-butadiene, and the like.

Gel-Forming Primer Layer

M. Sato, N. Miyagawa and J. Kobayashi; U.S. Patent 4,112,190; September 5, 1978; assigned to Mitsubishi Rayon Co., Ltd., Japan describe a metallized plastic molding comprising a plastic substrate, a primer layer of a gel percentage of at least 80% and a swelling degree of not more than 20% on the substrate and a metallic film layer on the primer layer. The metallized plastic molding is prepared by coating a primer-forming material which contains at least one compound having a molecular weight of 150 to 2,000 and containing at least two polymerizable unsaturated groups in one molecule on the surface of a plastic substrate, curing the material so that it has a gel percentage of at least 80% and a swelling degree of 20% or less to form a primer layer and then depositing a metallic film on the primer layer by a dry tape method.

Suitable plastic substrates used in this process include thermoplastic resins such as ABS resins, AS resins, polystyrene resins, polyacrylic resins, polyvinyl chloride resins, polycarbonate resins, polypropylene resins, polyethylene resins, polyester resins, polyamide resins and the like; and thermosetting resins such as phenol resins, urea resins, melamine resins, unsaturated polyester resins and the like.

The unsaturated compounds which are used for forming a primer layer are those which have a molecular weight of 150 to 2,000 and which contain at least 2 polymerizable unsaturated groups in one molecule. Specific examples of the compounds include ethylene glycol diacrylate and dimethacrylate, diethylene glycol diacrylate and dimethacrylate, triethylene glycol diacrylate and dimethacrylate, tetraethylene glycol diacrylate and dimethacrylate, propylene glycol diacrylate and dimethacrylate, butylene glycol diacrylate and dimethacrylate, etc.

Preferred primer-forming materials for the preparation of the primer layer include mixtures of a compound (A) which has a molecular weight of at least 150 but less than 450 and which contains at least two polymerizable unsaturated groups in one molecule and a compound (B) which has a molecular weight of 450 to 2,000 and which contains at least two polymerizable unsaturated groups in one molecule in a weight ratio (A)/(B) of 1/9-9/1.

Primer layers which are formed from such primer-forming materials adhere excellently to the plastic substrate and the properties of the resulting film are excellent. Moreover, the primer layer has a high crosslink density. Therefore, when a metallic film layer is formed on the primer layer, metallized plastic moldings are formed which exhibit excellent adhesivity between the primer layer and the metallic film. Consequently, the laminated product does not exhibit whitening, haze, the rainbow phenomenon and cracks do not develop in the laminated product.

The deposition of the metallic film layer can be accomplished, preferably by a vacuum evaporation method, a sputtering method, an ion plating method, or the like which are all conventional dry type methods for the deposition of metallic films. Suitable metals useful for the deposition of metallic films include various metals such as antimony, aluminum, tin, iron, nickel, zinc, chromium, copper, and the like and alloys of these metals may be used. Examples of the preparation of unsaturated compounds follow.

Example 1: *Preparation of Compound (A)* – 207 parts of phthalic anhydride and 321 parts of neopentyl glycol were charged into a reactor equipped with a stirrer, a thermometer, a partial condenser, a total condenser and a nitrogen gas introducing pipe, and the reaction was continued at 210°C for 5 hours while nitrogen gas was passed therethrough. The water of condensation was removed until a material having an acid value of 0.5 was obtained. The polyester obtained had a hydroxyl value of 376.

To 418 parts of the polyester were added 253 parts of acrylic acid, 4.4 parts of hydroquinone, 1.5 parts of conc sulfuric acid and 98 parts of toluene and the mixture was reacted in a reactor equipped with a stirrer, a thermometer and a Dean Stark type trap at 95° to 100°C under reduced pressure for 5 hours while the water of condensation was removed. The reaction was discontinued when no further water of condensation was evolved. Unreacted acrylic acid, hydroquinone and conc sulfuric acid were removed by washing the reaction product with alkali and water and then toluene was removed whereby an unsaturated compound (A) having 2 acryloyloxy groups in 1 molecule which had a number average molecular weight of 446 as determined by the end-group determination method was obtained.

Example 2: *Preparation of Compound (B)* – 330 parts of bisphenol A type epoxy resin (epoxy equivalent: 300 to 375; Epon 864), 72 parts of acrylic acid, 4 parts of tributylamine and 0.5 part of hydroquinone monomethyl ether were blended and were reacted at 95°C for 6 hours whereby an unsaturated compound (B) having 2 acryloyloxy groups in 1 molecule was obtained. Compound (B) had a number average molecular weight of 844 as determined by the end-group determination method.

Two-Layer Protective Surface

K. Manabe, J. Masumi, T. Tochitani and K. Shinoda; U.S. Patent 4,104,432; August 1, 1978; assigned to Toyoda Gosei KK, Japan describe molded plastic articles having on the surface thereof a thin metal film coated in any order of succession with protective layers of (a) a coat derived from 100 pbw of an acrylic copolymer having a hydroxyl number of 10 to 150, 10 to 150 pbw of an antiyellowing polyisocyanate and 2 to 10 pbw of an ultraviolet ray-absorbing agent and (b) a coat cured by the action of ultraviolet rays.

Such molded plastic articles are manufactured by applying the protective layers (a) and (b) in any order of succession onto a thin metal film covering a molded plastic article.

Plastic articles used in the process are usually molded to have an appropriate shape according to the intended purpose and are provided on the surface thereof with a metal film by a suitable conventional means such as vacuum evaporation or plating. Examples of plastic substances utilizable for manufacturing the molded plastic articles include various plastic resins such as ABS resins, styrene resins, vinyl chloride resins, acrylic resins, polyolefins such as polyethylene and polypropylene subjected to a surface activation treatment, FRP, polyamides and polyesters.

In case a thin metal film is formed on the surface of a molded plastic article by means of a conventional vacuum evaporation treatment, a base coat such as a urethane paint (a urethane paint of a one component system or two component system, whichever is adequate) is first applied onto the surface of the molded plastic article and then cured. A thin metal film is then formed on the surface of the cured base coat by means of vacuum evaporation.

The thickness of a metal film formed on the surface of plastic articles varies according to the method adopted and the intended purpose of the product but is usually within a range of 0.03 to 10 μ, preferably 0.05 to 1 μ. The thickness of a metal film may be defined by the weight of the metal per unit area of the metal film, for example, in terms of micrograms of the metal per square centimeter of the metal film.

A combination of specific protective layers is then applied according to the process onto the surface of the film to furnish it with excellent resistive properties to abrasion and weathering actions. The thin metal film is first coated with a paint (middle coat) composed of 100 pbw of an acrylic copolymer having a hydroxyl number of 10 to 150, 10 to 150 pbw of an antiyellowing polyisocyanate and 2 to 10 pbw of an ultraviolet ray-absorbing agent. This paint is then cured by heating and is further coated with a paint (top coat) curable by the action of ultraviolet rays. The plastic article thus treated is finally subjected to irradiation of ultraviolet rays whereby the product having on the surface thereof a thin metal film coated with a combination of the specific protective layers is obtained.

Example: A linseed oil-modified paint was applied as base coat onto the surface of a molded plastic article made of ABS resin until the thickness of the base coat became 10 μ. The base coat was dried and cured at 70°C for 60 minutes. Aluminum metal was then applied onto the surface of the base coat by means of vacuum evaporation so as to form an aluminum film of 0.2 to 0.8 μ in thickness. A paint consisting of 100 parts of methyl methacrylic polyester polyol having a hydroxyl number of 50, 25 parts of hexamethylene diisocyanate, 7 wt % of an UV-absorbing agent, Tinuvin 328, of the general formula:

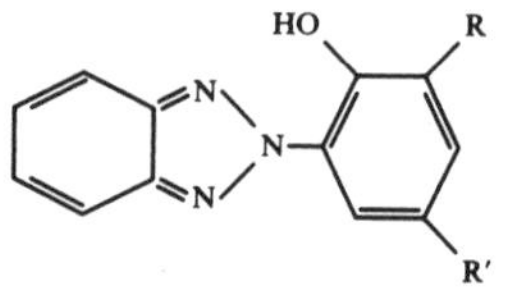

wherein R and R' each stand for an alkyl group, 2.0 parts of ethylbenzene, 20 parts of methyl ethyl ketone and 40 parts of methyl isobutyl ketone was applied as middle coat onto the aluminum film until the thickness of the middle coat became 10 μ. The coated article was then heated at 80°C for 60 minutes to effect curing of the intermediate coat. A paint consisting of 40 parts of hexamethylene diisocyanate, 100 parts of an unsaturated polyester of fumaric acid series containing 0.5% by weight of benzoin methyl ether, 50 parts of styrene, 10 parts of methyl ethyl ketone, 10 parts of isopropyl alcohol and 30 parts of methyl isobutyl ketone was applied as top coat onto the cured middle coat until the thickness of the top coat became 20 μ. The coated article was finally irradiated for 45 seconds with ultraviolet rays having an intensity of 600 W/m^2 to effect curing of the top coat. A combination of the protective layers was thus formed on the surface of the aluminum film.

Copper Deposited from an Acidic Bath

T.F. Davis; U.S. Patent 4,143,186; March 6, 1979; assigned to AMP Incorporated describes an electroless plating of copper in an acidic bath whereby copper by a chemical reduction process is continuously being plated to a desired thickness on a substrate. The process is especially applicable to plating of substrates which are susceptible to alkaline solutions. For example, high temperature substrates such as polyimides, polyparabanic acids, polyimides-amides, and polyhydantoins are especially adaptable for plating with an electroless copper solution of the process.

With reference to the copper salts, these are divalent salts and are added as soluble salt to an aqueous solution. Generally, the salt is $CuSO_4 \cdot 5H_2O$. As a concentrate, the amounts are stated for adding to a 1 liter solution.

For copper sulfate, it is found that the range should be from 4 g/ℓ to 50 g/ℓ. In addition, a complexing agent which buffers the solution to the desired pH range is added to the solution. Complexing agents suitable for this purpose are tartrates, acetates, glycolic acid and ethylenediaminetetraacetic acid. The following complexing agents are also useful: nitrilotriacetic acid, glyconic acid, glyconates, or triethanolamines.

Dimethylamine borane is used as the preferred reducing agent. The deposit of copper contains only traces of boron which is easy to overplate by an electrolytic process.

Still further, stabilizing agents are added to the bath which prevent decomposition. Suitable stabilizing agents are thioureas in an amount of 0.01 to 1.0 ppm; antimony, arsenic and bismuth alone or in combination, in an amount of 1.0 to 20 ppm.

The abovementioned stabilizers such as antimony, arsenic, or bismuth are preferably in a lower oxidation state such as a divalent or trivalent state although the higher valent states will also work but not as well.

In order to adjust the pH of the bath within the desired limits, it is generally adjusted with ammonium hydroxide; potassium or sodium hydroxides are equally useful. The operating temperature of the bath is generally from 70° to 160°F.

In summarizing the above, the general operating conditions for the bath are as follows:

(1) Cu^{++} added as any soluble salt, such as sulfate, nitrate, chloride, etc. 0.05 to 0.15 M

(2) Dimethylamine borane as a reducing agent

(3) A complexing agent that buffers in the pH range desired (tartrates, acetates, glycolic acid)

(4) Stabilizing agents to prevent decomposition [thiourea (0.01 to 1.0 ppm), antimony, arsenic and bismuth (alone or in combination, 1 to 20 ppm)]

Appropriate bath compositions are as follows:

	Bath Composition	Grams per Liter
(A)	$CuSO_4 \cdot 5H_2O$	4-24
	Dimethylamine borane	8-24
	Ethylenediaminetetraacetic acid	12-72
	KH_2PO_4 (as a pH buffer)	20-60
	Operating pH = 5.0 to 7.0; temperature = 140° to 180°F	
(B)	$CuSO_4 \cdot 5H_2O$	25
	$K_4P_2O_7$	100
	Dimethylamine borane	5
	Operating pH = 6.0 to 7.5; temperature = 70° to 100°F	

After the electroless deposition, the deposit may be overplated electrolytically with copper or with various other metals in various combinations as is needed, all as well known in the art.

Unbroken Hydrous Layer

V. Luft; U.S. Patent 4,154,869; May 15, 1979; assigned to Honeywell Inc. has found that the method of electroless plating on a nonconductive substrate or surface, wherein the substrate is sequentially contacted with a tin salt solution, a palladium salt solution, and an electroless plating bath containing the metal can be improved in the following manner.

Specifically, the improvement comprises preparing the nonconductive surface to form an unbroken layer of water adhering thereto upon dipping the surface in deionized water. This is done by cleaning the surface, followed by contacting the surface with a tin salt and then by contacting the sensitized surface with a palladium salt to catalyze the surface. It may be necessary to repeat these three steps to obtain the unbroken layer of water which indicates proper preparation of the surface. The next step includes electrolessly plating the surface with a metal for sufficient time to plate a visible amount of metal thereon, inspecting the surface for plating imperfections, removing the plating, and recleaning the areas of imperfections, followed by recleaning, resensitizing, recatalyzing, and finally, electrolessly plating the surface with the desired amount of metal.

A preferred tin salt solution would be 0.0147 g of stannous chloride, 0.010 ml of 37% hydrochloric acid, and sufficient deaerated deionized water to make up 1 ℓ of solution.

A preferred palladium salt solution comprises 5.0 g of palladium chloride, 7.0 ml of 37% hydrochloric acid and sufficient deaerated deionized water to make up 1 ℓ of solution.

In order to form a continuous, uniform, adherent metal film, it is necessary that the substrate be cleaned and prepared so that an unbroken layer of water will adhere to the substrate when immersed and withdrawn from deionized water. The cleaning step may be accomplished by physical means, such as ultrasonic vibration cleaning or by chemical means. Examples of cleaning solutions are potassium carbonate solutions, sodium hydroxide solutions, and the various chromic/sulfuric acid solutions which are conventionally employed as cleaning agents. The most important consideration is to use a cleaning method which will not undesirably etch or otherwise alter the surface of the substrate. Once the substrate has been cleaned, it should be rinsed thoroughly with deionized water.

Following cleaning of the substrate, the substrate is immersed and agitated in a tin salt solution such as that described hereinabove. Normally, it is necessary to immerse the substrate in the tin salt solution for at least 15 seconds so as to ensure complete contact of the substrate with the tin ions. Following the contact with the tin salt solution, the substrate is then immersed in a palladium salt solution and allowed to remain for at least 15 seconds or more with no agitation. It is preferred that no rinsing of the substrate be done between the immersion in the tin salt solution and the immersion in the palladium salt solution.

At this point, once the palladium salt solution has been contacted on the substrate, the substrate is rinsed again in deionized water, and the sensitizing/catalyzing sequence is repeated one or more times, until an unbroken layer of water is formed. Once this step (or steps) is complete, the substrate is placed in an electroless plating solution and is initiated by a catalyst.

One useful example of a plating solution is Cuposit NL-63, which is a nickel/phosphorus plating solution. Normally, the metallization need take place for only a short period of time until a visible amount of metal has been formed. Typically, 15 to 30 seconds of electroless plating at 180°F will produce a visible amount of metal on the nonconductive surface.

After the light amount of metal has been plated on the surface, it should be inspected to determine if complete metal coverage has been achieved. Any residual contaminants, such as adhesive, used to bond the material for one reason or another, or other impurities which tend to mask and/or poison the sensitizing or catalyzing agents will be revealed by the thin layer of metal since metallization only occurs on the surface where it has been both sensitized and catalyzed. The second advantage to precatalyzing the surface with the minor amount of metal deposited thereon is to increase the wettability of the surface since many materials sought to be metallized are relatively hydrophobic. Surfaces which have been lightly metallized, cleaned, and remetallized appear to be more easily wetted by the sensitizing solution and catalyst solution, although the reason for this is not precisely understood. Metal films formed adhere more strongly and uniformly.

Once the inspection of the visible thin metal has been completed and areas of imperfection have been noted, the thin layer of metal may be dissolved or otherwise removed, such as, for example, by immersion in a nitric acid solution diluted with equal parts of deionized water. Normally, approximately 1 minute of time is necessary to remove the plated metal.

At this point, the sensitizing through the use of a tin salt solution and the catalyzing through the use of a palladium salt solution are repeated, followed by contacting the surface with an electroless plating bath for sufficient time to plate the desired quantity of metal thereon. After plating, the finished product may be rinsed in deionized water and blown dry with dry nitrogen or other oxygen-free gases. In a preferred embodiment, it is desirable to heat the metallized surface at a temperature above the boiling point of water for a short period of time.

The process is useful on a variety of nonconductive surfaces, such as glass, quartz, and other nonporous materials, as well as on plastic or porous materials such as epoxies and other thermosetting resins, thermoplastics, fiber glass cloth, paper, cotton cloth, and synthetic cloth.

Liquid Seeder Preparation for Catalysis

F.J. Nuzzi and D.F. Vitellaro; U.S. Patent 4,160,050; July 3, 1979; assigned to Kollmorgen Technologies Corporation describe liquid seeders and catalyzation processes for the electroless deposition of metal on substrates. The seeders comprise the admixture of sources of cuprous, hydrogen and halogen ions, organic solvent(s), and an agent to convert cupric ions to cuprous ions. The processes include the steps of contacting a substrate with a certain seeder fixing the seeder to the substrate with water, and catalyzing the fixed seeder by further water treatment, use of strong reducing agent, or both.

The seeders may be used with any type of surface that will withstand the necessary contacting with the liquid seeders, including, for example, such materials as glass and ceramics, thermoplastic resins and thermosetting resins, and laminates such as phenolic/paper and epoxy/fiber glass.

The purpose of the organic solvent constituent(s) varies somewhat with the desired application of the liquid seeder being formed. If the seeder is being used basically for its catalyzation capability alone, then the primary purpose of the organic component is to act as a solvent for the copper salt and permit the generation of cuprous ions. If, however, the seeder is to be used to promote adhesion between the catalytic material and, for example, a resinous substrate, then another important function of the organic solvent, or at least one of them where the seeder incorporates two or more organic solvents, is to swell or attack the resinous substrate that is to be metallized. The preferred choice of solvent(s) will vary with the particular substrate being used and the application.

For seeding thermosetting resinous laminates where adhesion promotion is desired, a very polar solvent such as dimethylformamide is a preferred constituent. Where thermoplastic resins are to be so treated, such as acrylonitrile-butadiene-styrene polymers, lower chain alkyl glycols and glycol ethers, such as ethylene glycol, dipropylene glycol and ethylene glycol monoethyl ether are useful organic solvent constituents of the seeder.

Liquid copper seeders of the sort described are preferably utilized in the following fashion: at least the portion(s) of the substrate to be provided with a layer of electroless metal are contacted with a liquid seeder comprising the admixture of cuprous ions, halogen ions and at least one organic solvent. The work piece is then briefly (on the order of 30 seconds to 1 minute) contacted with water, resulting in the formation of a precatalytic coating over whatever surfaces were contacted with the catalyst solution. More precisely, the water treatment fixes the catalytic agent itself, which is still in a precatalytic state, to the substrate. Following this step, the practitioner may follow one of three routes: further water treatment, or contacting with a reducing agent, or both.

Since the agent fixed on the treated substrate surface(s) by the initial water treatment is still in a precatalytic form, it must then be rendered catalytic to the deposition of metal from an electroless metal deposition bath. Water treatment for a period longer than that necessary to initially fix the catalytic agent to the substrate will result in the coating becoming catalytic to the deposition of electroless metal. When cuprous chloride is used in the seeder, for example, the initial fixed coating has a whitish appearance and is essentially noncatalytic. If the water treatment is extended to about 5 minutes, the coating evolves to a pale green color and is then catalytic to electroless metal deposition. The substrate may then be plated in an electroless metal deposition bath.

Alternatively, the substrate may be treated with a strong reducing agent immediately following the initial brief water treatment in order to render it catalytic. With a cuprous chloride-containing catalyst, the previously described whitish precatalytic coating turns gray or black following treatment with a strong reducing agent and is then catalytic to electroless metal deposition.

Still a third alternative is to prolong the initial water treatment past the point necessary to merely fix the precatalytic agent and then treat the substrate in a strong reducing agent.

Strong reducing agent, as used herein, includes any reducing agent that will reduce cuprous ions to elemental copper without detrimental effect on the other process steps. Preferred strong reducing agents include the boranes and borohydrides in aqueous solution, such as dimethylamine borane and sodium borohydride. Aqueous solutions of hydrazine hydrate are also useful for this purpose.

Example 1: A piece of acrylonitrile-butadiene-styrene (ABS) is cleaned by immersion in methanol and dried with high-pressure air. A seeder liquid is prepared by admixing:

Ethylene glycol monoethyl ether, 950 ml
Dimethylformamide, 75 ml
HCl (37%), 50 ml
Sodium hypophosphite, 15 g
CuCl, 40 g

The ABS is treated for 10 minutes in the above seeder, rinsed for 5 minutes in running water, and plated in the following electroless copper bath:

N,N,N'N'-tetrakis(2-hydroxypropyl)ethylenediamine, 0.058 mol/ℓ
Cupric sulfate pentahydrate, 0.036 mol/ℓ
Sodium hydroxide, 0.36 mol/ℓ
Formaldehyde, 0.27 mol/ℓ
Sodium cyanide, 0.0002 mol/ℓ
Wetting agent, 0.001 g/ℓ
Deionized water, to make 1 ℓ
Temperature, 30° to 34°C

The entire surface of the ABS is covered with an adherent layer of bright electrolessly deposited copper.

Example 2: A piece of copper-clad epoxy/fiberglass laminate with holes drilled through it is immersed for 7 minutes in a liquid seeder formed by admixing:

Ethylene glycol, 400 ml
Dimethylformamide, 400 ml
HCl (37%), 200 ml
Deionized water, 200 ml
CuCl, 80 g
Sodium hypophosphite, 20 g

The laminate is then rinsed in running water for 2 minutes and then treated in a reducing agent prepared from 1 g of dimethylamine borane and 1,000 ml of deionized water for 5 minutes. Another water rinse for 2 minutes is followed by plating for 30 minutes in the electroless copper deposition bath of Example 1. The surface and hole walls of the laminate are covered with adherent, bright electroless copper.

GLASS AND CERAMIC SUBSTRATES

GLASS

High Speed Coating of Glass Ribbons

R. Leclercq, P. Capouillet and A. Van Cauter; U.S. Patent 4,123,244; Oct. 31, 1978; assigned to BFG Glassgroup, France describe a process for forming a metal or metal compound coating on a face of a glass substrate by contacting such face while it is at elevated temperature with a gaseous medium containing a substance in gaseous phase, which substance undergoes chemical reaction or decomposition to form the metal or metal compound on the face.

The superiority of the process appears to be particularly evident when attempting to build up coatings rapidly, e.g., at a rate of at least 700 Å of coating thickness per second. The process therefore promises to be of special importance when coating glass ribbons, in the course of continuous production at high speeds, e.g., speeds in excess of 2 m/min and even in excess of 10 m/min such as are often attained by the float process.

Example: Coating apparatus as represented in Figures 5.1a and 5.1b was used for coating a ribbon of glass **1** traveling in the direction indicated by arrow **2** from a float tank (not shown) in which the glass ribbon was formed by a float process on a bath of molten tin. The glass ribbon had a speed of 15 m/min and was supported at the coating station by rollers **3**.

The coating station is located in a compartment **4** of a horizontal gallery having a refractory roof **5**, a refractory sole wall **6** and refractory side walls **7** and **8**, the ends of the compartment being formed by displaceable refractory screens **9**, **10**. The coating apparatus may be disposed within a part of the gallery in which the glass ribbon is annealed, or at a position between the float tank and the annealing gallery. The coating apparatus comprises a vessel **11** containing a gas mixture and having a feed channel **12** extending across substantially the full width of the glass ribbon. The feed channel intrudes into the entry end of a shallow flow passage **13** defined in part by a shroud **14** and in part by the top face of the glass ribbon.

Figure 5.1: High Speed Coating of Glass Ribbons

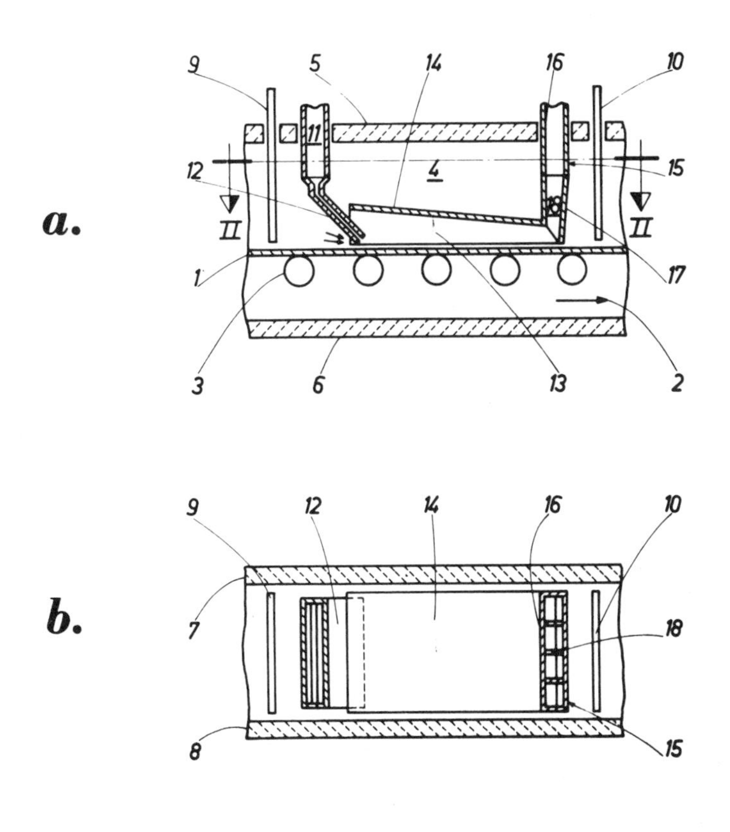

(a) Diagrammatic side view, partly in vertical section, of one coating apparatus embodying the process
(b) Cross-sectional view taken along the line **II–II** of Figure 5.1a

Source: U.S. Patent 4,123,244

The top of the shroud **14** is inclined slightly downward in the gas flow direction so that the gas flow passage **13** slightly decreases in height toward its gas exit end. The length of the passage is 50 cm and its height varies from 25 mm at its entry end to 10 mm at its gas exit end. At such exit end the shroud is connected to a chimney **15**. The front wall **16** of the chimney is vertical but the lower portion **17** of its rear wall slopes upward and rearward relative to such front wall.

The chimney is internally subdivided by partitions such as **18** into a plurality of exhaust passageways distributed in side-by-side relationship across substantially the whole width of the gas exhaust path.

A vapor mixture containing $SnCl_4$ and $SbCl_5$ was generated from a liquid phase containing such ingredients in a volume ratio of 100:1 and this vapor mixture, entrained in a stream of nitrogen, was delivered from vessel **11** through the feed channel **12**.

The temperature of the glass ribbon at the region beneath the entry end of the flow passage **13** was of the order of 585°C.

The rate of delivery of the vapor mixture into the flow passage and the draught forces through the chimney **15** were regulated so as to establish along the passage a substantially turbulent-free flow of vapor mixed with air induced into the passage by the discharge of the vapor stream, as suggested by the arrows beneath the feed channel. Such regulation was, moreover, such that a coating composed essentially of SnO_2 together with a small quantity of Sb_2O_5 serving as doping agent, and having a thickness of 2500 Å, was formed on the traveling glass ribbon. Regulation of the draught forces can be achieved, e.g., by using a regulatable fan in the chimney.

The coating on the glass had a green tint viewed by reflected light. The coated glass had a very high visible light transparency but reflected a significant proportion of incident radiation in the far infrared spectral region.

The emissivity of the coating was 0.4; its diffuse luminous transmission was practically nil.

Examination of the coating showed that it had a homogenous structure and had uniform thickness and optical properties.

Spray Coating Apparatus

R. Van Laethem; U.S. Patent 4,125,391; November 14, 1978; assigned to BFG Glassgroup, France describes a process for forming a metal or metal compound coating on a face of a glass substrate by contacting such face, while the temperature is elevated, with droplets comprising a metal compound which by pyrolysis forms the coating metal or metal compound on the face.

An embodiment of the process, selected by way of example, is illustrated in Figure 5.2 which is a diagrammatic side view partly in vertical section of part of a flat glass manufacturing plant incorporating coating apparatus according to the process.

The coating apparatus is located in an annealing chamber **1**, having a roof wall **2** and a sole wall **3**, through which chamber the glass ribbon **4** is conveyed from a ribbon-forming section of the plant. The chamber may, for example, be part of the annealing lehr of a Libbey-Owens type sheet glass drawing machine, or it may be associated with a float tank in which the glass ribbon is formed by the float process.

The glass ribbon is supported by rollers **5** and travels through the chamber in the direction indicated by arrow **6**. Above the path of the glass ribbon the chamber is provided with displaceable refractory screens **7** and **8** which define between them a compartment in which the metal or metal compound coating is formed on the upper face of the glass ribbon as it travels through the chamber.

Figure 5.2: Spray Coating Apparatus

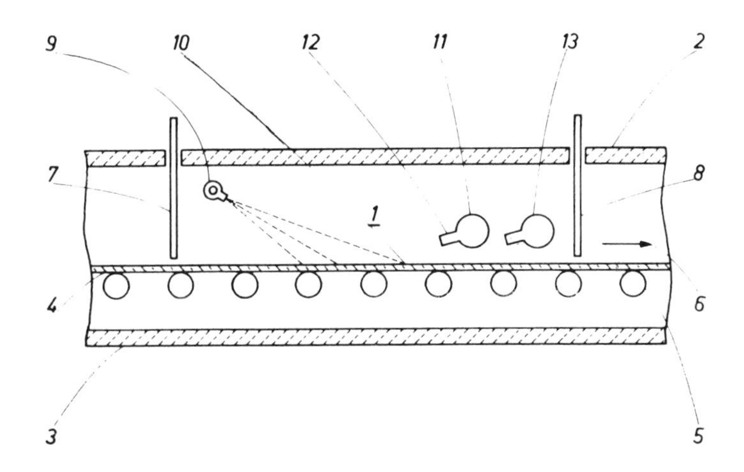

Source: U.S. Patent 4,125,391

A spray gun **9** is mounted above the horizontal path of the glass ribbon and is connected to mechanism (not shown) for displacing such gun to and fro along a horizontal path normal to the direction of the ribbon displacement. The vertical distance between the spray gun and the top face of the glass ribbon is from 15 to 35 cm. The spray gun is orientated so that droplets are discharged in a conical spray whose mean angle α of inclination to the ribbon is from 25° to 35°, the cone angle being 20°.

At a distance of the order of 10 to 30 cm downstream from the downstream boundary **10** of the zone of impingement of the droplet stream on the glass ribbon, there is an exhaust duct **11** which is connected to means (not shown) for maintaining suction forces in the duct. The duct extends transversely across the ribbon path and has a nozzle **12** defining a slotlike gas inlet passage. The entry orifice of the nozzle is at a height of from 1 to 20 cm above the glass ribbon.

In this particular embodiment, a second exhaust duct **13** is provided which is spaced downstream from the duct **11**.

When the apparatus is used, the discharge from the spray gun and the suction forces which serve to aspirate gases into the exhaust ducts **11** and **13** are adjusted so that in the zone upstream of the spray cone the atmosphere is substantially quiescent and unpolluted by vapor droplets or vapors of the sprayed substance, and so that the paths of the droplets from the spray gun to the glass ribbon are substantially unaffected by the suction forces. Moreover, the atmosphere above the zone of impingement of the droplets on the ribbon remains clear. The spray gun is continuously moved to and fro transversely across the ribbon path and the continuously exerted suction forces are such that the atmosphere above any coated zone across the ribbon becomes completely cleared in the period of time taken for the spray gun to complete one movement cycle comprising a to and fro movement across the ribbon path.

By way of modification, the spray gun **9** could be replaced by a series of stationary spray guns mounted in side-by-side positions across the ribbon path so that they together apply coating substance over the full width of the ribbon path, or by a stationary atomizer having a droplet discharge head extending across such path.

Example: Coating apparatus as described with reference to Figure 5.2 was employed for coating a ribbon of glass 3 m in width in the course of its production by a Libbey-Owens type drawing process, the speed of the glass ribbon being of the order of 1 m/min. The coating apparatus was installed at a position such that the temperature of the glass at a zone of impingement of the droplet stream was of the order of 600°C.

The spray gun was of a conventional type, and was operated at a pressure of the order of 4 kg/cm^2. The gun was displaced to and fro across the ribbon path at a height of 30 cm above the glass ribbon, so as to complete nine reciprocations per minute. The spray gun was directed so that the axis of the spray was at 30° to the plane of the glass ribbon.

The suction forces in the exhaust ducting were adjusted to maintain a suction of the order of 100 mm of water in the suction nozzle of each of the ducts **11** and **13**, which nozzles were 20 cm above the glass ribbon.

The spray gun was fed with an aqueous solution of tin chloride obtained by dissolving in water 375 g/ℓ of hydrated tin chloride ($SnCl_2 \cdot 2H_2O$), and adding 55 g/ℓ of NH_4HF_2.

The rate of delivery of the coating solution was 20 ℓ/hr, in an amount of 10 Nm^3/hr of carrier gas. A coating of tin oxide doped by fluorine ions and having a thickness of 7500 Å, was formed on the glass ribbon.

Examination of the coating showed it to be of uniform thickness and optical properties and to have a homogeneous structure. The coating had a neutral tint viewed by reflected light. The coating possessed a high visible light transmissivity and possessed an appreciable reflective power in respect to infrared rays in the wavelength range 2.6 to 40 μ. The emissivity of the coating was 0.1. The diffuse luminous transmission of the coating was very small.

Similar results were obtained in a process in which the same coating procedure was followed for coating a ribbon of float glass as it traveled from the float tank.

Antibonding Film

L.A. Conant, W.M. Bolton and J.E. Wilson; U.S. Patent 4,125,640, Nov. 14, 1978 describe a method of metal armoring glassware by chemical vapor deposition, utilizing an antibonding film to prevent cracking or other defects from developing in the glassware as a result of metal bonding to the glassware.

The method is based on the use of a thin film coating applied to the surface of the glassware, which separates the outer metallic armor from direct contact with the glass surface. This film may be described as an antibonding film since it prevents bonding or adherence of the metal to the glass surface, particularly to defects, including those of microscopic size, which generally exist on the surface of commercial glass. Such defects may be pits, scratches, abrasions, roughened areas, also intentional insignia etch marks and graduation marks frequently used on laboratory and process glassware, where the depositing metal can penetrate and mechanically interlock or bond.

Since the metal coating is applied at elevated temperatures, and since the expansion coefficient of nickel for example is four times greater than borosilicate glass, a high contraction of the nickel will occur upon cooling. If during this contraction process the nickel coating cannot freely move because it is strongly adherent to the glass surface, or mechanically locked to pits, etc., high stress concentrations may develop which can easily cause cracking or fracture of the glassware. It is the function of the intermediate antibonding film to allow the metal to freely contract and conform to the geometry of the glassware on cooling, whereupon it forms a tight shrinkfit to the glassware which is placed in strong compression.

Types of Antibonding Films: These films may be broadly grouped in the following categories:

(1) Dispersion types where fine solid particles, generally colloidal, are suspended in a vehicle that is completely volatilized, leaving the particles lightly bonded to the glass surface.

(2) Dispersions where the particles are suspended in a vehicle composed of a polymer dissolved in a solvent. After curing, the film consists of a plastic matrix with a dispersion of the fine solid particles, or reaction products. This type of film is generally well bonded to the glass surface. The dispersion particles in types (1) and (2) may be graphite, molybdenum disulfide, etc., for low friction and improved thermal conductivity, and/or metal particles like aluminum or silver for improved thermal conductivity, electrical conductivity, enhanced inner vessel reflectivity where this is desired. A black coating of course would not be reflective to the inside of a glass vessel.

(3) Polymer films, generally composed of a plastic or resin with low viscosity for spraying, or that can be thinned in suitable solvent, are coated on the glass surface and cured. These films are usually well bonded to the glass.

(4) Emulsion types are like the suspensions except the dispersion is composed of colloidal liquid droplets, usually a polymeric material. These also may be strongly or weakly adherent to the glass surface depending upon whether or not they are resin bonded.

It is preferred to use antibonding films of the type (1) group, specifically dispersions of colloidal graphite, molybdenum disulfide, fluorocarbons, in vehicles that are easily applied by spraying techniques. The liquid should have a rather fast evaporation rate, e.g., isopropyl alcohol, Freon, etc. These particulate films are not only easy to apply but are easily removed, if required (for example where windows will be placed), economical, require no cure (only drying or a bake, to thoroughly remove liquid). They have good thermal conductivity, some electrical conductivity like graphite, good lubricity (low friction), and high thermal stability, particularly the graphite and molybdenum disulfide. These films are usually applied to the thickness range of 1 to 25 μ (usually about 2 to 10 μ). Good antibonding films that may be classified as type (2) result when fluorocarbon resins and suitable modifiers are dispersed in solvents. These coatings require curing at elevated temperatures (600° to 800°F) and are considerably more expensive.

In practice the coating is applied by gas plating, otherwise known as chemical vapor deposition. It is a preferred method since it can be applied at an elevated temperature, and strongly compresses the glass object upon cooling. This exerts a strengthening effect on the object.

Coating with nickel from nickel tetracarbonyl is particularly versatile since deposition can easily be accomplished between 350° and 550°F or from 150° to 550°F with the use of catalysts. Gas plating nickel from biscarbonyl is well known, although the strengthening effect on glassware by this coating method is not really too obvious.

Aluminum coating is another good method particularly from the compound diethyl aluminum hydride. A deposition temperature of about 750°F is used. Of the two metals, nickel is probably the best because of its superior corrosion resistance, and ease of application.

The method of gas plating is as follows: The object to be coated is placed in an air tight chamber, and purged of air with an inert gas. The object such as a glass flask is then heated by a resistance heater, placed inside it, and controlled by a Variac. The object may also be heated by a fluid such as heat exchange fluid which is piped to it and circulates within its interior. It may also be heated externally by an infrared lamp or other exterior radiating heat source. In this case the sides of the chamber are transparent.

When the object reaches the desired coating temperature, the coating compound (i.e., nickel carbonyl) in the form of a fluid, is metered through a flowmeter to a vaporizer, which converts it to a vapor. The vapor is blended with carrier gas, metered through a flowmeter and flows to the plating chamber where it impinges on the heated object. The compound then thermally dissociates into a metal and a gas. In the case of nickel carbonyl it decomposes to Ni and carbon monoxide as shown by the following chemical reaction: $Ni(CO)_4 \rightarrow Ni + 4CO$. The nickel is deposited as a dense, continuous coating and the carbon monoxide is exhausted and burned or collected.

The nickel is contained in a pressurized cylinder, and metered to the vaporizer through a flowmeter. The inert purging gas is metered through a flowmeter. The coating in the case of nickel may be of any desired thickness, depending upon the application, ranging from less than 5 mils to 100 mils or more. A preferred range found suitable for most applications, is from 5 mils to 20 mils.

Metal Acetyl Acetonate Coating

E. Plumat and R. Posset; U.S. Patent 4,129,434; December 12, 1978; assigned to Glaverbell, Belgium describe a process for forming a metal oxide coating on a substrate wherein a solution of at least one metal compound is applied to the substrate and the compound is converted in situ to leave a coating of at least one metal oxide. The solution used for coating the substrate is a solution of a metal acetyl acetonate or a mixture of metal acetyl acetonates in an aprotic solvent, a substituted or unsubstituted monocarboxylic acid solvent, an amine or diamine solvent, or a mixture of two or more solvents selected from solvents of these classes.

By applying this process, it is possible to form metal oxide coatings which are of substantially uniform thickness and composition, as is required for example in the case of very thin coatings having a substantial light transparency. The process is therefore very suitable for forming optical coatings on vitreous bodies or articles for modifying their light-transmitting and/or light-reflecting properties, e.g., for giving the bodies or articles a tinted appearance when viewed by transmitted or reflected light.

Preferably the substrate is preheated to a sufficiently high temperature to provide the heat required to effect the conversion of the metal acetyl acetonate. By preheating the substrate it is possible to bring about substantial evaporation of the solvent and conversion of the metal compound or compounds immediately on contact of the solution with the substrate.

In general, the optimum temperature of the substrate at the time that it is coated is in the range of 300° to 700°C. The temperature should in general preferably be chosen as high as possible consistent with avoiding impairment of the substrate. For coating vitreous substrates, the recommended temperature range is 450° to 650°C. By working within this range, very uniform coatings can be formed and moreover a very strong adherence of the coating to the vitreous substrate can be achieved, this adherence also being influenced by the temperature of the substrate on coating.

The solution of metal acetyl acetonate(s) is preferably applied in the form of droplets to achieve the required results. Use can be made of an inside-mixing atomizing gun fed separately with compressed air and the solution of metal acetyl acetonate(s), both at the same pressure, which may, e.g., be of the order of 1.5 kg/cm^2 above atmospheric pressure. The acetyl acetonate solution itself can be at ambient temperature or any higher temperature provided that there is no undue premature evaporation of the solvent or decomposition and oxidation of the acetyl acetonate(s), and provided that the substrate is not subjected to a harmful thermal shock.

As examples of suitable aprotic solvents, the following are cited: dimethylformamide, dimethylacetamide, tetramethylurea, dimethyl sulfoxide, acetonitrile, nitrobenzene, ethylene carbonate, tetramethylene sulfone, hexamethylphosphoramide. Particular preference is given to dimethyl formamide.

Suitable solvents also include substituted and unsubstituted monocarboxylic acids. Preference is given to aliphatic substituted and unsubstituted monocarboxylic acids. Particularly good examples include acetic acid (CH_3COOH), butyric acid (CH_3CH_2COOH), acrylic acid ($CH_2CHCOOH$), thioglycolic acid ($HSCH_2COOH$), formic acid ($HCOOH$).

The third specified class of solvents comprises amine and diamine solvents. Preference is given to alkyl and alkylene amino and diamino solvents in which the amino group or groups is or are unsubstituted. Particularly good examples include: ethylene diamine, propylene diamine, butyl amine, propyl amine. These solvents do not decompose to yield oxygen and the process can be performed using such a solvent or solvents to obtain a suboxide coating, e.g., vanadium dioxide (VO_2).

Example: Various filmogenic solutions were prepared for use in forming on vitreous substrates, coatings respectively comprising an oxide of chromium, titanium, iron, zirconium, cobalt and zinc. The solutions were prepared by dissolving the acetyl acetonate of the corresponding metal in the solvents identified in the table set out below.

The concentration of each metal acetyl acetonate solution corresponded to 40 g of the corresponding metal oxide per liter of solution.

The solutions were sprayed onto pieces of flat glass 4 mm in thickness, preheated to a temperature of the order of 580°C.

The radiation transmitting, reflecting and absorbing characteristics of the metal oxide layers obtained are indicated in the table in which the term energy denotes total radiation energy over the whole spectrum including infrared and ultraviolet light.

Solvent	TiO_2 Formic Acid	Fe_2O_3 Glacial Acetic Acid	ZrO_2 Ethylene Diamine	Co_3O_4 Dimethyl Sulfoxide	ZnO Propylamine	Cr_2O_3 Dimethyl Formamide
Coating thickness, Å	450	510	600	920	300-350	600
Tint by transmitted light	grey	yellow-amber	grey	brown	grey	greenish grey
Energy transmission, %	68.1	56.0	76.3	35.5	77.4	68.2
Energy reflection, %*	27.1	31.6	16.9	38.2	14.0	19.8
Energy absorption, %*	4.8	12.4	6.8	26.3	8.6	12.0
Light transmission, %	64.6	41.2	78.6	26.8	79.3	64.9
Light reflection, %*	34.5	44.2	20.2	39.9	15.2	24.8

*At coated side.

Stannic Oxide Coating

W.A. Larkin; U.S. Patent 4,130,673; December 19, 1978; assigned to M&T Chemicals Inc. describes a method for reducing the coefficient of friction and susceptibility to scratching exhibited by glass surfaces, the method consisting essentially of the following sequence of operations:

(1) Maintaining the glass surface at a temperature of between 450° and 600°C while applying to the glass surface vaporized or finely divided butyltin trichloride; and

(2) Maintaining the glass surface at a temperature of 350°C or below while applying to the glass surface a coating of a natural wax or synthetic polymer.

The process can be more clearly understood by referring to Figure 5.3 which is a diagrammatic flow sheet depicting the various steps involved in the present method for coating glass articles.

Figure 5.3: Flow Sheet for Glass Coating Process

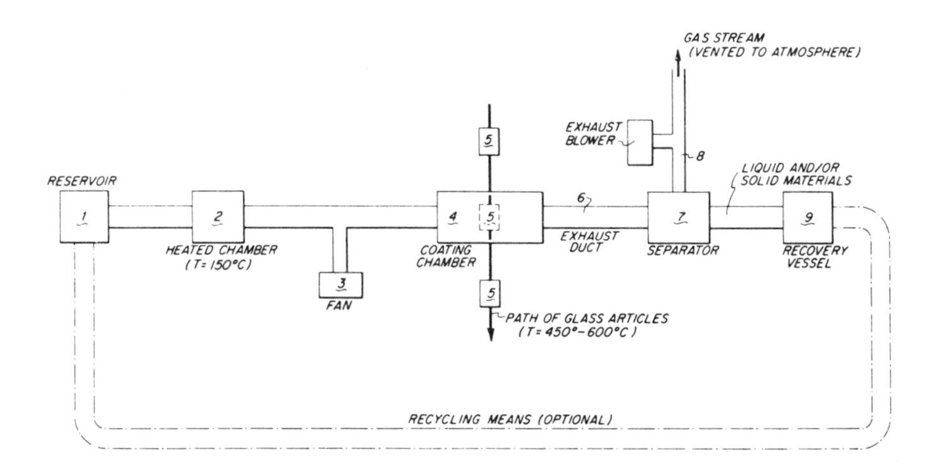

Source: U.S. Patent 4,130,673

Liquid butyltin trichloride, which is stored in a suitable airtight reservoir **1** flows into a heated chamber **2** wherein vaporization of the halide occurs. The flow of organotin compound to the heated chamber is controlled by means of a suitable valve pump or gate (not shown). The temperature within the heater is maintained at approximately 150°C. The spray of finely divided organotin compound is entrained in an airstream generated by a fan **3** and enters the coating chamber **4**. The flow rate of the airstream is between 25 and 50 cfh. A portion of the butyltin trichloride contacts the heated glass container **5** or other articles in the coating chamber. The glass articles are transported into and out of the coating chamber by means of a suitable conveyor (not shown).

The surface of the glass article is at a temperature of between 450° and 600°C. At these temperatures the tin compound which contacts the surface of the glass is converted to stannic oxide. A mixture of air and the unpyrolyzed butyltin trichloride is drawn out of the coating chamber by means of a blower (not shown) through an exit duct **6** which contains an electrostatic precipitator **7** or other suitable device for separating gasses from liquid and solid materials. The gas stream passes through an exhaust passage **8** and is discharged into the atmosphere. The solid and liquid components leaving the coating chamber are recovered in a suitable receiver **9**. Means for returning the recovered material to the reservoir may optionally be included, as indicated by the broken lines in the drawing.

The glass articles are transported through the coating area by means of a suitable conveyor. Preferably, the glass articles are coated shortly after being shaped and prior to being placed in the annealing lehr. The temperature of the articles in the coating area is above the pyrolysis temperature of butyltin trichloride, preferably between 450° and 600°C. The residence time of each article in the coating area is sufficient to deposit an oxide film that is about 500 mμ in thickness. Thicker coatings are undesirable since they cause the glass to assume an iridescent appearance.

It has been found that when a coating of natural wax or synthetic organic polymer is applied over a film of stannic oxide prepared in accordance with the present method, the coefficient of friction exhibited by the coated glass is significantly lower than can be achieved by applying the organic coating to a stannic oxide film prepared using other organic or inorganic tin compounds.

Among the synthetic polymer coatings that can be applied to the glass articles are polyethylene, oxidized polyethylene, copolymers of a polyoxyalkylene or etherified polyoxyalkylene and a dialkylpolysiloxane such as dimethylpolysiloxane, copolymers, including those described in U.S. Patent 3,554,787, wherein the major constituent is a vinyl ester such as vinyl acetate or an ester of an unsaturated acid, such as ethyl acrylate; vinyl alcohol polymers, including those disclosed in U.S. Patent 3,352,707; polyurethanes, including those described in U.S. Patent 3,407,805; and maleic anhydride copolymers, including those described in U.S. Patent 3,598,632.

The organic coating composition is usually applied as the glass articles emerge on a conveyor from the cold end of the annealing lehr using any suitable means such as traversing spray nozzles. The coating is applied as a solution or emulsion at a rate of between 0.5 to 5 qt for each 1,000 ft^2 of conveyor that passes in front of the spray nozzles.

Example: A film of liquid butyltin trichloride was applied to one surface of heated glass plates measuring 2" x 3" (5 x 8 cm) by exposing each plate for 3 seconds to a finely divided spray of the organotin compound emanating from the orifice of an atomizer. The plates were located at a distance of 1 ft (30 cm) from the orifice. The atomization was accomplished using compressed air at a flow rate of between 7,500 and 8,000 cc/min. Just prior to being sprayed the plates were heated for ten minutes in a furnace maintained at a temperature of 900°F (480°C). The coated plates were heated to a temperature of 250°C for five minutes to convert the butyltin trichloride to stannic oxide.

Before the plates had an opportunity to cool to ambient temperature (about 21°C) they were sprayed with an aqueous emulsion prepared by diluting with 980 parts of water a mixture containing 4.2 parts of an oxidized polyethylene (Polyethylene 629), 0.75 part of mixture of fatty acids commonly referred to as tall oil fatty acids, 0.75 part of morpholine and 14.3 parts of water. The emulsion was applied using the same procedure described for applying the butyltin trichloride, with the exception that each plate was exposed to the spray for 5 seconds. After cooling to ambient temperature the coated plates were immersed for two minutes in a 4% by weight aqueous solution of sodium hydroxide maintained at a temperature of 60°±4°C. This treatment is equivalent to the washing cycle used commercially for glass soft drink bottles.

Semireflecting Glass Products

The following are prior art processes for producing semireflecting windows by the spraying of organometallic salts in solution and pyrolysis of these salts:

> Using acetylacetonates dissolved in aromatic or chlorinated solvents which gives a quality product, produced at low cost, but with all the drawbacks associated with the use of these solvents; or
>
> Using organometallic salts with longer hydrocarbon chains which makes it possible to eliminate these drawbacks, but leads to more costly product and which, in certain cases, shows less adherence although with a greater homogeneity of the deposit.

D. Philibert; U.S. Patent 4,120,679; October 17, 1978; assigned to Saint-Gobain Industries, France describes a process to eliminate these drawbacks by replacing the abovementioned organometallic salts with a complex molecule which may be dissolved in aliphatic solvents and which will make it possible to produce, under normal conditions, i.e., by spraying of the solution onto a glass product, such as a window, at a suitable temperature and pyrolysis of the complex molecule, a metal oxide deposit adhering perfectly to its support and showing satisfactory homogeneity at a cost of the same order as that associated with conventional processes using acetylacetonates dissolved in aromatic or chlorinated solvents.

Briefly stated, the process comprises applying to at least a portion of one surface of the product a solution of at least one charge-transfer complex in a nonaqueous solvent therefor, the product having been heated prior to the application to a temperature sufficient to evaporate the solvent and cause pyrolysis of the charge-transfer complex. The charge-transfer complex has the general formula MeX_mY_n where Me is a transition metal or metalloid, X is at least one β-diketone ester or alcohol radical having no more than 10 carbon atoms, and Y is at least one β-diketone ester or alcohol radical having no more than 10 carbon atoms and different from X; or one of X or Y can be at least one halogen radical; and m plus n equal the valence of Me. Complexes containing an alcohol radical will better dissolve in alcohols and those containing an ester radical in esters.

The particular temperature of the glass sheet, of course, depends on the charge-transfer complex and the solvent used, but it generally ranges from 580° to 640°C as in the conventional processes.

The preferred metals are transition metals such as chromium, iron, cobalt, nickel, titanium and vanadium, and the metals of Groups III-A, IV-A and V-A of the Periodic Table. As to the metalloids, silicon is preferred. The most useful elements considered as a group, are Ti, V, Cr, Mn, Fe, Co, Ni, Cu, Zn, Zr, Cd, Al, In, Sn, Sb, Bi and Ce, as well as Si.

As to the β-diketones, it is preferred to use at least one acetylacetone radical and at least one isovaleroylacetone radical. In fact, it is preferred X or Y be at least one acetylacetone radical and that the other a C_{6-10} β-diketone radical or radicals. As noted, X or Y can also be a halogen such as chlorine, fluorine, or bromine with fluorine being preferred.

The solvent used is a nonaqueous, organic solvent and can be selected from any known aliphatic or aromatic solvents and the solvent can be chlorinated, an alcohol, ester, ketone, or similar type, or combination thereof.

It was noted quite unexpectedly that when two distinct charge-transfer complexes of the process each have a low solubility in a solvent, they may become jointly soluble in this same solvent. This unusual and unexpected property makes it possible to increase the quantity of acetylacetonate. The result is that when the acetylacetonate content increases, the industrial price decreases, the solubility in aliphatic solvents decreases, but the amount of pyrolysis increases for the molar mass decreases and the nonvolatile organic part decreases. For a given solvent showing little toxicity, it is, therefore, possible to thus push back further the limit of solubility.

The charge-transfer complexes can be prepared by any of the known procedures by using, for example, distilled acetylacetone and a C_{6-10} β-diketone, such as isovaleroylacetone in excess with respect to a metal salt such as $FeCl_3$, $CrCl_3$, $Cr(NO_3)_3$, $CoCl_2$, and the like. The mixture of these compounds is vigorously shaken and distilled and the distilled solution is then washed with water to eliminate the remaining inorganic salts. The complex is concentrated to dryness and then dissolved in an appropriate organic solvent. The organic phase is solubilized and then dried on $CaCl_2$. All traces of inorganic salts are eliminated by filtration and the solvent is removed by evaporation.

Example: A chromium charge-transfer complex containing an acetylacetone radical and two isovaleroylacetone radicals was synthesized as previously described. The valence of the chromium was 3 and its coordination number was 6. This complex was perfectly stable under normal conditions and its metal content was 11.75% by weight. A 3% chromium solution by weight in ethyl acetate was then prepared from this molecule.

This solution was sprayed onto glass as it left a vertical tempering oven at a temperature ranging from 600° to 620°C.

The thus treated glass had a grey metallic tint when viewed by reflection and a yellowish brown tint when viewed by transmission.

Silicon/Metal Oxide Coated Glass

H.E. Donley; U.S. Patent 4,100,330; July 11, 1978; assigned to PPG Industries, Inc. describes a method for coating a glass substrate with a first film of silicon and a second film of metal oxide by pyrolytic coating techniques. The resultant coated article is more durable than a silicon coated article and has better solar energy control properties than a metal oxide coated article.

The preferred commercially practicable embodiment of the method involves moving a hot glass substrate, preferably a continuous ribbon of float glass at a temperature of at least about 400°C, in relation to a first coating apparatus through which a silane-containing gas is released toward the hot glass surface in a nonoxidizing atmosphere to deposit a coating of silicon, and a second coating apparatus through which a solution of an organometallic coating reactant is released toward the hot glass surface in an oxidizing atmosphere to deposit a coating of metal oxide.

In the first coating step, the temperature of the substrate is preferably between about 500° and 700°C. In a typical float glass operation, where a ribbon of glass is being advanced along a bath of molten metal over which a protective atmosphere is maintained, the silane-containing gas is preferably released in a hot zone where the temperature is between about 600° and 670°C and within the protective atmosphere of nitrogen and hydrogen typically present. The rate of movement of the glass ribbon is generally dictated by glass making considerations. Therefore, the coating parameters are adjusted in relation to the rate of advance of the glass substrate. The concentration of the silane in the gas is controlled to provide the desired coating thickness, while the gas flow rate is regulated to insure uniformity of the coating.

The silane-containing gas preferably comprises from about 0.1 to 20% by volume of a silane, for example, monosilane (SiH_4), up to about 10% by volume of hydrogen, and from about 70 to 99.9% of an inert gas such as nitrogen. The temperature of the gas is maintained well below 400°C prior to contacting the hot glass surface to prevent premature decomposition of the silane. A protective atmosphere of 90 to 95% nitrogen and 5 to 10% hydrogen is preferred. A detailed description of apparatus and method for carrying out this first coating step is found in Belgian Patent 830,179. The silicon coating typically has a thickness of about 250 to 600 Å, preferably from about 300 to 450 Å.

After the silicon coated glass ribbon has advanced beyond the nonoxidizing atmosphere, but while it is still at a temperature sufficient to pyrolyze the organometallic coating reactant, preferably between about 250° and 650°C, the silicon coated glass surface is contacted with a coating composition containing an organometallic coating reactant which thermally decomposes on contact with the hot glass surface in an oxidizing atmosphere to form a metal oxide coating.

The preferred organometallic coating reactants include metal diketonates, such as cobalt, iron, nickel, and chromium acetylacetonates; metal carboxylates, such as cobalt, iron, nickel and chromium 2-ethylhexanoates; and quaternary metal carboxylates, such as cobalt, iron, nickel, chromium, vanadium and titanium neodecanoates, neohexanoates and neoheptanoates. The coating reactants are preferably dissolved in an organic solvent at a concentration of about 0.1 to 10% by weight of the metal.

Various aromatic and aliphatic solvents are suitable, particularly highly volatile solvents such as benzene, toluene and heptane. Preferred solvents, for safety reasons, include halocarbons and halogenated aliphatic hydrocarbons such as trichloroethane or methylene chloride. The quaternary metal carboxylates may be solubilized in water with ammonia.

A solution of coating reactant may be vaporized and transported to the hot glass surface where the coating reactant pyrolyzes to deposit a metal oxide film by chemical vapor deposition in an oxidizing atmosphere. However, it is preferred to spray a solution of coating reactant in liquid form onto the hot glass surface. A detailed description of apparatus and method for carrying out this second coating step is found in U.S. Patent 3,660,061. The metal oxide coating typically has a thickness of about 300 to 600 Å, preferably about 450 to 500 Å.

Example 1: Clear float glass 6 mm in thickness is maintained in an atmosphere consisting of about 93% by volume nitrogen and about 7% by volume hydrogen while the glass surface, at a temperature of about 610°C, is contacted with a gaseous mixture of about 5% by volume monosilane and about 95% by volume nitrogen. A silicon coating approximately 350 Å thick is deposited.

The silicon coated glass appears brown by transmission with a luminous transmittance of 34% and silver in reflection with a luminous reflectance of about 56% and a dominant wavelength of reflected light of about 480 nm. The shading coefficient is 0.52 for a monolithic coated sheet and 0.47 for a double glazed unit.

The silicon coating is completely removed from the glass by 20 strokes of uniform force with a slurry of cerium oxide or pumice. The silicon coating is also completely removed by immersion for 30 seconds in a solution of 200 g of sodium hydroxide in 1,000 ml of water at 200°F.

Example 2: Clear float glass 6 mm in thickness and at a temperature of about 590°C is contacted in air with a spray of a solution containing 32.0 g of cobalt acetylacetonate, 7.0 g of iron acetylacetonate, 8.5 g of chromium acetylacetonate, 12 cc of meta-cresol, and 189 cc of solvent which is equal parts by volume methylene chloride and trichloroethylene. A metal oxide coating approximately 500 Å thick is deposited.

The metal oxide coated glass has a luminous transmittance of about 39% and luminous reflectance of about 35% with a dominant wavelength of reflected light of 556 nm. The shading coefficient for a monolithic coated sheet is 0.60.

The metal oxide coating is not damaged by 20 strokes of uniform force with a slurry of cerium oxide or pumice, nor is it removed from the glass surface by immersion for 30 seconds in a solution of 200 g of sodium hydroxide in 1,000 cc of water at 200°F.

Example 3: Clear float glass is coated with silicon as in Example 1, then with metal oxides as in Example 2 to a final luminous transmittance of 26%.

The silicon/metal oxide coated glass has a luminous reflectance of 58%, similar to the silicon coated glass, but the dominant wavelength of reflected light is 556 nm, similar to the metal oxide coated glass. Moreover, the shading coefficient for a monolithic coated sheet is 0.46, lower than either the silicon or metal oxide coated sheets, and comparable to the silicon coated sheet in a double glazed unit.

The silicon/metal oxide coating is not damaged by 20 strokes of uniform force with a slurry of cerium oxide or pumice, nor is it removed by immersion for 30 seconds in a solution of 200 g of sodium hydroxide in 1,000 cc of water at 200°F.

CERAMIC

Ceramic Shell Molds

I.R. Walters and H.G. Emblem; U.S. Patent 4,102,691; July 25, 1978; assigned to Zirconal Processes Limited, England describe a method of producing a shaped refractory object which comprises the steps of coating a shaped pattern with a slurry made by combining a refractory powder, an aqueous solution of a zirconium salt which when dissolved in water yields an aqueous solution which is acidic, and a gellation-delaying agent selected from the group consisting of magnesium acetate, magnesium lactate, ammonium lactate, glycine, betaine, fructose and polyhydric alcohols and thereafter contacting the slurry with ammonia vapor thereby causing the zirconium salt to gell.

The ammonia vapor is preferably ammonia gas but the use of a mist of liquid ammonia droplets is also envisaged, as is the use of a mist of an aqueous solution of ammonia. The preferred aqueous solution of ammonia has a specific gravity of about 0.91 corresponding to about 25% NH_3 and desirably has a specific gravity of about 0.88 corresponding to about 35% NH_3.

Preferably more than one coating step is performed each with a subsequent contacting with ammonia vapor. A further preference is that after each coating step but prior to the vapor contacting step the unset slurry should be dusted, i.e., lightly covered, with refractory granules.

In the preparation of molds for coating metals or alloys according to the process a monolithic ceramic shell mold may be prepared by applying several coatings to the expendable pattern, with coarse refractory material being dusted on to each wet coating. Each coating is hardened with ammonia before the next coating is applied. Usually six to eight coatings and dustings are suitable.

Alternatively, a solid block mold may be prepared by first applying only one coating to the expendable pattern, dusting coarse refractory material on to the wet coating and hardening the coating by the action of ammonia, then surrounding the coated pattern with a slurry of a powdered refractory material in a suitable binding agent. The refractory materials envisaged are silica, the aluminum silicates such as mullite, sillimanite, calcined kaolin and calcined fireclay grog, together with zircon and zirconia.

Example 1: *Ceramic Shell Mold Production* – A solution of magnesium acetate crystals ($4H_2O$) in zirconium acetate solution (containing zirconium corresponding to a ZrO_2 content of 22% nominal) can be used as follows in the preparation of ceramic shell molds. A suitable solution contains 10 g magnesium acetate crystals ($4H_2O$) per 100 ml of zirconium acetate solution (containing zirconium corresponding to a ZrO_2 content of 22% nominal).

A slurry containing 200 g of Molochite–120 grade per 100 ml of the magnesium acetate/zirconium acetate solution was prepared. This slurry was used to coat a wax pattern. Tabular alumina 28 to 48 grade was dusted on to the wet coating, which was hardened by exposure to ammonia vapor. This operation of dipping, dusting and exposure to ammonia vapor was repeated a further three times, giving a total of four dips and four dustings. The coatings were allowed to dry and harden overnight. When drying was complete a strong shell was obtained. To build up a large shell, more than four coatings will be required. A coarser grade of tabular alumina is desirable for dusting the later coatings.

An alternative procedure is to carry out two of the dipping, dusting and exposure to ammonia vapor sequences described above, then complete the forming of the ceramic shell mold by known methods using a slurry in which the binder is a silica aquasol or a hydrolyzed ethyl silicate solution.

Molochite is a china clay calcination product.

Example 2: *Solid Block Mold Production* – A solid block mold may be prepared by first applying only one coating to the expendable pattern, dusting coarse refractory material on to the wet coating and hardening the coating by the action of ammonia vapor, then surrounding the coated pattern with a slurry of a powdered refractory material in a suitable binding agent.

A suitable coating composition is a slurry containing 200 g of Molochite–120 grade per 100 ml of a solution containing 10 g of magnesium acetate ($4H_2O$) crystals per 100 ml of zirconium acetate solution (containing zirconium corresponding to a ZrO_2 content of 22% nominal). In some cases it may be advantageous to carry out two of the dipping, dusting and exposure to ammonia vapor sequences before surrounding the coated pattern with a slurry of a powdered refractory material in a suitable binding agent.

Differential Cooling Applied to Metal-Ceramic Interlayer

W.D. Marscher; U.S. Patent 4,109,031; August 22, 1978; assigned to United Technologies Corporation describes a process which utilizes differential cooling for the production of metal-ceramic composite articles. The articles have a mixed metal-ceramic interlayer between the metal and ceramic components and this interlayer has a characteristic softening temperature at which the interlayer material becomes plastic and deforms freely upon the application of load. Differential cooling is applied to the metal component to cool it quickly below the softening temperature of the intermediate layer while the intermediate layer is above its softening temperature and while the ceramic component is above the softening temperature of the intermediate layer.

After the preferential cooling of the metal component the article is allowed to cool at a uniform rate. The effect of the preferential cooling step is to cause the metal component to contract while the intermediate layer is still in a plastic condition and the ceramic component is at an elevated temperature. Upon subsequent cooling, the intermediate layer abruptly regains most of its strength but, due to the nonuniform temperature gradient remaining from the preferential cooling, the residual stress is different from that which would result from that obtained by the conventional uniform cooling which has previously been employed.

The process will be described with reference to the specific application to gas turbine seals.

Figure 5.4 shows a schematic cross section of a gas turbine air seal.

Figure 5.4: Cross Section of Gas Turbine Air Seal

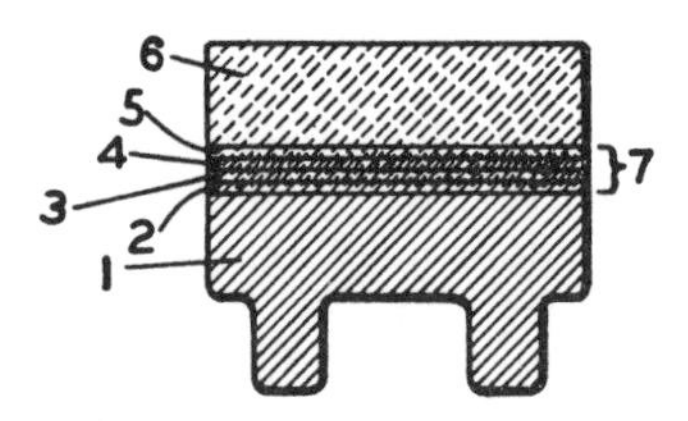

Source: U.S. Patent 4,109,031

The air seal consists of a metal backing plate **1** and a ceramic seal component **6** separated by interlayer **7** of mixed metal-ceramic. The backing plate is attached to the turbine engine case while the inner ceramic component is in close proximity to the rotating blades. The function of the seal is to minimize gas leakage around the blades. In a specific embodiment, the backup layer may be a cobalt superalloy such as MAR-M509 and the ceramic component may be based on zirconia.

The interlayer may be comprised of several layers **2**, **3**, **4** and **5** which may have different compositions, for example, if the interlayer is comprised of four layers as shown in Figure 5.4, layer **5** might be 80% ceramic, 20% metal; layer **4** might be 60% ceramic, 40% metal; layer **3** might be 40% ceramic, 60% metal; and layer **2** might be 20% ceramic, 80% metal. Of course other combinations of metal and ceramic layers, including more or fewer layers, or a continuously graded arrangement might be used. While seals have been made using MAR-M509 and zirconia, other materials could of course be substituted. MAR-M509 is a cobalt base superalloy which has a nominal composition of 0.6% C, 24% Cr, 10% Ni, 7% W, 7.5% Ta, 0.2% Ti, 1% Fe, balance essentially cobalt.

Other nickel and cobalt base superalloys could be substituted for the MAR-M509. Likewise, other ceramics such as alumina might be substituted for the zirconia. These ceramics may be modified, as for example, additions may be made to zirconia to stabilize the crystal structure. The metal powder in the intermediate layer need not be of the same composition as the backup layer. Of course, if other materials are substituted, their mechanical and thermal properties must be taken into consideration.

In the operation of the gas turbine engine, the seals are constrained to operate under conditions such that the intermediate layer is always below its softening temperature, despite the fact that the surface of the ceramic component may be in excess of 2000°F. Sufficient cooling air is applied to the metal backup component to maintain the intermediate layer at a temperature below its softening point. Consequently, there is always a degree of prestress in conventionally produced metal-ceramic seals during operation. This prestress in the ceramic will be compressive since the metal will contract more than ceramic as the temperature of the seal drops below 1600°F.

This process involves differential cooling so that the degree of prestress in the seal can be modified to be other than that which would be found under an equilibrium situation. In particular, by preferentially cooling the metal backup element while maintaining the intermediate layer and the ceramic component at an elevated temperature and then allowing the complete seal to cool uniformly, the compressive prestress in the ceramic portion of the seal may be reduced. Another way to consider the process is to say that the stress free temperature has been reduced. The reduction of stress free temperature is a consequence of the preferential cooling since the intermediate layer still becomes hard at the same temperature.

In summary, the process involves heating the complete seal assembly to a temperature above the softening temperature of the intermediate layer and preferentially cooling the metal component while maintaining the ceramic and the intermediate layer at an elevated temperature with the temperature of the intermediate layer staying above its softening point. After the preferential cooling step, the whole seal assembly is then allowed to cool substantially uniformly as, for example, by convective cooling in a gaseous environment. For example, the preferential cooling step may be applied by an air jet or equivalent cooling jet of another gas or heat transfer media.

Heat Treatment of Cuprous Oxide Coated Article

J.S. Breininger and C.B. Greenberg; U.S. Patent 4,170,461; October 9, 1979; assigned to PPG Industries, Inc. describe a wet chemical method for the direct deposition of cuprous oxide onto a transparent nonmetallic substrate such as glass wherein the electrolessly deposited cuprous oxide film is heated to effect a change in the color of transmitted light.

In more detail, large sheets of glass, preferably soda-lime-silica glass about 7/32" thick, are cleaned by conventional procedures, preferably a blocking operation carried out with rotating felt blocks which gently abrade the glass surface with an aqueous slurry of a commercial cleaning compound, preferably cerium oxide. A suitable continuous line apparatus for washing, rinsing and sweeping the surface is shown in U.S. Patent 3,723,158.

The surface to be coated is contacted with a dilute aqueous solution of a sensitizing agent, preferably 0.01 to 1.0 g/ℓ of stannous chloride. The sensitized surface is preferably activated by depositing on it a thin catalytic silver film, preferably by contacting the sensitized surface with an alkaline solution of ammoniacal silver nitrate and a solution of a reducing agent, preferably dextrose. The thickness of the silver film preferably reduces the luminous transmittance of the sheet to about 40 to 80%.

It is important that the silver film be substantially free from silver oxide as the presence of the oxide appears to favor deposition of copper in the subsequent coating step. Therefore, if it is likely that the silver film has undergone significant oxidation prior to the cuprous oxide coating step, it is preferred to rinse the activated surface with a dilute solution of an oxide inhibitor such as sodium borohydride, ammonium polysulfide, formaldehyde, or preferably sodium thiosulfate.

The silver activated surface is contacted with an electroless plating bath comprising a copper salt, a complexing agent, a reducing agent and sufficient alkali to raise the pH above about 12.9. The preferred copper salt is copper sulfate and the preferred complexing agent is Rochelle salt (sodium potassium tartrate) although other complexing agents such as gluconic, citric, malic and lactic acids and their alkali metal salts may be used. Formaldehyde, particularly a 37% aqueous solution is a preferred reducing agent, although other common reducing agents such as dextrose or hydrazine sulfate can be used. The preferred alkali is sodium hydroxide.

In a most preferred embodiment, a silver activated glass substrate is contacted for several minutes at ambient temperature with an aqueous solution comprising per liter, about 2 to 5 g copper sulfate, 15 to 40 ml formaldehyde solution, and 1 to 5 g Rochelle salt and sufficient sodium hydroxide to maintain the pH of the solution above about 12.9. Contacting a silver-activated glass substrate having a luminous transmittance of about 50% with such a solution for sufficient time to deposit a cuprous oxide film of sufficient thickness to lower the luminous transmittance to about 10 to 40% results in a cuprous oxide coated article which appears blue-green at the film surface, yellow-green at the glass surface and greenish by transmission. The most preferred coated glass sheets have, in addition to a luminous transmittance of about 10 to 40%, a luminous reflectance of about 10 to 40% from the film surface and a luminous reflectance of about 30 to 60% from the glass surface. The coated sheets are then heated, preferably at a temperature of at least about 300°F, and more preferably, above about 600°F, for sufficient time to effect a change in the color of transmitted light and a significant reduction in the reflectance from the uncoated glass.

Example: Flat sheets of clear soda-lime-silica glass are cleaned and sensitized with an aqueous solution of 0.5 g/ℓ stannous chloride. The sensitized surface is contacted for 25 seconds at ambient temperature with an alkaline aqueous solution containing 2.1 g/ℓ silver nitrate, 10 ml/ℓ ammonium hydroxide (28% aqueous solution), and 0.32 g/ℓ sodium hydroxide, and a solution of 2.6 g/ℓ dextrose to deposit a thin catalytic silver film which lowers the luminous transmittance of the glass sheet to about 50%.

The silver-activated surface is rinsed with a solution of 0.1 g/ℓ sodium thiosulfate and contacted for 3½ minutes at ambient temperature with a solution containing 3.8 g/ℓ copper sulfate, 29 ml/ℓ formaldehyde (37% aqueous solution), 3 g/ℓ Rochelle salt and 25 g/ℓ sodium hydroxide to deposit a cuprous oxide film.

The cuprous oxide coated sheet appears blue-green at the film surface with a reflectance of about 30%, yellow-green at the glass surface with a reflectance of about 44%, and greenish by transmission with a luminous transmittance of about 24%.

The cuprous oxide coated sheet is heated to 900°F for 15 minutes in order to simulate the temperature conditions of a glass edged multiple glazed unit fabrication method. After such heat treatment, the reflectance from the uncoated glass surface is decreased to 23% and the article appears brown by transmission.

ELECTRICAL APPLICATIONS

PRINTED CIRCUITS

Substrate of Laminated Anodized Aluminum

J.J. Grunwald, E.D. D'Ottavio, H.L. Rhodenizer and M.S. Lombardo; U.S. Patent 4,110,147; August 29, 1978; assigned to MacDermid Incorporated describe a process by which plastic parts are formed against an anodically treated aluminum surface by lamination under heat and pressure, whereby the surface of the plastic part, after removal of the aluminum, has a high energy level and is peculiarly receptive to adherent coatings of metal film deposited thereon by standard additive or semiadditive electroless plating processes. The process finds particular application in the production of electronic circuit boards.

Two distinct methods of manufacture of printed circuit boards for electronic equipment have, in general, been employed previously. One is termed the subtractive method, while the other is called the additive procedure. The manufacture of a printed circuit board by the subtractive method starts with a composite consisting of a sheet of insulating (e.g., plastic) material as a base or substrate, having adhered on one or both of its faces a thin copper foil of from 0.001" to 0.003" thick. This foil is secured to the insulating base by means of an appropriate adhesive and the application of heat and pressure in forming the composite blank circuit board. The substrate or insulating base used to support the conductive foil is generally made in the form of a flat sheet of compression molded epoxy, phenolic, polyester or other thermosetting resin material, with or without glass fiber or other reinforcement.

In preparing a printed circuit board by the subtractive method, a suitable mask is applied over the foil-covered surface or surfaces of the composite to define the desired printed circuit configuration, and the board is then introduced into a spray etching machine or equivalent in order to chemically dissolve away unmasked portions of the copper foil from the surface of the plastic. The resulting circuit board containing the desired circuit configuration is then treated in various ways to prepare it for mounting of accessory electronic components, etc.

The retained portions of the original metal foil thus serve as the conductor paths of the finished circuit board.

The alternative, i.e., the additive process of printed circuit board manufacture differs from the subtractive process in that the circuit boards are prepared by directly depositing conductive metal in the desired circuit configuration directly on a suitably prepared plastic substrate. That is, conductor metal is applied to the substrate only where conductor paths are wanted.

A laminate suitable for use in the additive circuit board process is prepared, for example, by placing a B-stage thermosetting epoxy-coated and impregnated glass fabric sheet in a laminating press on top of a foil of aluminum having an anodically treated surface with that surface confronting the resin substrate, and then further curing the thermosetting resin under the influence of heat and pressure. If a double-faced laminate is desired, it can be prepared in the same manner by placing sheets of aluminum foil above and below the prepreg plies in the laminating press in such a way that the anodically treated surfaces confront the substrate. Where the finished laminate is clad on one side only, a sheet of aluminum foil (unoxidized) is utilized to prevent adherence or sticking of the resin substrate to the platen of the laminating press.

The actual bonding of the B-stage prepregs to the anodically treated aluminum surface(s) is accomplished by simultaneously pressing the laminating components together and baking at a temperature of about 250° to 450°F, and preferably at 300° after 400°F, at a pressure of from 5 to 1,000 psig for a period of time ranging from about 5 to 30 minutes. During the laminating process it may be necessary to water-cool the laminates under the pressure applied in order to promote temperature control of the resin during the curing cycle.

The metal foil thickness may be varied widely although, preferably, it will be from about 0.001 to 0.003" in thickness. In a like manner, the thickness of the thermosetting or thermoplastic resin substrate utilized may vary from about 0.015 to 0.125", or more, as by increasing the number of plies of prepreg in the laminating step.

Example: A sheet of aluminum foil, having a thickness of about 0.002", is immersed in an alkaline soak cleaner bath for 5 minutes at a temperature of 190°F to remove surface grime and oils. The clean aluminum foil is then preferably etched in ammonium bifluoride at room temperature for 3 minutes, then treated anodically in an electrolytic bath containing 10% by weight phosphoric acid for 10 minutes at a current density of 10 A/ft^2 and at a temperature of 110°F.

The anodically treated aluminum foil is then placed in a laminating press on top of a stack, e.g., eight to ten, epoxy B-stage prepregs each having a thickness of about 0.003". A sheet of cellophane is placed between the undersurface of the stack of prepregs and the platen in order to prevent sticking during the curing operation. The press, preheated to a temperature of 350°F is closed and the laminate components are heated at a pressure of about 5 psig for about 30 seconds, after which the pressure is raised to 250 psig and curing is continued at the same temperature for about 15 minutes. The result is a laminate in which the aluminum foil is firmly attached on the surface to a cured, hard, infusible thermoset resin substrate.

Amino-Alkanoxy-Substituted Silane Bonding Layer

M.S. Lombardo, E.F. Jacovich, E.D. D'Ottavio and J.J. Grunwald; U.S. Patent 4,100,312; July 11, 1978; assigned to MacDermid Incorporated describe a method of producing consistently higher peel strengths between a metal conductor film and plastic substrate, to be able to do this over a wider range of operating conditions in the preparation of the laminate and thus provide greater tolerance for variables which inherently and unavoidably arise under commercial production operations, and especially to materially improve the thermal shock resistance of the final laminate product.

It has been found that applying to the anodized sacrificial metal foil, prior to lamination of it to the substrate, a film of a suitable silicon derivative, more especially an organic silicon derivative of the class comprising the amino-alkanoxy-substituted silanes, can substantially improve the adhesion of the conductor metal to the substrate both before and after soldering.

The silicon derivatives can be applied by both aqueous and nonaqueous solutions. The concentration of the silicon derivatives in solution can be quite small, indicating that no more than perhaps a monomolecular layer is retained on the sacrificial metal foil prior to lamination. Surprisingly, it appears that the silicon derivative in its mono layer form is so tenaciously held at the plastic-metal interface that its effect is not diminished or destroyed in the process of laminating the metal to the polymer substrate or in the subsequent chemical stripping step.

A wide variety of organic silicon derivatives is available but apparently not all are useful in the process. The best results are obtained by use of silane-type materials, and more particularly there is preferred a rather specific type of silane having the general formula $R \cdot Si(R_1)_3$, where R is a lower alkyl (up to 6 carbons) amino-substituted radical, and R_1 is a lower alkanoxy (up to 3 carbons) radical.

Example: A glass-epoxy aluminum foil laminate is first prepared by a method similar to that described in U.S. Patent 3,620,933, Example 1. This comprises taking a sheet of aluminum foil of approximately 2 mils thickness and immersing it in an alkaline soak cleaner bath for 5 minutes at a temperature of about 190°F to remove surface grime and oil. The clean aluminum foil is then preferably etched slightly in ammonium bifluoride solution at room temperature for 3 minutes preliminarily to anodic treatment in an electrolytic bath containing phosphoric acid (10% by volume) at 115° to 120°F for about 3 minutes at a current density of 25 to 30 A/ft^2.

The anodized aluminum foil is placed in a solution of N-β-(aminoethyl)-γ-aminopropyltrimethoxysilane in isopropanol for about 3 minutes at room temperature. The concentration of the silane is 4 ml/ℓ in this solution. The foil is then dried for about 2 minutes at 300°F and laminated to a plastic substrate.

The substrate consists of stacked plies (e.g., eight in number) of glass fiber reinforced epoxy B-stage resin (e.g., Precision No. 1528) and the composite of resin and foil is placed in a laminating press using a release strip, such as a sheet of cellophane, between the aluminum foil and platen in order to prevent sticking during the curing operation. The laminating press is preheated to a temperature of about 350°F, is then closed and the laminate components are preheated at a pressure of about 5 psi for 30 seconds, after which the pressure is raised to 250 psi and the curing is continued at that same temperature and pressure for about 15 minutes. The resulting laminate is a hard, infusible resin substrate having the aluminum foil permanently adhered to its surface.

This aluminum clad laminate is optionally cleaned of any surface grime and is immersed, sprayed or otherwise contacted with an etchant solution capable of dissolving away all visible traces of the aluminum foil. As described in Example 7 of the abovementioned U.S. Patent 3,620,933, any of the usually employed aluminum etchant solutions, such as hydrochloric acid (10 to 40% by volume), or alkali metal hydroxide (5 to 20% by weight), is effective.

Typical treatment conditions comprise a solution temperature of 80° to 180°F, preferably 100° to 130°F, for a period of 2 to 30 minutes, but normally about 5 minutes at the preferred temperature. When the substrate is free of aluminum foil, it is dipped in an aqueous phosphoric acid bath containing 50% by volume of phosphoric acid, for about 7 minutes at 160° to 170°F, after which it is again thoroughly water rinsed.

The substrate is ready for metal plating. In this example the procedure employed is the so-called one-step activation technique described in U.S. Patent 3,523,518, Example 1. This comprises immersing the substrate in a palladium-stannous-chloride hydrosol activator solution, prepared in accordance with the above patent, for about 3 minutes at room temperature; carefully rinsing and then immersing the substrate in an accelerating solution of fluoroboric acid; rinsing again and placing the substrate in a commercial electroless copper plating solution (e.g., Metex 9030, or equivalent) for a period of about 20 minutes at room temperature; and finally rinsing and electroplating an additional copper deposit to a thickness of about 1 mil.

The plated substrate is dried and then subjected to an oven bake at 300°F for about 1 hour. The adhesion of the plated metal deposit to the plastic substrate of this sample was checked by the standard technique of measuring the pull of a 1" wide strip of metal peeled from the surface and pulled at 90° to that surface. The average adhesion value was found to be 20 to 22 lb/in.

A small sample of the same plated board was floated for 10 seconds on the surface of a solder pot filled with solder at a temperature of 510° to 530°F. After cooling, the adhesion test gave a value of 15 to 18 lb/in.

For purposes of comparison, a second plated board is prepared using identically the same procedure described above except that the step of immersing the anodized aluminum foil in the silane bath is omitted. The adhesion of this is found to be about 3 to 4 lb/in before thermal shock test, and 1 to 3 lb/in after such test.

One Hundred Percent Pattern Plating

W.P. Dugan and J.A. Muhr; U.S. Patent 4,135,988; January 23, 1979; assigned to General Dynamics Corporation describe a process for manufacturing an electrical printed circuit board having plated through-holes for connecting circuits on opposite surfaces of the board comprising the steps of:

(a) laminating a copper foil having a thickness of about 0.00014" on opposite surfaces of an epoxy glass substrate board, the substrate board being substantially flat and having a thickness of about 0.062";

(b) drilling a plurality of apertures through the laminated substrate board at the desired locations of the plated throughholes;

(c) sensitizing the exposed surfaces of the substrate board in the apertures by a catalyst;

(d) electroless copper plating the surfaces of the apertures and the opposite surfaces of the substrate board to establish an initial copper thickness;

(e) laminating at an elevated temperature, a dry film photoresist having a thickness of approximately 0.0018" to opposite surfaces of the drilled assembly;

(f) photographically exposing and developing the photoresist to define nonpolymerized portions over the apertures and the desired circuit images and polymerized portions over the remainder of the photoresist;

(g) chemically removing the nonpolymerized portions of the photoresist to expose the apertures and the desired circuit images;

(h) removing the electroless copper plating;

(i) reapplying electroless copper plating over the surfaces of the apertures and the desired circuit images to establish an initial copper thickness;

(j) electroplating with copper the surfaces of the apertures to a thickness of approximately 1 mil and the circuit image to a thickness of approximately 1.5 mils;

(k) electroplating with solder the copper plating to a thickness of 0.0003";

(l) removing the polymerized photoresist from the board; and

(m) chemically etching the exposed copper foil from the substrate.

By use of this process, generally referred to as one hundred percent pattern plating, it is possible to reduce the conductor width and the space between conductors to 5±1 mils. Excellent chemical adhesion is achieved between the plated copper and the thin layer of copper clad which is laminated securely to the epoxy glass substrate. The extremely thin copper clad having a thickness of 0.14 mil is readily removed without adversely affecting the finished product. Precise control of the width of the circuit leads is readily achieved by containing the copper plating within the boundary of the walls of the polymerized photoresist.

Straight vertical copper circuit walls are also obtainable with this process. This is due to the fact that minimum undercutting of the plated circuit path occurs. The reason for this is that there is no electrolytic copper build-up on the copper clad which is later removed by etching. In this manner, higher density packaging of electronic components is readily achieved.

Flame-Retardant Dielectric Sheeting

S.A. Miller; U.S. Patent 4,140,831; February 20, 1979; assigned to Minnesota Mining and Manufacturing Company describes a flame-retardant metal-clad dielectric sheeting comprising a nonwoven web and an electrically conductive metallic layer adhered to at least one side of the web, the web comprising at least 10 pounds per ream of a fiber blend of discontinuous fibers compacted and held together with a matrix of film-forming high-molecular weight polymeric binder material that accounts for 5 to 75% by weight of the web.

Substantially all of the fibers in the fiber blend should generally have a softening point at about 450°F. At least 5% by weight of the fibers resist distortion when exposed for 10 seconds to a molten solder bath heated to 500°F, at least 30% by weight of the fibers exhibit a tensile elongation of at least 20% and a tenacity of at least 3.5 g/den, and at least 40% by weight of the fibers exhibit a moisture regain of less than about 3% by weight (which is the amount of weight gain in a previously dried sample when the sample is exposed at 65% relative humidity and 70°F for one day).

The polymeric binder material is solid and nontacky at room temperature, at least initially softens and flows slightly under the heat and pressure of the compacting operation, and in film form is foldable upon itself without cracking and resists distortion when heated for 10 seconds on a molten solder bath heated to 500°F.

The metallic layer is adhered to the web by a polyester/diepoxide adhesive system comprising a polyester formed from an acid selected from the group consisting of azelaic, adipic, and sebacic acids and neopentyl glycol and end-capped with a moderately brominated bisphenol A-epichlorohydrin epoxy; a curing agent comprising a polyanhydride selected from the group consisting of polyazelaic, polysebacic and polyadipic anhydrides; a sufficient quantity of highly brominated epoxy resin to provide at least about 14% by weight bromine in the adhesive system; and about 5% by weight of antimony trioxide.

This metal-clad dielectric sheeting generally exhibits no more than a 2 mil/inch change in dimensions upon etch-removal on part or all of the electrically conductive layer. Also, sheeting of the process exhibits less than a 10 mil/inch change in dimension, substantially no blistering upon exposure for 10 seconds to a solder bath heated to 450°F, and is flame-retardant.

Preferred high-temperature-resistant fibers used in the nonwoven webs of sheeting of the process are from the class of aromatic polyamides, such as described in U.S. Patents 3,094,511 and 3,300,450. Users of printed circuitry often desire temperatures of 500° to 550°F in their molten solder baths, and sheeting based on aromatic polyamides has been found to provide the best dimensional stability at such temperatures. These aromatic polyamides have the formula:

$$(-NR_1-Ar_1-NR_1-CO-Ar_2-CO-)_n$$

in which R_1 is hydrogen or lower alkyl, and Ar_1 and Ar_2 are divalent aromatic radicals. Among the preferred polymers are those in which R_1 is hydrogen and Ar is a m- or p-phenylene radical.

Another useful class of high-temperature-resistant fibers is the class of acrylic fibers, preferably those that are homopolymers of acrylonitrile, but also including copolymers of acrylonitrile (which generally include at least 85% by weight acrylonitrile) and any additional monomer that does not detract from the high-temperature-resistant properties of the fibers. Glass fibers in forms which can be handled on air-laying equipment are another useful high-temperature-resistant fiber.

Additionally, to provide strength to the web, at least 30% of the fibers should exhibit a tensile elongation of at least 20% and a tenacity of at least 3.5 g/den. Preferred tough high-tensile-strength fibers, capable of providing the elongation and tenacity requirements, are polyester fibers of the formula:

$$(-O-A-O-CO-Ar-CO-)_n$$

where A is a divalent straight-chain or cyclic aliphatic radical, Ar is a divalent aromatic radical, for example, m-phenylene and n is the index of polymerization. These polyesters are prepared in a known manner from difunctional alcohols, e.g., ethylene glycol and 1,4-cyclohexanedimethanol, and difunctional carboxylic acids (or esters thereof), e.g., terephthalic acid, isophthalic acid, and mixtures thereof. For best strength properties, the fibers are drawn, that is, stretched or oriented, causing them to be crystalline in structure. The aromatic polyamide fibers described above are also often tough, high-strength fibers, as are nylon fibers.

General ranges for the amounts of fibers as listed above in nonwoven webs used in the process are as follows: for drawn polyester fibers, 30 to 95% by weight, preferably at least 50% by weight, and more preferably at least 70% by weight; for aromatic polyamide fibers, 5 to 60% by weight, and preferably 10 to 30% by weight; and for acrylic fibers, up to 70% by weight, and preferably up to 40% by weight.

Particularly useful polymeric binder materials are reactive acrylic-based resins, generally comprising copolymers having a major portion of lower-alkyl (generally 1 to 8 carbon atoms) esters of acrylic or methacrylic acid, such as ethyl acrylate, butyl acrylate and 2-ethyl hexyl acrylate, and a minor portion of acrylic or methacrylic acid. Polyurethane binder materials may also be used. Other useful binder materials are thermoplastic polymers, such as polymers based on vinyl chloride (such as Vinylon resins).

The binder material is generally included in nonwoven fibrous webs in sheeting of the process in an amount between 5 and 75% by weight of the backing, and preferably in an amount less than 35% by weight of the backing.

Suitable conductive layers include foils of copper, aluminum, nickel, silver, gold or suitable transition metals. The thickness of the metal foil is usually about 0.02 to 0.05 mm. Conductive layers can also be provided in a laminate of the process by electroless plating processes.

Deposition of Colloidophobic Material

W.J. Baron and J.T. Kenney; U.S. Patent 4,100,037; July 11, 1978; assigned to Western Electric Company, Inc. describe a method of selectively depositing a metal on a surface of an electrically nonconducting substrate.

The method comprises selectively coating portions of the surface with a colloidophobic material to render the portions colloidophobic and to delineate an exposed surface pattern capable of retaining a colloidal species thereon. The selectively coated surface is treated with a sol comprising a colloidal species selected from the group consisting of (a) a colloidal species capable of reducing an activating metal ion to an activating metal, and (b) a colloidal activating metal species capable of participating in an electroless metal deposition to deposit the colloidal species on the exposed surface pattern.

Referring to Figure 6.1, a suitable substrate **70** is selected. For the production of electrical circuit patterns, suitable substrates are those which are generally electrically nonconductive. Substrate **70** may then be cleaned or degreased employing techniques well known in the art. A suitable colloidophobic material is selectively applied to portions of surface **72** of the substrate to form a colloidophobic coat or surface **73** which delineates an exposed surface pattern **74**, including the walls of through-holes **71**.

Figure 6.1: Deposition of Colloidophobic Material

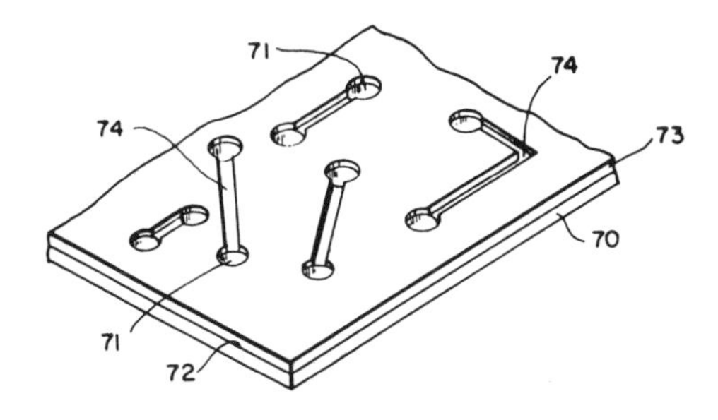

Source: U.S. Patent 4,100,037

Suitable colloidophobic materials are those which lower the surface energy or tension of a substrate surface treated therewith or present a surface having a surface energy which has a relatively low value as compared to untreated portions of the substrate and as compared to the colloidal species-containing medium, e.g., a sol. Particularly effective colloidophobic materials are compositions comprising (1) polytetrafluoroethylene

$$\left[-\underset{F}{\overset{F}{\underset{|}{\overset{|}{C}}}}-\underset{F}{\overset{F}{\underset{|}{\overset{|}{C}}}}-\right]_n$$

where n is about 1,000; (2) polyethylene (molecular weight of 1,500 to 100,000); (3) polysiloxanes having a structural formula of

$$CH_3-\underset{CH_3}{\overset{CH_3}{\underset{|}{\overset{|}{Si}}}}-O-\left[\underset{CH_3}{\overset{CH_3}{\underset{|}{\overset{|}{Si}}}}-O\right]_x-\underset{CH_3}{\overset{CH_3}{\underset{|}{\overset{|}{Si}}}}-CH_3$$

where x is the number of repeated units and having a viscosity ranging from 20 to 100,000 cp; (4) polyfluoroalkyl esters such as (a) polyperfluorooctyl methacrylate or (b) polyperfluorolauryl methacrylate, both dissolved in a suitable solvent vehicle such as hexafluoroxylene; (5) perfluoroepoxy resins such as (a) the fully cured reaction product of the diglycidyl ether of 1,3-bis[2-hydroxyhexafluoro-2-propyl]-5-heptafluoropropylbenzene having the structural formula of

$$\underset{\text{epoxide}}{CH_2\text{–}CH}\text{–}CH_2\text{–}O\text{–}C(CF_3)_2\text{–}C_6H_3(CF_2CF_2CF_3)\text{–}C(CF_3)_2\text{–}O\text{–}CH_2\text{–}\underset{\text{epoxide}}{CH\text{–}CH_2}$$

and 1,4-bis(aminomethyl)cyclohexane, having the structural formula

$$H_2NCH_2\text{–}C_6H_{10}\text{–}CH_2NH_2$$

and (b) the fully cured reaction product of the diglycidyl ether of 1,3-bis[2-hydroxyhexafluoro-2-propyl]-5-pentadecafluoroheptylbenzene, having the structural formula of

$$\underset{\text{epoxide}}{CH_2\text{–}CH}\text{–}CH_2\text{–}O\text{–}C(CF_3)_2\text{–}C_6H_3[(CF_2)_6CF_3]\text{–}C(CF_3)_2\text{–}O\text{–}CH_2\text{–}\underset{\text{epoxide}}{CH\text{–}CH_2}$$

and 1,4-bis(aminomethyl)cyclohexane and of perfluorotetradecane; (6) polyfluoroethanes; and (7) colloidal silicas having chemically bonded to the surface thereof amounts of from 0.01 to 30% by weight based on the weight of the silica of a disilazane treating material, such as described in U.S. Patent 3,627,724. The disilazanes useful as treating materials with this process have the general formula

$$R_4\text{–}\underset{R_5}{\overset{R_3}{Si}}\text{–}\overset{H}{N}\text{–}\underset{R_6}{\overset{R_2}{Si}}\text{–}R_1$$

where R_1, R_2, R_3, R_4, R_5 and R_6 are alike or unlike alkyl radicals having from 1 to 5 carbon atoms.

Representative examples of the disilazane compounds which may be employed herein are hexamethyldisilazane, hexaethyldisilazane, hexapropyldisilazane, hexabutyldisilazane, trimethyltributyldisilazane, tripropyltributyldisilazane, dimethyltetrapropyldisilazane, tetrabutyldiethyldisilazane, and the like.

The colloidal silicas are well known in the art and include nonporous silicas prepared by pyrogenic and precipitation processes, having an average ultimate particle diameter of less than about 0.5 micron and preferably less than 0.1 micron.

Typically, the colloidophobic material is in the form of an ink composition comprising a suitable liquid vehicle which is applied to surface **72** of substrate **70** (Figure 6.1) using conventional printing techniques and dried or cured, if necessary, using conventional drying or curing techniques.

Upon application of the colloidophobic material to surface **72**, the colloidophobic material may be further treated, e.g., as by heating to effect a full cure of the material, whereby colloidophobic coat or surface **73** is incapable of retaining a catalyst species thereon upon exposure thereto. Exposed solution to pattern **74** retains its original capability of retaining (relative to surface **73**) a colloidal species and upon exposure thereto will retain such colloidal species, e.g., colloidal stannous hydrous oxide particles contained in a hydrosol.

The resultant substrate **70** having colloidophobic surface **73** and colloidophilic surface **74** is treated with a suitable sol containing a colloidal species capable of reducing a precious metal, e.g., palladium, platinum, silver, gold, from a salt solution thereof.

Upon treatment or contact with the sol, the colloidal species contained therein, e.g., colloidal hydrous oxide particles of tin (Sn^{+2}), are deposited on exposed surface **74** to form a film or coat thereon (not shown) thereof. Substrate **70** is then treated, e.g., rinsed, with a suitable inert rinsing agent, e.g., water, whereby excess sol is removed from the surfaces of substrate **70** including colloidophobic surface **73**.

The colloidal species deposited on substrate **70** is then activated, i.e., is exposed in a conventional manner, e.g., by immersion, to an activating solution, e.g., an aqueous $PdCl_2$ solution, containing an activating metal ion, e.g., Pd^{+2}, wherein the activating metal ion, e.g., Pd^{+2}, is reduced to the metal, e.g., Pd, and deposited on area **74** of the substrate **70** in the form of a catalytic activating metal pattern.

The patterned, activating metal-deposited substrate **70** may then be water-rinsed and is then immersed in a conventional electroless metal deposition solution wherein an electroless metal ion, e.g., Cu^{+2}, Ni^{+2}, is reduced to the metal, e.g., Cu°, Ni° and deposited on surface **74** of substrate **70** to form an electroless metal deposit. The electroless metal deposit may be built up to a desired thickness by prolonged immersion in the electroless metal deposition solution or alternatively may be further built up by being electroplated in a standard electroplating bath.

Example: A colloidophobic ink was prepared by mixing 22 g of a screening medium comprising 8% by weight ethylcellulose and 92% by weight β-terpineol and 9 g of a commercially obtained colloidal silica having chemically bonded to the surface thereof amounts of from 0.01 to 30% by weight based on the weight of silica of hexamethyldisilazane. The resultant ink was then conventionally screen printed and dried at 70°C for 30 minutes in a pattern on a surface of each of the following substrates: (1) an epoxy-glass laminate; (2) an epoxy-coated metal laminate; (3) a rubber-modified epoxy base; to form a colloidophobic inked surface pattern thereon which delineated an exposed colloidophilic surface pattern.

A colloidal sensitizing sol was prepared by dissolving 10 g of $SnCl_2$, 10 ml of concentrated hydrochloric acid (37% by weight aqueous HCl) in 1 liter of deionized water. The inked substrates were immersed in the resultant sensitizing sol for 1 minute at 25°C and then rinsed with deionized water at 25°C for 1 minute. The thus-sensitized substrates were then activated in a 0.05% by weight aqueous $PdCl_2$ solution (pH = 2.2) by immersion therein at 25°C for 2 minutes.

The activated substrates were then immersed for 10 minutes at 25°C in an electroless copper plating solution comprising 15 g/ℓ of solution of cupric sulfate, 3 g/ℓ of a solution of $NiSO_4 \cdot 6H_2O$, 9 g/ℓ of solution of formaldehyde, 30 g/ℓ of solution of sodium potassium tartrate, 8 g/ℓ of solution of NaOH and 1 ppm $Na_2SO_3 \cdot 7H_2O$ wherein an electroless copper pattern corresponding to noninked, exposed areas of each substrate surface, having a 10-microinch thickness was obtained. There was no metal deposition on the inked areas of any of the substrate surfaces.

Copper Film of High Elongation

H. Morishita, M. Kawamoto, M. Wajima and K. Murakami; U.S. Patent 4,099,974; July 11, 1978; assigned to Hitachi, Ltd., Japan describe an electroless copper solution capable of providing an electroless deposited copper film having high elongation.

This preparation is characterized by adding either 2,2'-dipyridyl or 2,9-dimethyl-1,10-phenanthroline, and polyethylene glycol to a plating solution containing a copper salt, a complexing agent, a reducing agent and a pH-adjusting agent as main components. From 5 to 300 mg/ℓ of 2,2'-dipyridyl is added to the solution. Alternatively, 2,9-dimethyl-1,10-phenanthroline is added to the solution in a range of 1 to 50 mg/ℓ. As far as an effect upon the improvement of the elongation of deposited film is concerned, it is preferable to add 2,2'-dipyridyl rather than 2,9-dimethyl-1,10-phenanthroline.

As to the polyethylene glycol to be used together with either 2,2'-dipyridyl or 2,9-dimethyl-1,10-phenanthroline, polyethylene glycol having molecular weight in a range of 200 to 6,000 are used. In view of the effect upon the improvement of elongation, solubility in the copper solution, etc., it is preferable to use polyethylene glycol having molecular weights of 400 to 2,000.

The amount of polyethylene glycol to be added depends upon the molecular weight, and thus is hard to determine, but at least 1 g/ℓ of polyethylene glycol must be added to the solution. In the case of less than 1 g/ℓ, the elongation fails to reach 3%. A preferable amount of the polyethylene glycol is at least 3 g/ℓ, if the elongation and depositing rate are taken into account, though the amount depends also upon the amount of 2,2'-dipyridyl or 2,9-dimethyl-1,10-phenanthroline added. However, in the case of more than 100 g/ℓ, the depositing rate is decreased to less than 3 μm/hr in terms of the deposited film thickness.

According to this procedure, a deposited copper film having elongation equivalent to that of the electrodeposited copper film can be obtained by a combination of the additives, but such effect cannot be obtained by using the individual additives alone.

Conductive Noble Metal Composition

R.S. Kosiorek; U.S. Patent 4,098,949; July 4, 1978; assigned to Hercules Incorporated describes a conductive metal composition adapted to be applied to and fired on a heat-resistant substrate to form thereon an electrically conductive, adherent film which does not adversely affect the overall strength of the substrate, the composition comprising, by weight (a) from 50 to 88% of at least one finely divided, electrically conductive metal; (b) from 2 to 40% of a crystallizable glass frit binder comprising by weight 40 to 70% of SiO_2, 10 to 31% of Al_2O_3, 3 to 20% of Li_2O, 2 to 15% of B_2O_3, 0 to 4% of As_2O_3, 0 to 5% of Na_2O, 0 to 5% of K_2O, 0 to 6% of Bi_2O_3, and at least one oxide component selected from the group consisting of 4 to 19% of ZrO_2 and 1 to 10% of TiO_2; (c) from 5 to 48% of an inert liquid vehicle; and (d) from 0 to 20% of an inert filler or pigment.

The composition, as indicated, contains a major amount which is from 50 to 88% by weight of the composition of at least one electrically conductive metal such as the noble metals, gold, silver, platinum, palladium, rhodium and iridium in particulate form. Silver is the preferred noble metal and most preferably will be present in an amount of from 60 to 86% by weight of the composition.

The particulate metal is generally present in its elemental form and the particles thereof are usually in powder or flake form. Preferably, the metal is a powder which will pass through a 200 mesh or smaller screen. Particulate compounds of the metal can also be used provided the compound can be reduced to the elemental metal during processing such as by carrying out the firing in the presence of a reducing atmosphere.

The liquid vehicle portion of the composition functions to hold the conductive metal particles and the frit binder together and to permit the facile application of the composition to the substrate. Particularly suitable are pine oil, turpentine, mineral oils, glycols, clean burning heavy bodied oils and the like. The method of application and the thickness of the desired coating will, of course, influence the proportion and type of liquid vehicle in the composition.

The metallizing composition can also contain, if desired, a small amount and preferably up to 20% by weight of the composition of inert fillers and/or pigments. The inclusion of fillers such as silica or alumina usually results in films having improved abrasion durability.

The compositions can be applied to any substrate which retains its integrity during the firing operation such as ferrous and nonferrous metal substrates and ceramic substrates but is particularly suited to the ceramic substrates such as glass, china, porcelain and glass-ceramics. The method of application of the composition to the substrate is conventional and typically includes such well-known techniques as spraying, brushing, dipping, banding or screen or stencil printing.

Following application of the composition to the substrate in the desired pattern or design, the coating is preferably, but not necessarily, dried prior to firing. Usually, firing at a temperature of at least 590°C and which is within the tempering cycle of the substrate will fuse the frit and effect a firm bond between the composition and the substrate. Preferably, firing is carried out for 2 to 15 minutes at from 600° to 900°C.

The compositions of the process are particularly useful in the production of electrical components since the fired-on metallic films, in an electrical pattern or design on ceramic objects, are easily connected in electrical circuits. Among the many uses are conductors, resistors and other components in printed circuits and other electronic applications, capacitors, glass electrodes, electroconductive heating elements on aircraft, architectural and automobile windows to remove moisture or ice deposits, ornamental metallic coatings or designs, and the like.

Reducible Salt/Radiation-Sensitive Reducing Agent Photosensitizer

B.S. Madsen; U.S. Patent 4,133,908; January 9, 1979; assigned to Western Electric Company, Inc. describes a method of depositing a metal on a surface which includes coating the surface with a sensitizing solution comprising at least a reducible salt of a nonnoble metal and a radiation-sensitive reducing agent for the salt to form a sensitized surface. The sensitized surface is exposed to a source of light radiant energy to reduce the metal salt to a reduced metal salt species, wherein the sensitizing and/or exposing steps are restricted to a selected pattern on the surface to form a catalytic real image capable of directly catalyzing the deposition of a metal thereon from an electroless metal deposition solution. The catalytic real image is treated with a stabilizer comprising a reducing agent for the nonnoble metal ions of the reducible salt, a complexing agent and an accelerator to at least stabilize the catalytic layer.

Typical suitable substrates include bodies comprising inorganic and organic substances, such as glass, ceramics, porcelain, resins, paper, cloth and the like. For printed circuits, among the materials which may be used as the bases, are dielectric coated metal or unclad insulating thermosetting resins, thermoplastic resins and mixtures of the foregoing, including fiber, e.g., fiber-glass-impregnated embodiments of the foregoing.

Porous materials, comprising paper, wood, fiber glass, cloth and fibers, such as natural and synthetic fibers, e.g., cotton fibers, polyester fibers, and the like, as well as such materials themselves may also be metallized in accordance with the process.

The reducible metal salt can comprise, in general, a cation selected from the metals of Group VIII and I-B of the Periodic Table of the Elements. The anion associated in such metal salts can vary widely and can comprise organic and inorganic anions such as halides, sulfates, nitrates, formates, gluconates, acetates and the like. The cations in such salts will include copper, nickel, cobalt and iron, in any of the usual degrees of oxidation, e.g., both cuprous and cupric, ferrous and ferric, etc., will serve. Some typical salts include cupric formate, cupric gluconate, cupric acetate, cupric chloride, cupric nitrate, nickel chloride, cobalt chloride, ferrous sulfate and cobalt chloride.

The surface of the substrate, if necessary, is cleaned. A sensitizing solution of a reducible metal salt composition, e.g., cupric formate, and a light radiant energy-sensitive reducing agent contained in a suitable solvent, e.g., water, an alcohol, mixtures of water and an alcohol (ethanol, butanol, etc.), dimethylformamide, dimethyl sulfoxide, is applied to the surface to form a sensitizing solution layer. The coated surface is typically dried and then exposed through a positive or negative of an original pattern or photograph, to form the real image on selected portions of the surface. The real image comprises reduced metal salt species nuclei, e.g., copper metal nuclei.

The radiant energy-sensitive compound used in association with the reducible metal has the property of decomposing to a compound which will exercise a reducing action on the exposed metal salt. Preferred radiation-sensitive compounds are anthraquinone and derivatives thereof, such as 9,10-anthraquinone, β-chloroanthraquinone, β-phenylanthraquinone, 1,2-benzanthraquinone, anthraquinone-2-sulfonic acid, anthraquinone-2,6-(or 2,7-)-disulfonic acid and salts thereof, and anthraquinone-2,6-disodium sulfonate, anthraquinone-2,7-disodium sulfonate, anthraquinone-2,7-dipotassium sulfonate, and the like. Particularly preferred are the anthraquinone disulfonic acids and the salts thereof.

A preferred additional ingredient in the treating composition is a secondary reducer, such as an organic, oxygen- or nitrogen-containing compound. Especially preferred as secondary reducing compounds are alcohols or polyols. Among the organic oxygenated compounds can be mentioned glycerol, ethylene glycol, pentaerythritol, mesoerythritol, 1,3-propanediol, mannitol, 1,2-butanediol, pinacol, sucrose, dextrin, polyethylene glycols, lactose, starch, gelatin, and the like. Also included are compounds such as triethanolamine and propylene oxide.

Compounds which are also useful as secondary reducers are amino compounds, polyethers, certain dyestuffs and pigments. Among these may be mentioned aldehydes, such as formaldehyde, benzaldehyde, acetaldehyde, n-butyraldehyde, polyamides, such as nylon, albumin and gelatin; leuco bases of triphenylmethane dyes, such as 4-dimethylaminotriphenylmethane; leuco bases of xanthene dyes, such as 3,6-bisdimethylaminoxanthane and 3,6-bisdimethylamino-9-(2-carboxyethyl)xanthene; polyethers, such as ethylene glycol diethyl ether, tetraethylene glycol dimethyl ether; alizarin, erythrosin, phthalocyanine blue, zirconium silicate and the like. A preferred secondary reducer is sorbitol.

Usually to insure complete reduction, a substantial excess of the radiant energy-sensitive compound (based on the reducible metal ions) will be present. The metal salt concentration in solution can vary over wide limits, e.g., from 0.5 to 100 grams or more per liter can be used but it is most economical and convenient not to use more than about 25 g/ℓ and preferably less than about 15 g/ℓ. The radiant energy-sensitive compound can comprise from 1 to 10 or more equivalents, based on the metal salt. The amount of the secondary reducer, e.g., glycerol, sorbitol, pentaerythritol, dyestuff or the like, can likewise vary over a wide range, e.g., from 0.5 to 500 g/ℓ.

After exposure to the light radiant energy source, the real image is treated with a suitable stabilizer. A suitable stabilizer typically comprises a solution comprising (a) a reducing agent for the nonnoble metal ions of the metal salt contained on the surface, (b) a complexing or chelating agent, (c) an electroless metal accelerator, and (d) a basic compound, where the real image comprises a copper species. Typical reducing agents include, under alkaline aqueous conditions paraformaldehyde and formaldehyde, and under acid aqueous conditions ($pH<7$) hypophosphite species, e.g., sodium hypophosphite, potassium hypophosphite, etc. The reducing agent is present in an amount sufficient to reduce the nonnoble metal ions contained in the real image which were not reduced by the radiant exposure. Typically, the reducing agent is present in an amount ranging from 0.7 to 1.4% by weight of the resultant stabilizer solution.

Suitable complexing or chelating agents include any conventional complexing agent employed in an electroless metal deposition solution. Some suitable complexing agents include ethylenediaminetetraacetic acid and the sodium mono-, di-, tri-, and tetrasodium salts thereof, N-hydroxyethylenediaminetriacetate, triethanolamine and N,N,N',N'-tetrakis(2-hydroxypropyl)ethylenediamine. Typically, the complexing agent is present in an amount ranging from 2.8 to 4.7% by weight of the resultant stabilizer solution. Suitable accelerators are the sodium, potassium and ammonium ferricyanides and ferrocyanides. Typically, the accelerator is present in amount ranging from 0.19 to 0.75% by weight of the resultant stabilizer solution.

Where a reducible copper salt is employed to obtain the real image, the stabilizer must contain in combination with the reducing agent, the complexing agent and the accelerator a basic compound. Any basic compound which is chemically compatible with the other components of the stabilizer can be employed. Typically, alkali metal compounds, e.g., NaOH, Na_2CO_3, KOH, ammonium hydroxide and amines are employed. The amount of basic compound employed is not critical and typically is present in an amount ranging from 0.9 to 1.2% by weight of the resultant stabilizer solution. However, in the case of a real image containing a copper species, the pH of the resultant stabilizer solution should be above 7 and preferably 12.5 to about 12.8 whereby optimum bleed stabilizer is obtained.

Example 1: For comparison purposes, a substrate comprising a steel core with a fully cured diglycidyl ether of bisphenol A coating thereon was selected. The substrate comprised about 200 through-holes having a diameter of about 0.050". The substrate was immersed in a solvent bath comprising a mixture of 90% by volume methyl ethyl ketone and 10% by volume methanol for 10 minutes at 25°C. The substrate was water rinsed for 1 minute at 25°C and then etched in an aqueous solution comprising 300 g CrO_3 and 250 g H_2SO_4 in 1,000 ml of water, maintained at 25°C for 10 minutes. The etched substrate was then water rinsed at 25°C for 10 minutes.

A sensitizing solution was prepared by dissolving 10 g of cupric acetate, 4 g of 2,6-anthraquinone disulfonic acid disodium salt, and 50 g of sorbitol in a solvent comprising 950 ml of H_2O. The etched substrate was immersed in the sensitizing solution for 1 minute at 25°C, removed therefrom and dried at 50°C for 2 to 4 minutes. A surface of the dried substrate was selectively exposed to a high-pressure mercury discharge lamp (30 W/cm^2 surface at 3660 A) for 90 seconds to form a real image. The imaged surface was then water rinsed at 25°C for 0.5 minute. The surface was then immersed in a conventional electroless metal deposition solution comprising cupric sulfate, formaldehyde, sodium cyanide, alkali and ethylenediaminetetraacetic acid for 15 minutes at 35°C. A smeared, 50 μinch-thick electroless copper pattern which extended in part beyond the boundaries of the real image, i.e., a deposit which bled or ran from the real image, was obtained.

Example 2: For comparison purposes, the procedure of Example 1 was repeated except that after the real image was formed, the imaged surface was first treated by immersion for 5 minutes at 25°C in an aqueous stripping solution comprising 5% by weight acetic acid of the resultant solution. The stripping solution treated surface was then water rinsed at 25°C for 0.5 minute. Upon treatment with the electroless metal deposition solution, a smeared electroless copper deposit as in Example 1, was obtained.

Example 3: For comparison purposes, the procedure of Example 1 was repeated except that after the real image was formed, the imaged surface was immersed in an aqueous solution comprising 200 ml of a 40% by weight aqueous solution of ethylenediaminetetraacetic acid, 80 ml of a 37% by weight aqueous solution of formaldehyde, 64 ml of a 10 N aqueous NaOH solution, and 1,636 ml of water. The surface was immersed in the solution at 38°C for 4 minutes whereafter it was water rinsed at 25°C for 0.5 minute, and then treated with the electroless metal deposition solution. A smeared 50 μ-inch-thick electroless copper pattern, similar to that obtained in Example 1, was obtained after 15 minutes of immersion in the electroless copper solution.

Example 4: The procedure of Example 3 was repeated except that the aqueous solution comprised (in addition to the formaldehyde, the ethylenediaminetetraacetic acid and the NaOH), 20 ml of a 20% by weight aqueous solution of $K_4Fe(CN)_6$. The real imaged surface was immersed in the resultant stabilizing solution at 38°C for 4 minutes, water rinsed at 25°C for 0.5 minute and then treated with the electroless metal deposition solution. A 50 μ-inch-thick electroless copper pattern, deposited on the real image was obtained after 15 minutes of immersion in the electroless copper solution at 35°C. The copper pattern did not exhibit any smearing, running or bleeding outside the original boundaries of the real image as evidenced by examination with a microscope at a 30-fold magnification.

Example 5: The procedure of Example 4 was repeated except that after the real imaged surface was treated with the stabilizing solution and water rinsed, it was then immersed in an aqueous stripping solution comprising 5% by weight acetic acid for 5 minutes at 25°C. The surface was then water rinsed at 25°C for 0.5 minute and immersed in the electroless copper deposition solution. A 50 μ-inch-thick electroless copper pattern which did not spread from the boundaries of the real image was obtained, as in Example 4.

Sensitizing Solution of Nonnoble Metal and Alcohol

D. Dinella, J.A. Emerson and T.D. Polakowski, Jr.; U.S. Patent 4,098,922; July 4, 1978; assigned to Western Electric Company, Inc. describe a method of depositing a metal on a surface which includes coating the surface with a sensitizing solution comprising at least a reducible salt of a nonnoble metal dissolved in a solvent comprising water and an alcohol having a structural formula of

$$R_1-\underset{R_2}{\overset{R_3}{\underset{|}{\overset{|}{C}}}}-OH$$

where R_1, R_2 and R_3 are members selected from the group consisting of an alkyl group having 1 to 7 carbon atoms and the hydrogen atom, where the alcohol has a total of 2 to 8 carbon atoms. The coated surface is then treated to reduce the metal salt to metallic nuclei to form a catalytic layer thereon capable of directly catalyzing the deposition of a metal on the nuclei from an electroless metal deposition solution.

Typical suitable substrates include bodies comprising inorganic and organic substances, such as glass, ceramics, porcelain, resins, paper, cloth and the like. Metal-clad or unclad substances of the type described may be used. For printed circuits, among the materials which may be used as the bases, may be mentioned metal-clad or unclad insulating thermosetting resins, thermoplastic resins and mixtures of the foregoing, including fiber, e.g., fiber glass, impregnated embodiments of the foregoing.

Referring to Figure 6.2, a surface **12** of substrate **10** is selectively deposited with an electrically nonconductive layer or real image **13** comprising nuclei of a metal which is capable of catalyzing the deposition of electroless metal from an electroless metal deposition solution with which it is destined to be exposed or treated.

Figure 6.2: Cross-Sectional View of Substrate with Nonconductive Layer

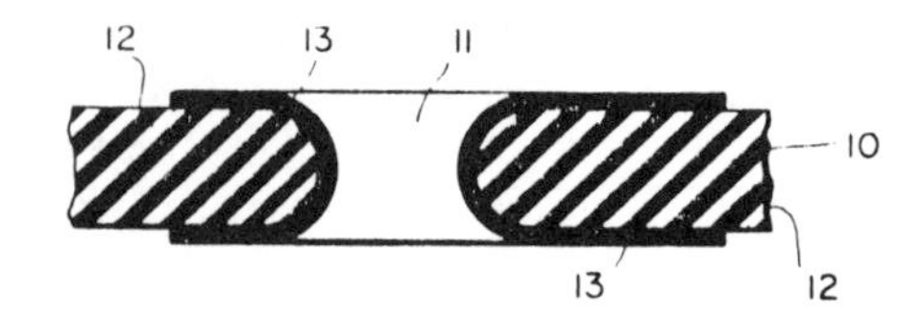

Source: U.S. Patent 4,098,922

Real image **13** comprises metallic nuclei in which the metals are selected from Groups VIII and I-B of the Periodic Table of Elements. Preferred metals are selected from Period 4 of Groups VIII and I-B; iron, cobalt, nickel and copper. Especially preferred for the production of the real image is copper. If desired, the surface can be coated with an adhesive before being coated with the compositions.

In producing the real image, the metal is reduced from its salt or a composition of the salt in situ in selected areas on the surface of the substrate by application of radiant energy, e.g., thermal or light, such as ultraviolet light and visible light, x-rays, electron beams, and the like, or by treatment with a chemical reducing agent. The catalytic surface is then electrolessly metal deposited. If it is desired to pattern the metal deposit, subtractive techniques can be used, such as masking and etching.

Example: A substrate comprising a steel core with a fully cured diglycidyl ether of bisphenol A coating thereon was selected. The substrate comprised about 200 through-holes having a diameter of about 0.050". The substrate was immersed in a solvent bath comprising methyl ethyl ketone for 10 minutes at 25°C. The substrate was water rinsed for 1 minute at 25°C and then etched in an aqueous solution comprising 360 g CrO_3, 250 g H_3PO_4 and 180 g H_2SO_4 in 1,000 ml of water, maintained at 25°C for 10 minutes. The etched substrate was then water rinsed at 25°C for 10 minutes.

A sensitizing solution was prepared by dissolving 21.5 g of cupric formate, 16 g of 2,6-anthraquinone disulfonic acid disodium salt and 66 g of sorbitol in a solvent comprising 1,000 ml of water. The etched substrate was immersed in the sensitizing solution for 1 minute at 25°C, removed therefrom and dried at 90° to 100°C for 3 minutes. A surface of the dried substrate was selectively exposed to a high-pressure mercury discharge lamp (30 W/cm^2 surface at 3660 A) for 90 seconds to form a real image.

The exposed surface was water rinsed for 1 minute and then immersed in a conventional electroless metal deposition solution comprising cupric sulfate, formaldehyde, sodium cyanide, alkali and EDTA, to obtain a 1.4-mil electroless copper-deposited pattern corresponding to the real image. The resultant electroless copper pattern was discontinuous on the walls of the through-holes.

Hydrazine-Alkali Hydroxide Photosensitizer

A. Shirk and J.P. Redmond; U.S. Patent 4,112,139; September 5, 1978; assigned to AMP Incorporated have found that when polyimide film of polyimides such as disclosed in U.S. Patent 3,436,372 is treated with hydrazine or substituted hydrazines or mixtures of same of the formula:

$$H_2N{-}N\begin{matrix} \diagup R \\ \diagdown R_1 \end{matrix}$$

where R and R_1 are hydrogen, an alkyl of 1 to 4 carbons or mixtures of these radicals in admixture with an alkali hydroxide, the treated polyimide substrate becomes photolytic and when exposed to ultraviolet light, heat or chemical treatment, the photolytic surface can be destroyed thereby providing a method for desensitizing the photolytic surface of the selected areas. It is emphasized that a hydrazine-alkali solution must be used.

By exposing the polyimide polymer, for example, to ultraviolet light through a masking pattern, a high degree of resolution can be achieved because the photolytic action affects molecular configurations. Hence, the deposited metal pattern, which is obtained in the nonphotolytic area when the polyimide is introduced in an electroless bath, will plate a metal in excellent pattern quality. Moreover, it is a noteworthy fact that exposed areas can be reactivated numerous times after exposure to ultraviolet light by immersion in a solution of 1% sodium hydroxide and other masking patterns can be provided in the treated area to provide subsequent metallization schemes.

Generally, the alkali hydroxide can range from 5% alkali hydroxide, by weight, to 95% hydrazine, by weight, to 95% alkali hydroxide to 5% hydrazine, by weight, the balance being water. Hydrazine is generally available as a solution, typically as a hydrazine hydrate solution of 85%, by weight, balance water.

Example: A polyimide polymer film, such as Kapton, is immersed from 1 to 3 minutes in an aqueous solution of 70% hydrazine hydrate by volume and 30% of 25% sodium hydroxide by volume then followed by a cold water rinse. The film is then dried by warm air at about 140°F. After drying the previously treated polyimide polymer film is exposed to an ultraviolet light source preferably about 2537 Å at about 25 mW-sec/cm^2 or more directed through a quartz glass with a suitable complementary pattern thereon. After proper time/exposure cycle the film is then immersed in a solution of 0.1 to 1 g/ℓ palladium chloride and hydrochloric acid of 5 to 10 ml/ℓ. Areas that have been shielded from the ultraviolet light will accept the $PdCl_2$ and areas exposed will reject such catalyst. Plating will then occur only in the areas where the Kapton has been shielded from ultraviolet light exposure. When a plating bath such as electroless nickel is employed the reduction of metal salts to metal is achieved by the hypophosphite in such baths. When it is desirable to plate electroless copper a reducing agent such as sodium hypophosphite is required prior to immersion in the electroless copper to reduce metal salts to metal.

At no time is a photosensitive material added in the subject process; rather the polyimide is made photosensitive by the abovedescribed method of hydrazine-alkali hydroxide treatment.

Dispersion Imaging Material

H. Kobayashi, T. Arakawa, T. Shiga, K. Ohmura and S. Ito; U.S. Patent 4,121,007; October 17, 1978; assigned to Asahi Kasei Kogyo KK, Japan describe a method for preparing a printed circuit board.

The method comprises applying a polymer having a softening point or a melting point of 50° to 300°C onto a base at least on one side thereof to form a polymer layer on the base; applying onto the polymer layer a dispersion imaging material which is dispersible in response to energy absorbed by the material in an amount exceeding the threshold value of the dispersion imaging material to form a dispersion imaging material layer coated on the polymer layer; applying energy to the dispersion imaging material through a circuit pattern mask until the absorbed energy exceeds the threshold value of the material to cause dispersion of the material at its exposed portions, thereby forming an imaged board. The imaged board is subjected to an electrically conductive metal plating to deposit the electrically conductive metal on the dispersion imaging material at its unexposed portions, forming a pattern circuit.

As specific examples of the polymer employable in this process, there can be mentioned polyurethane, polyvinylidene chloride, polyvinyl acetate, polyvinyl butyral, a copolymer of vinylidene chloride and vinyl acetate or acrylonitrile, a copolymer of vinyl chloride and vinyl acetate, a polycarbonate, a polyamide, a copolymer of styrene and acrylonitrile, a copolymer of acrylonitrile, butadiene and styrene, an epoxy resin and a blend thereof.

The thermohardening resins are employed after being suitably blended with an initiator for hardening reaction or a crosslinking agent. The epoxy resin may be preferably employed in the form of a blend of an epoxy resin and a polyurethane or polyvinyl butyral. In this case, the adhesion characteristics between the base and the polymer undercoat layer and between the polymer undercoat layer and the deposited metal are by far improved. The thickness of the polymer layer is not so critical but advantageously within the range of 0.5 to 100 μ and more advantageously from 1 to 50 μ.

As specific examples of the dispersion imaging materials, there can be mentioned iron, cobalt, nickel, copper, silver, gold, bismuth, antimony, tin, aluminum, cadmium, zinc, lead, selenium, indium, tellurium, alloys of these atoms and compositions including these atoms. These materials all have a slight solubility in a plating bath, but the iron family metals such as iron, cobalt and nickel are preferable, especially nickel, which is the most suitable in view of its good adhesion characteristics to the dispersion imaging material. The thickness of the dispersion imaging material layer is advantageously within the range of 50 to 2500 Å, and more advantageously 100 to 1000 Å.

The material of the base employable for this process may be selected from a wide class of materials so long as they have insulating properties. Stated illustratively, both an inorganic matter such as a silicate glass, ceramics, mica, etc., and an organic matter such as cellulose acetate, polystyrene, polyethylene, polypropylene, etc. may be employed. Further, a laminate of phenolic resin, polyester resin or epoxy resin with paper or glass cloth substrate incorporated therein, or a film of polyester or polyimide may be suitably employed. These have excellent heat resistance property. The epoxy resin laminate is especially desirable because of dimensional stability and chemical resistance, and the polyimide film is desirably employed as a flexible base of excellent heat resistance.

The polymer layer may be formed in such a manner as ordinarily employed for application of coating materials, e.g., brushing, dipping or spraying. The dispersion imaging material layer is formed on the base which is coated with the polymer layer by vacuum deposition, sputtering, ion plating, etc., by using an appropriate apparatus. In this connection, it is noted that when through-holes are preliminarily formed in the base and the dispersion imaging material is applied also to the inner walls of the through-holes by e.g., ion plating having good throwing power, before the steps of exposure and conductive metal deposition, there can be produced a printed circuit board with necessary through-holes at the same time.

Specific examples of the sources of energy are a flashlamp, infrared lamp, laser beam, electron beam, etc. The irradiation is desired to be as short as possible in view of the resolution of the image obtained and the possible decomposition of the underlying polymer layer so that a pulse of less than several msec is preferable. The shorter the irradiation time, the more is improved the resolution of the image obtained and the more is reduced or minimized the decomposition of the polymer layer. To meet this requirement, a xenon flashlamp or a pulse laser is optimum.

Example: National Lite R-1610 (an epoxy resin laminate with a substrate of glass cloth incorporated therein, produced by Matsushita Electricworks, Ltd., Japan) having a thickness of 2 mm was subjected to a surface treatment with an aqueous dichromic acid-sulfuric acid mixture (containing sulfuric acid in an amount of more than saturation) for 5 minutes. Hamakollan VK-1080 (polyurethane produced by Yokohama Rubber Co., Ltd., Japan and having a softening point of 91° to 103°C) was coated on the laminate using as a solvent N,N-dimethylformamide and then dried to form a polymer layer of 10 μ in thickness. Copper was sputtered on the coat in a thickness of 300 Å.

The thus-obtained material was subjected through a circuit pattern mask to flashlight of 1 msec pulse width and a quantity of 0.8 joule/cm^2 emitted from a xenon flashlamp to obtain a circuit pattern. The exposed portion had an extremely high electrical resistance and was substantially insulated. Then, the material was subjected to copper plating at a temperature of 50°C for 30 minutes using TCP-701 (electroless plating solution, Shipley) which is mainly composed of copper sulfate, formalin, lithium hydroxide and a complexing agent. Copper was deposited on the unexposed portions in a thickness of 5 μ, thereby to form a pattern circuit of copper having a sharp electrical contrast. The obtained copper pattern circuit conformed, in high precision, to the used circuit pattern mask and satisfied the characteristic requirements for a circuit board.

The adhesive properties of the copper pattern circuit were excellent and passed a Scotch tape test based on the standard of ANSI pH 1.28. In a thermal shock test, no change such as a blister or peeling was observed in three cycles. The thermal shock test was conducted in such a way that the sample was held in either one of temperature controlled baths of -30°C and 70°C for 1 hour, and moved into another bath within 30 seconds to be held there for another 1 hour, repeating these operations three times to observe a change of the sample.

Use of Enzymes for Patterning

J.H. McAlear and J.M. Wehrung; U.S. Patent 4,103,073; July 25, 1978; assigned to Dios, Inc. describe a process whereby micropattern devices, such as electronic microcircuits, are produced by establishing on a substrate base a film of resist material, such as a polymeric film, containing dispersed therethrough a substantial proportion of an enzyme and then producing a pattern of a metal by reactions depending upon presence of the enzyme.

Enzymes suitable for use according to the process are the transferases which are effective to catalyze the transfer of an inorganic radical containing the desired metal from a source compound to a compound from which the metal can be derived in elemental form, and the oxidases.

The film-forming materials most suitable for use as a carrier for the enzyme or enzymes are those polymeric materials which can be polymerized in situ on the substrate base under conditions which will not inactivate the enzyme. Typical of such materials are the vinyl monomers, especially the methacrylates, including, for example, methyl methacrylate, ethyl methacrylate and butyl methacrylate, capable of polymerization to the solid state at low temperature under the influence of ultraviolet radiation.

Microsubstrates are prepared according to the process by agitating the enzyme, in dry (lyophilized) form, in the liquid monomer or monomers at low temperature, so as to obtain a dispersion of the enzyme in the monomeric material, with the enzyme essentially in molecular solution. The resulting dispersion is spread evenly on the face of the substrate base and polymerized, with or without a polymerization initiator, under the influence of ultraviolet radiation under controlled temperature conditions such that the temperature does not exceed the highest temperature at which the enzyme remains active. For production of electronic microcircuits, the substrate base can be a silicon wafer covered by an insulating film of silicon dioxide. The enzyme-containing resist film can be 0.1 to 5 μ thick and can be established, e.g., by centrifugal spreading.

The amount of enzyme provided in the resist film should be tailored to the particular chemical reactions to be carried out with the aid of the enzyme and, ordinarily, should be at least 1% based on the weight of the film former used to produce the resist film. In most cases, higher proportions of the enzyme are advantageous and it is advantageous to introduce as much of the enzyme as can be uniformly distributed in the monomers. Amounts of the enzyme equal to as much as 50% of the total weight of the resist film are achievable.

Such microsubstrates can be further processed in a number of ways to produce a desired micropattern of metal or other functional material. Thus, the film can be removed mechanically, as by use of a stylus, in those areas which are to constitute the negative, so that the enzyme-containing film remains only in the positive areas, where the functional material is to be established. The entire surface of the substrate can then be treated with an aqueous solution of both a compound from which, e.g., metal-containing radicals are liberated by the enzyme and a compound which will react with the liberated radicals to yield a product from which the metal can be derived in elemental form.

Employing phosphatase as the enzyme, the aqueous treating solution can be of creatinine phosphate and lead nitrate, with the enzyme liberating phosphate radicals and with the lead nitrate reacting to form lead phosphate. The lead phosphate can be thermally reduced to deposit metallic lead in the positive area of the pattern, residual polymeric material being destroyed by the heat employed for the thermal reduction, so that the metallic lead deposits directly on the substrate base.

Though mechanical steps, matter beam etching and irradiation with electromagnetic radiation are all useful ways of tracing to establish the desired micropattern, it is particularly advantageous to write the pattern with a focussed electron beam. Electron beam writing can be employed in either of two modes; one using the beam at a dosage level adequate to depolymerize the polymeric material, so that the irradiated portions can be selectively removed by dissolution with the corresponding monomer, and the other using the beam at a higher dosage level adequate to deactivate the enzyme in the irradiated areas in which case treatment with the aqueous solution can proceed without requiring removal of any of the polymeric material.

In this connection, it will be understood that conventionally established films of polymeric resist material are pervious to aqueous solutions, and that an enzyme, such as phosphatase, supported in a 0.1 to 5 μ thick film of, e.g., polymethyl methacrylate will react with constituents of a solution in which the film is immersed.

Irradiation of the enzyme-containing resist film with a focussed electron beam at a dosage in the range of from 10^{-7} to 10^{-5} coulombs/cm^2 at 10,000 eV is effective to depolymerize most polymeric resist film formers, without deactivating the enzyme. Irradiation with a focussed electron beam at a dosage level of from 10^{-7} to 25 coulombs/cm^2 at 10,000 eV is adequate both to totally depolymerize the polymeric material and totally deactivate the enzyme in the irradiated area. Depending upon the micropattern to be formed, the electron beam can have a diameter of from 50 to 500 Å.

Example 1: 90 parts by volume of butyl methacrylate and 10 parts by volume of methyl methacrylate are blended and the blend purged of oxygen by bubbling nitrogen gas through the liquid monomers. The liquid blend is cooled to -10°C over dry ice and acetone, an amount of benzoyl peroxide equal to 1% of the combined weight of the two monomers is added, and the liquid thoroughly mixed. A known volume of commercially available lyophilized phosphatase is placed in an Erlenmeyer flask and an equal volume of the monomer blend is added, followed by ultrasonic agitation for 30 minutes to thoroughly disperse the enzyme through the monomer blend. The resulting material constitutes a uniform molecular dispersion of phosphatase in the liquid monomers.

A silicon wafer having a silicon dioxide surface is mounted in horizontal position, silicon dioxide face up, for rotation about a vertical axis. The phosphatase-containing monomer blend is then supplied by an eye dropper to the center of the upper face of the dish while the dish is rotated to spread a thin film of the liquid uniformly over the face of the wafer. The wafer is then placed face up under a nitrogen-filled glass container and subjected to ultraviolet light from a conventional laboratory black light overnight with the ultraviolet source 6" from the wafer and the temperature of the wafer maintained below 40°C. The amount of liquid supplied and the speed and time of centrifugal spreading are controlled to provide a film which, after polymerization is complete, is approximately 1 μ thick.

The finished film can be characterized as a solid matrix of copolymerized polybutyl methacrylate and polymethyl methacrylate through which the macromolecules of phosphatase are evenly distributed with the phosphatase retaining its enzymatic activity.

Example 2: A microsubstrate prepared according to Example 1 is employed to produce a supported micropattern of conductive lead, as in an electronic microcircuit. The microsubstrate is placed in a computer-controlled electron microscope and a negative of the desired micropattern is traced on the phosphatase-containing polymeric film of the microsubstrate with the electron beam of the microscope focussed to a diameter of approximately 100 Å and with an irradiation dosage of 10^{-6} coulomb/cm^2 at 10,000 eV. As a result, the polymeric material in the areas traced by the beam is depolymerized.

Monomeric methyl methacrylate is then flowed over the surface of the microsubstrate for 10 minutes to dissolve away the depolymerized polymeric material and the enzyme contained thereby, leaving the silicon dioxide layer of the substrate base exposed throughout the negative area. The microsubstrate is then washed with distilled deionized water and then immersed in an aqueous solution of creatinine phosphate and lead nitrate for 3 hours at room temperature, the active phosphatase retained by the polymeric material remaining on the positive areas of the micropattern causing the creatinine phosphate to break down into creatinine and phosphate radicals, with the phosphate radicals reacting with lead nitrate to yield lead phosphate.

The microsubstrate is then again washed with distilled deionized water. The cleaned microsubstrate is then placed in an oven and maintained for 5 hours at 320°C under a nonoxidizing atmosphere to reduce the lead phosphate to elemental lead and to destroy the polymeric material remaining in the positive area. The elemental lead so produced is deposited on the silicon dioxide film as a conductive body extending throughout the positive area of the micropattern, the negative area being constituted by exposed portions of the silicon dioxide film.

TiO_2-NiO-Sb_2O_5 Masking Material

M. Wajima, M. Kawamoto, K. Murakami, H. Morishita and H. Suzuki; U.S. Patent 4,151,313; April 24, 1979; assigned to Hitachi, Ltd., Japan describe a method for producing printed circuits which comprises forming a plating resist on a negative pattern on a surface (one side or both sides) of an insulating substrate including through-holes with a masking material, depositing an initiator for electroless metal plating, removing the initiator on the resist after acceleration, and forming a circuit of electroless metal only on a positive pattern on the surface of the insulating substrate, wherein an improvement comprises (1) the masking material containing a solid solution of oxides of titanium, nickel and antimony, and (2) removing the initiator on the plating resist through contact with a hydrochloric acid solution of ammonium persulfate.

The most preferable masking material of this process has the following composition:

(a) 100 parts by weight of epoxy resin;

(b) an effective amount of a curing agent and/or curing catalyst for the epoxy resin;

(c) 0.5 to 3 parts by weight of copolymers of acrylic acid esters comprising at least two kinds of acrylic acid monomers, having a molecular weight of 10,000 to 50,000;

(d) 2 to 40 parts by weight of solid solution of oxides of titanium, nickel and antimony; and

(e) a necessary amount of an organic solvent for mixing, dissolving and dispersing the components (a) through (d) to make a viscosity of the resulting solution 150 to 450 poises (20°C, B-type rotary viscometer, SC4-14 rotor, 100 rpm).

The masking material having the foregoing composition is coated on a negative pattern of an insulating substrate according to a screen printing method or others and cured at 100° to 160°C for 10 to 60 minutes, whereby a resist film is formed.

An initiator of known material is deposited on the insulating substrate having the resist thus formed according to the known procedure, and accelerated. One example of the initiator solution used in the pretreating step is a hydrochloric acid solution of stannous chloride ($SnCl_2 \cdot 2H_2O$) and palladium chloride ($PdCl_2$). The acceleration is carried out by an acid or alkaline solution. The acid solution includes, for example, hydrogen fluoride or oxalic acid, or other dissolved in a hydrochloric acid solution and diluted in water, and the alkaline solution includes, for example, an aqueous sodium hydroxide solution containing a metal chelating agent such as triethanolamine, ethylenediaminetetraacetate salt, etc. in a solution state.

The initiator deposited on the resist must be removed from the insulating substrate treated with the accelerating solution to prevent unwanted deposition of electroless metal on the resist. As the initiator-removing solution, ammonium persulfate dissolved in dilute hydrochloric acid is used. The initiator deposited on the resist can be readily removed by contacting the substrate with the solution and then washing the substrate with water.

The preferable composition can be obtained by dissolving 0.5 to 20 g of ammonium persulfate in 50 to 200 ml of 35% hydrochloric acid, and diluting the resulting solution to 1 liter with water. When the insulating substrate is made to contact such inorganic acid solution, for example, by dipping, it is sufficient to dip it at a liquid temperature of 15° to 25°C (room temperature) for 3 to 15 minutes.

As the electroless metal plating solution, for example, an electroless copper plating solution contains the following components: copper sulfate, copper acetate, copper carbonate, copper formate, etc. as a metal salt; Rochelle salt, ethylenediaminetetraacetic acid, etc. as a complexing agent; formalin, paraformaldehyde, etc. as a reducing agent; and sodium or potassium hydroxide as a pH controller.

In addition to these basic components, for example, sulfur compounds, cyanide compounds, pyridines, alcohols or glycols can be added thereto to improve the stability of the plating solution or physical properties of the resulting plated film. The desired plated film can be obtained by subjecting such electroless copper plating solution to plating operation at pH 12 to 13 (20°C) and a plating temperature of 60° to 80°C for 5 to 36 hours.

Example: Phenolic resin-modified nitrile rubber-based, thermosetting adhesive (NB-3033, Sale Tilney, Ltd., Japan) was applied to a paper-based phenolic resin laminate (LP-43 N, produced by Hitachi Kasei Kogyo KK, Japan) according to a curtain coat method, and cured by heating at 165°C for 80 minutes. Then, the laminate was provided with holes at necessary locations, and a masking material containing solid solution of oxides and having the following composition (viscosity at 20°C, 200 poises) was formed on a negative pattern thereof according to a stainless steel printing method:

	Parts by Weight
Epoxy resin:	
Epikote 1007 (epoxy equivalent 2200)	65
Epikote 828 (epoxy equivalent 188)	35
Filler:	
Silicon oxide	2
Calcium carbonate	4
Zirconium silicate	2
Pigment:	
Phthalocyanine blue	2
Solid solution of oxides:	
TiO_2-NiO-Sb_2O_5	
(particle sizes: 0.8 to 1.2 μm)	20
Solvent:	
Butyl Carbitol	40
Methyl Cellosolve	10
Curing agent:	
Dicyandiamide	5
N,N,N',N'-tetramethylbutanediamine	1.1

The masking material was then cured at 145°C for 40 minutes to form a resist film on the laminate. The laminate was dipped in a chromium-sulfuric acid mixed solution (prepared by dissolving 50 g of CrO_3 in 200 ml of concentrated sulfuric acid and making the resulting solution 1 liter with water) at 40°C for 7 minutes to roughen the positive pattern, and washed with water and neutralized. The laminate was then dipped in a solution consisting of 400 ml of 35% HCl and 600 ml of water for 1 minute, then dipped in aqueous hydrochloric acid solution containing stannous chloride and palladium chloride in a solution state as an initiator solution (sensitizer HS101B, Hitachi Kasei Kogyo, Japan) for 5 minutes, then washed with water for 1 minute.

The laminate was then dipped in an aqueous alkaline solution containing a metal chelating agent as an accelerating solution (adhesion promoter ADP101, Hitachi Kasei Kogyo, Japan) for 5 minutes and washed with water for 1 minute, after which the laminate was dipped in an inorganic acid solution having the following composition for 4 minutes and washed with water for 1 minute: 3.0 g of ammonium persulfate and 80 ml of 35% hydrochloric acid with sufficient water to make the entire solution 1 liter.

A plated film having a thickness of about 35 μm was deposited onto the positive pattern of the substrate with an electroless copper plating solution having the following composition at 70°C over a period of 11 hours, and no deposition of electroless metal was found on the resist containing the solid solution of oxides of metals printed on the negative pattern on the surface of the substrate.

Metal salt:	
Copper sulfate	10 g
Complexing agent:	
Ethylenediaminetetraacetic acid	30 g
Reducing agent:	
37% formalin	4 ml
pH controller:	
Sodium hydroxide in an amount to make pH 12.9 (20°C)	
Additive:	
Polyethylene glycol (molecular weight 400)	30 ml
Water to make the entire solution 1 liter	

Addition of Catalytic Promotors

N. Feldstein; U.S. Patent 4,151,311; April 24, 1979 describes a process of preparing a catalytic composition for electroless plating or image intensification processes, which comprises the steps of (a) forming a principal catalyst consisting of a colloidal dispersion and then (b) admixing the primary catalyst with a promotor. When necessary, the above is then treated with an activating composition.

The catalytic compositions of the process are applicable to the metallic plating of a wide variety of dielectric substrates including thermoplastic, thermosetting resins and especially in the fabrication of printed circuitry arrays.

More specifically, the process is comprised of the following sequence of steps followed by electroless plating.

(1) immersing a dielectric substrate (preferably previously etched) in a catalytic composition comprising the admixture of the principal catalytic agent of a nonprecious metal, preferably selected from at least one member of the group consisting of copper, nickel, cobalt, iron and mixtures thereof; and the catalytic promoting agent preferably selected from the group consisting of manganese, chromium, tungsten, molybdenum, zirconium and vanadium and mixtures thereof;

(2) rinsing the substrate with water to remove excess catalytic composition;

(3) (optional) immersing (or exposing) the substrate after rinsing to an activator, e.g., ultraviolet light;

(4) optionally rinsing the substrate with water prior to electroless plating; and

(5) electroless metal deposition.

Double-Sided Circuit Arrangement

K.C.A. Bingham, A.G.A. Gillingham and C. Baldwin; U.S. Patent 4,118,523; October 3, 1978; assigned to International Computers Limited, England describe a method of forming a printed circuit arrangement having conductive areas provided on opposite surfaces of a flexible substrate and through-holes with electrically conductive walls interconnecting the areas. The method also includes the steps of forming the hole in the substrate by a chemical process, removing portions of the conductive areas to expose ringlike regions of the substrate including the periphery of each end of each hole, metallizing the walls of the holes and the exposed regions of the substrate, and producing the required conductor patterns in the conductive areas.

Referring to Figure 6.3a, a 25-μm-thick polyimide film **1** is coated on both surfaces with a thin layer **2** of chromium having a thickness of about 600 Å and a layer of copper having a thickness of 0.1 to 0.5 μm. In order to be sure that there are no pinholes in the resultant sandwich of the substrate **1** and the layers **2**, the layers are coated with layers **3** of copper having a thickness of approximately 5 μm. These additional layers **3** also help to reinforce the sandwich.

The metallized layers **3** are then coated with layers of photoresist **4** which can have a thickness of 3 μm (Figure 6.3b). These two layers **4** are simultaneously exposed to ultraviolet light through two mirror-imaged, but otherwise identical, masks which define the required positions **5** of holes to be formed in the substrate. These masks (not shown) are conveniently adjusted by means of alignment marks on the masks which are held in accurate registrations by a semiconductor mask alignment device (not shown). The photoresist layers are then developed so as to expose the locations **5** of the required holes on the surface of the layer **3**.

Following the development of the photoresist masks, the substrate **1** and the metallization layers **2** and **3** are subjected to an etching operation in which both sides of the substrate and metal coatings sandwich are simultaneously subjected to etchants.

Figure 6.3: Double-Sided Circuit Arrangement

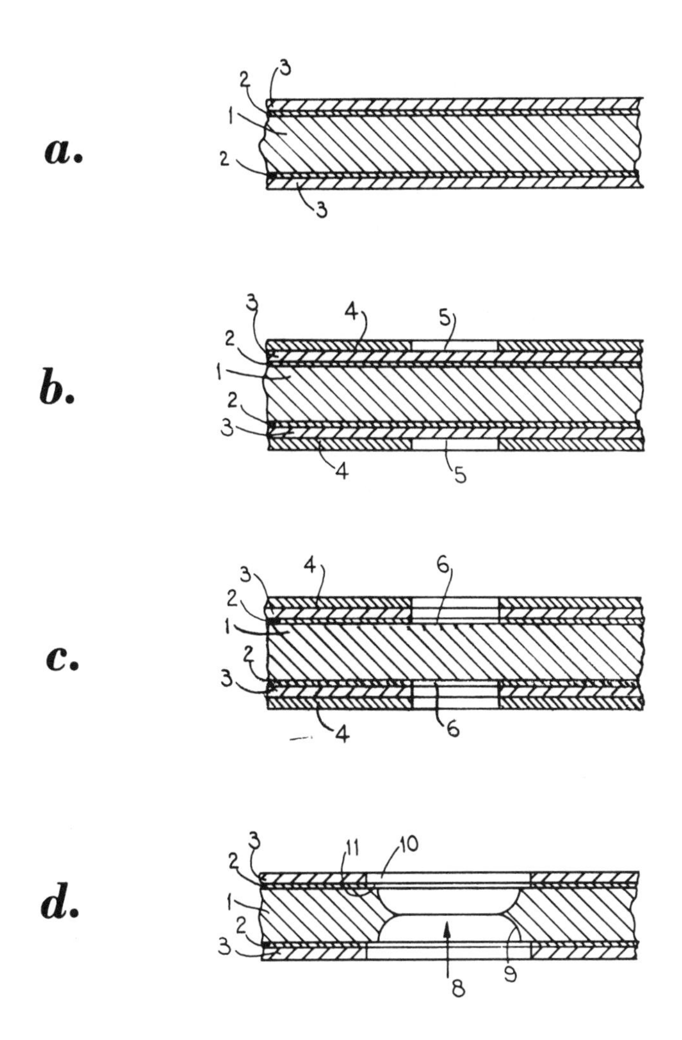

Source: U.S. Patent 4,118,523

The exposed areas of copper and chromium are respectively etched in ferric chloride (or ammoniacal etchant) and alkaline ferricyanide solutions to expose the surface regions **6** of the polyimide substrate **1**. This is shown in Figure 6.3c.

The sandwich, which is contained in an oscillating jig (not shown) to cause turbulence at the exposed regions **6** of the polyimide substrate, is then etched in a hydrazine-hydrate solution. This material will penetrate the substrate at a rate of approximately 25 μm per hour. Since the etching material acts sideways as well as inwards into the substrate center, the etching produces wider diameter regions at the ends of the holes being formed and a narrower diameter central region. Furthermore, since the metal is etched to a lesser extent there is a circumferential overhang.

A further photoresist mask is applied to each of the faces of the sandwich, comprising the substrate, the metal coatings and the through-holes. The further photoresist masks each define a pattern of holes which corresponds to the locations of the already formed holes, but which is such that each hole has a greater diameter than that of the formed substrate holes. The photoresist mask is exposed to ultraviolet light and developed.

The sandwich is subjected to a further etching operation using the same etchants as previously mentioned which etches the metal exposed by the removal of the rings and in so doing etches away the metal overhang together with a further ring of metal **10** so as to expose an annular ring **11** of polyimide around each through-hole **8** in the polyimide substrate, after which the remainder of the photoresist material is removed. This position is shown in Figure 6.3d.

The walls **9** of the holes **8** and the annular rings **11** of exposed substrate are then electrolessly copper-plated in order to provide through-hole electrical continuity from the metal layer on one surface of the substrate to the metal layer on the other surface of the substrate. In addition, with a view to reinforcing the through-hole plating, a further layer of copper is deposited on the electroless layer by an electroplating technique.

A new photoresist layer (not shown) is applied to both sides of the sandwich and these layers are exposed to ultraviolet light through masks which define the pattern of conductors required on the faces of the substrate. The photoresist layers are developed and the exposed substrate regions are electroplated with copper (or nickel) to a thickness of 25 μm followed by a protective coating of tin-lead or gold to provide the conductors.

The last-mentioned resist layers are removed and the sandwich with the conductors is subjected to two etching operations which remove the 600 Å chromium film and the 5 μm copper film from regions between the conductor regions so as to separate the conductors. This final etch can be effected using ferric chloride 30 g/ℓ when the conduction protective coating is gold or ammoniacal etchant when the protection is tin-lead or gold.

SEMICONDUCTOR INTEGRATED CIRCUITS

Improved Insulating Layers

Y. Sumitomo and Y. Ohashi; U.S. Patent 4,123,565; October 31, 1978; assigned to Tokyo Shibaura Electric Co., Ltd., Japan describe a method for manufacturing a semiconductor device, comprising the steps of: (a) forming a first insulating layer and a first wiring layer on one surface of a semiconductor substrate, the first wiring layer being formed on at least a portion of the one surface of the semiconductor substrate where the first insulating layer is not formed and having substantially the same thickness as that of the first insulating layer; (b) forming an insulating film flat on the first insulating and wiring layers by coating and then burning a liquid insulating material on the first insulating and wiring layers so that the liquid material occupies grooves between the first insulating and wiring layers; (c) forming an intermediate insulating layer on the insulating film; and (d) forming a second wiring layer on the intermediate insulating layer.

A method for manufacturing a semiconductor device according to this process will be explained below with reference to Figures 6.4a through 6.4f.

An approximately 1-μ-thick SiO_2 insulating layer **13** is formed by the CVD (Chemical Vapor Deposition) method on one surface of a semiconductor substrate, for example, a silicon substrate **1** as shown in Figure 6.4a. It is to be noted that electrode areas, SiO_2 layer covering the electrode areas and openings through which the electrode areas are connected are formed on the surface of the semiconductor substrate, but they are omitted for brevity.

A resist layer **21** is formed on the SiO_2 layer **13** and the SiO_2 layer is etched by an etching solution such as NH_4F with the resist layer as a mask in order to form a subsequent wiring layer (Figure 6.4b). The formation of the resist layer is effected using Negaresist OMR-83 (Tokyo Oka KK, Japan) and an excellent result is obtained.

Etching of the SiO_2 layer proceeds first from that portion of the SiO_2 layer which contacts the resist layer, leaving a V-shaped groove. The resist layer overhangs the V-shaped groove in an eaveslike fashion. The overhanging structure plays an important part in using the lift-off method. Then, for example, aluminum is evaporated in a thickness of 1 μ on the resultant semiconductor structure to form first wiring layers **12** (Figure 6.4c). The first wiring layer opens or is broken due to the overhanging structure of the resist layer **21**. Next, the resist layer is removed together with that portion of the wiring layer on the resist layer using a resist layer stripping agent (Figure 6.4d).

Figure 6.4: Manufacture of Semiconductor

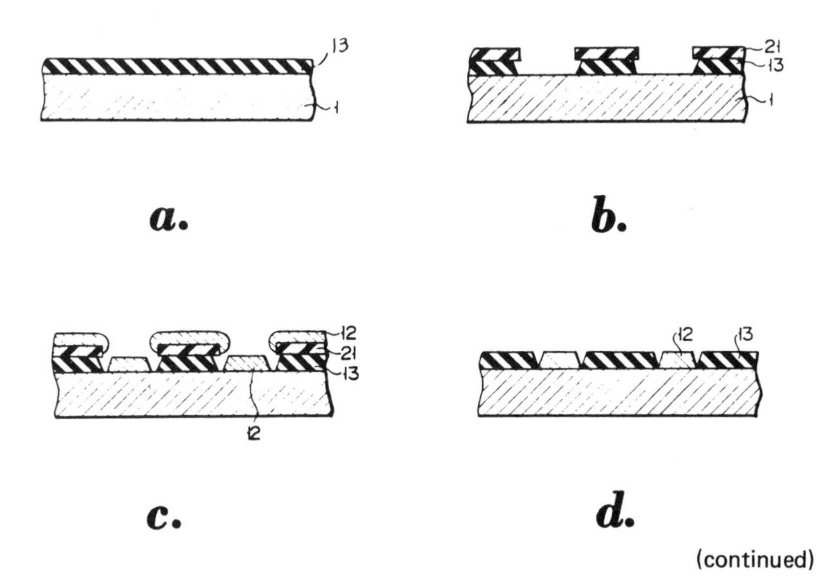

(continued)

Figure 6.4: (continued)

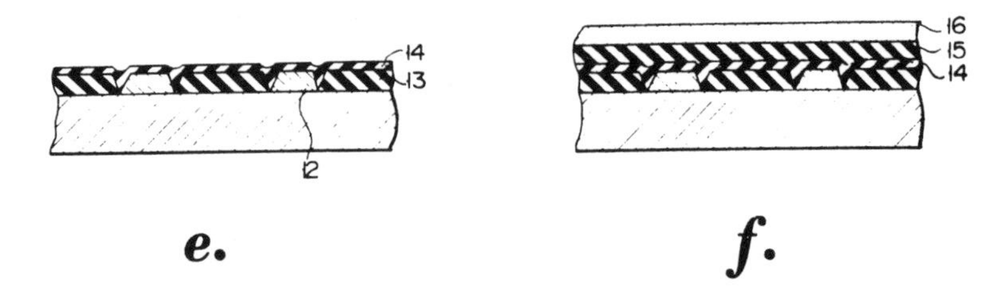

Source: U.S. Patent 4,123,565

As a result, the remaining wiring layer **12** is formed substantially flush with the insulating layer **13**. That is, the wiring layer is embedded in the insulating layer. A clearance or groove of 1 to 1.5 μ is left between the wiring layer and the insulating layer.

A liquid insulating material such as liquid silica is then coated on the resultant structure. As the liquid silica, OCD (Tokyo Oka KK, Japan) bearing 5.9% of SiO_2 may be used. A spinner method is used as a method for coating the liquid silica. Coating is effected by the spinner method at 4,200 rpm and the coated structure is burnt at 220°C in air for 10 minutes. The coating and burning processes are carried out once more.

As a result, a silica film **14** having a thickness of preferably 1000 to 1500 Å is formed as shown in Figure 6.4e. For a thickness below 1000 Å the silica does not completely fill the groove between the wiring layer **12** and the insulating layer **13** and for a thickness above 1500 Å the resulting silica film is liable to crack. The preferred coating thickness is effected to prevent a possible crack and to fill the groove between the wiring layer and the insulating layer with silica. As a result, a maximum height difference between that area of the silica film below which the groove is provided and the other area of the silica film is below 2000 Å.

Then, an intermediate insulating layer **15** is formed on the silica film **14** as shown in Figure 6.4f. As the intermediate insulating layer, PSG (Phosphor Silicate Glass) is preferred and its thickness is preferably about 1 μ. The intermediate insulating layer is etched to provide holes through which the first wiring layer is electrically connected to the next wiring layer. Where three or more wiring layers are provided, the corresponding wiring layers are formed by the lift-off method.

A second wiring layer **16** is then formed by a usual method. That is, aluminum is evaporated in a thickness of about 1.5 μ on the intermediate layer and photoetching is effected on the aluminum layer to form the second wiring layer **16** of a predetermined pattern (Figure 6.4f). In this way, the two-layer wiring structure is obtained.

Deposition of Nickel on Aluminum

F. Vratny; U.S. Patents 4,125,648; Nov. 14, 1978; 4,154,877; May 15, 1979;

and 4,122,215; October 24, 1978; all assigned to Bell Telephone Laboratories, Incorporated describes the selective deposition of nickel on aluminum metallized semiconductor devices in predetermined areas defined by openings in a suitable dielectric or photoresist. Figure 6.5 shows the method steps in an illustrative embodiment of the electroless deposition of nickel on aluminum.

Figure 6.5: Deposition of Nickel on Aluminum

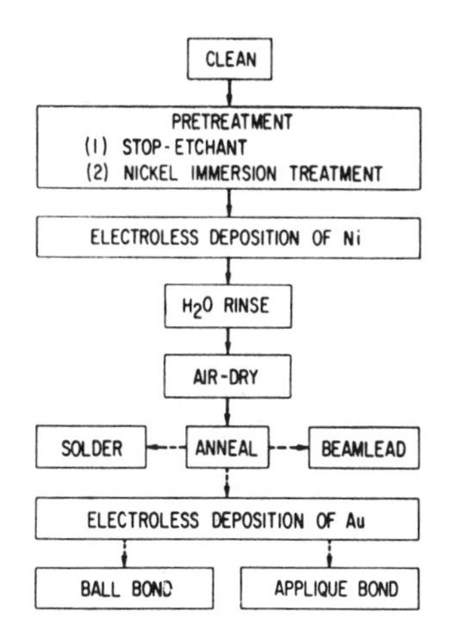

Source: U.S. Patent 4,125,648

The pretreatment encompasses two distinct steps which permit electroless deposition without deleterious side effects and confines deposition to the desired area if the substrate is masked. The first step in the pretreatment removes the aluminum oxide and simultaneously activates the entire surface. The second step activates the aluminum with nickel ions and, if patterned with a mask, deactivates the mask relative to the aluminum. A typical pretreatment for an aluminum metallized integrated circuit wafer having a silicon nitride mask is as follows:

Pretreatment: *Stop-Etchant* – The stop-etchant comprises buffered hydrofluoric acid in ethylene glycol, amyl acetate, ethyl acetate, ether or ethyl Cellosolve in a volume ratio of 1:2 to 4:1 at room temperature (18°C) for 0.25 to 3 minutes (depending on concentration).

The wafer is immersed in a solution containing per liter H_2O, 1.1 to 50 g of nickel sulfate, nickel chloride or nickel acetate (0.07-0.3 M), 3 to 40 g of ammonium chloride, ammonium citrate or ammonium acetate (0.05-0.75 M), 0.01 to 0.5 g of p-toluenesulfonic acid, and 0.01 to 10 ml buffered hydrofluoric acid at room temperature (18°C) for 15 to 60 seconds.

Following standard cleaning procedures, the substrate is first immersed in a buffered hydrofluoric acid stop-etchant. Buffered hydrofluoric acid, BOE (Buffered

Oxide Etchant), is a 6.7:1 (volume) mixture of 40% ammonium fluoride and 49% hydrofluoric acid. BOE when mixed with a nonaqueous solvent such as ethylene glycol, amyl acetate, ethyl acetate, ether, or ethyl Cellosolve acts as a stop-etchant since it dissolves the oxide at a much faster rate than the aluminum. Fluoride ions activate the substrate surface. Variation of the ratio of BOE to solvent (preferably between 1:2 to 4:1) varies the etch rate and is modified to suit the aluminum surface compositon.

Without rinsing, the wafer is transferred to the second step which is a nickel immersion treatment. Nickel ions exchange with fluoride ions on the aluminum surface and activate in a nondeleterious manner. The nickel complex is chosen by the amount of nickel ions one wants to produce. The chloride complex accelerates conversion to nickel ions while the acetate complex retards conversion relative to the sulfate complex.

The other major component produces a common ion effect and provides an ion to exchange with fluoride ions on the mask surface. For example, chloride ions in ammonium chloride exchange with fluoride ions on the mask surface to deactivate it relative to the aluminum. This confines nickel deposition to the desired area. The citrate and acetate complexes deactivate more slowly than the ammonium chloride complex. p-Toluenesulfonic acid (p-TOS) wets the surface but is an optional component of the bath. A small amount of BOE is also included to prevent the formation of aluminum hydrous oxide.

Without rinsing, the wafer is transferred from the nickel immersion treatment to the electroless plating bath. At this point, there are fluoride and nickel ions on the surface which can readily be replaced with nickel metal. The deposition of the nickel metal is self-propagating. A typical bath composition with suitable concentration is shown below. The bath is used under the following conditions: temperature 25° to 95°C with slight agitation and a pH of 3.5 to 7, at a rate of 0.1 to 5 μm/8 minutes.

Ingredients	Concentration*	Molarity
Nickel sulfate, g	15–45	0.05–0.2
Sodium acetate, g	5–85	0.04–0.5
Sodium hypophosphite	2.5–25	0.02–0.2
BOE, ml	trace–10	—
p-TOS, g	trace–0.15	—
Formaldehyde, ml	trace–50	—
Ethanol, ml	trace–150	—
Boric acid, g	trace–65	—

*Per 1.5 liter H_2O.

Concentration of the bath components is adjusted to accommodate various types of aluminum surfaces and to control deposit characteristics. Other reducible nickel salts, hypophosphites, or organic acid salt complexing agents may be used. The various buffers, stabilizers, and wetting agents affect deposit characteristics and bath controllability. The concentration of BOE requires control for quality deposits. A low molecular weight alcohol, such as methanol or ethanol, and p-TOS wet the substrate surface and reduce surface tension at the mask to aluminum interface. As an acid, p-TOS may also prevent formation of hydrous oxide on the substrate surface. Formaldehyde is a stabilizer. Boric acid stabilizes, buffers, and acts as a leveler to control particle size.

Time and temperature regulate the rate of deposit. Typically, 1 μm of nickel will be deposited in about 8 minutes at 72°C. To obtain thicker deposits, samples may be plated for longer time or the boric acid and BOE concentration can be reduced and/or sodium hypophosphite concentration can be increased. The nickel deposit contains 2 to 4% phosphorus which advantageously hardens the metal. Bath temperature can range from 25° to 95°C with maximum efficiency at approximately 72°C. High temperatures cause the bath to decompose more quickly and low temperatures excessively slow the rate and may allow the acid in the bath to etch into the aluminum. The pH can range between 3.5 and 7 with maximum efficiency at approximately 6.8. At pH 7, deposition is slow and particle size decreases. At pH 3.5, deposition is also slow and acid can attack the aluminum.

Subsequent to deposition, the substrate is rinsed with water, blotted to remove the excess, and allowed to air dry. It may be desirable to anneal the substrate in a reducing atmosphere such as forming gas (20% hydrogen and 80% nitrogen) at 200° to 425°C. Annealing assures bonding between aluminum and nickel.

In semiconductor processing, nickel pads may be directly soldered or with subsequent gold plating may be ball bonded, applique bonded, or subjected to other known procedures for providing leads or bonding to lead frames. As bonding pads, the thick nickel deposits spread laterally around the edges of the masked area and hermetically seal the contact area. This process also seals pinhole defects in the mask with nickel.

It may be desirable to plate the nickel deposit with gold or copper before further processing. A rinse with a mixture of BOE and ethylene glycol or some other nonaqueous solvent is recommended before electroless deposition of gold.

Silicon Layers from Si-Al-Zn Melts

H. Sigmund; U.S. Patent 4,113,548; September 12, 1978 describes a process for the production of semiconducting silicon layers on oxidized aluminum films by means of an isothermal silicon separation from silicon-aluminum-zinc melts. The silicon layer thicknesses in this regard are to amount to 10^{-3} to 4×10^{-3} cm.

This process includes the use of an aluminum-zinc melt which preferably contains 90 mol % aluminum and 10 mol % zinc up to 20 mol % aluminum and 80 mol % zinc. An aluminum film having an oxidized surface is used as a substrate, the thickness of the substrate film having approximately the thickness of the silicon layer to be applied thereto and the structure of the aluminum oxide being amorphous.

The temperature of the aluminum-zinc melt is below the melting point of the oxidized aluminum film, but it is high enough so that sufficient silicon may be dissolved in the aluminum-zinc melt. The melt saturated with silicon is contained in a container at the same high temperature. In a part of the container, solid silicon pieces are in contact with the melt. In a second part of the container, the oxidized aluminum strip is drawn across the surface of the melt, the aluminum-zinc melt saturated with silicon wetting the amorphous aluminum oxide surface. The aluminum-zinc melt saturated with silicon is recirculated in drawing direction or opposite to the drawing direction of the aluminum film.

In order to achieve a separation of silicon on the oxidized aluminum film, the film is cooled in the contact region of the melt in that heat is passed off the face of the oxidized aluminum strip remote from the melt by radiation and convection.

For achieving a planarlike crystal growth on the amorphous aluminum oxide layer, the aluminum oxide layer may contain a coat of silicon atoms or of atoms of a metal having a high surface energy rating. The coating density of the silicon or metal atoms in this regard does not exceed the surface density of the aluminum and oxygen atoms of the amorphous aluminum oxide layer.

Differential Prebaking of Photoresist

D.R. Dyer, C. Johnson, Jr. and R.R. Wilbarg; U.S. Patent 4,115,120; Sept. 19, 1978; assigned to International Business Machines Corporation describe a method for forming thin patterned films by lift-off techniques in which the deposition mask with apertures having the desirable negative slope or overhang are produced by a simplified efficient technique.

Figures 6.6a through 6.6f show the formation of a mask in accordance with the present process as well as the utilization of this mask for lift-off purposes. Referring to Figure 6.6a, a layer of photoresist material **10** is formed on a semiconductor substrate **11**. In the fabrication of integrated circuits, the substrate may be a semiconductor material or a semiconductor substrate having a surface layer of an electrically insulative material, such as silicon dioxide or silicon nitride. Layer **10** may be formed of a positive photoresist composition such as described in U.S. Patent 3,666,473. Other photoresist compositions such as those described in U.S. Patent 3,201,239 may also be used to form layer **10**. The photoresist composition is applied to substrate **11** by conventional spin coating at about 4,000 rpm.

Next, the photoresist is baked (Figure 6.6b) under the following conditions. The wafers **10** with the photoresist **11** coated thereon are mounted on a structure in which the semiconductor substrate or wafer surface coated with the photoresist is exposed while the other surface is seated on a cooling member **20**. Conveniently, the bath of wafers may be seated in recesses in the cooling member. In order to maintain the temperature differential on the photoresist coating **11** on the wafers while the wafer is being heated by conventional means, i.e., in an oven, the cooling member which may conveniently be made of a heat-conductive material such as metal and may be cooled in any conventional manner, e.g., by liquid being circulated through the coils in any standard liquid cooling scheme, thermoelectric cooling could also be used.

In order to partially cure photoresist coating **10** so that the exposed surface of the photoresist coating is partially cured to a greater extent than the lower surface of the coating at the interface with the substrate **11**, the heating or curing cycle should be relatively short, i.e., from 40 to 60 seconds as compared to more conventional photoresist prebake or preliminary curing cycles which are from 5 to 30 minutes.

With longer curing cycles it becomes more and more difficult to maintain the temperature differential between the upper and lower surfaces of the photoresist layer **10** which is necessary for the process.

Figure 6.6: Differential Prebaking of Photoresist

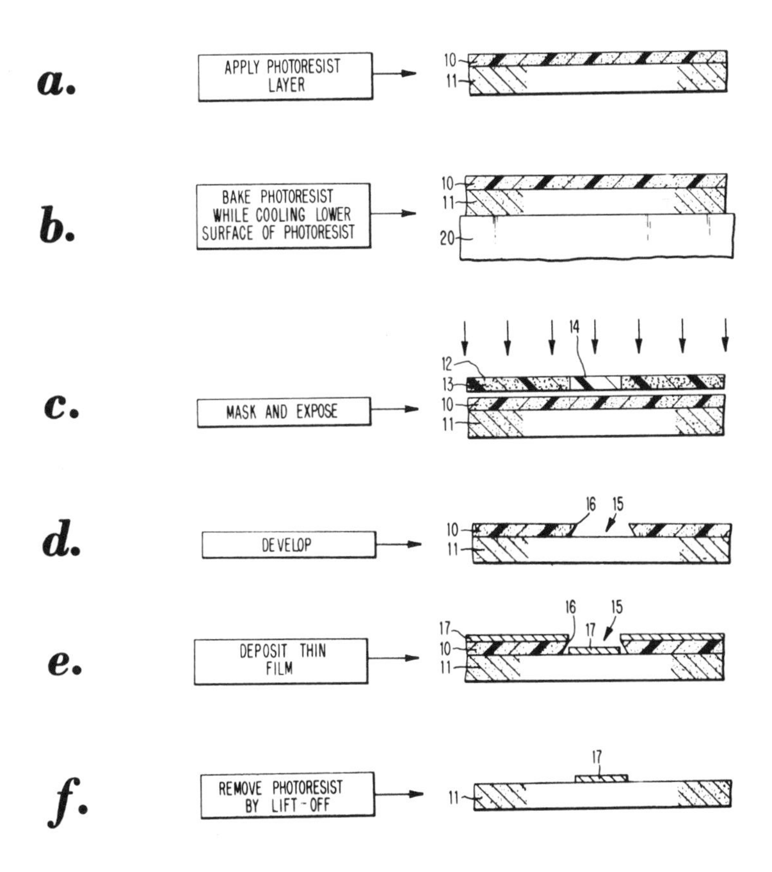

Source: U.S. Patent 4,115,120

For best results, the photoresist coated wafer **10** is baked at a temperature from 120° to 160°C and preferably at about 140°C for a period of from 30 to 60 seconds and preferably for 40 seconds, while the underside bottom of the wafer is maintained at a cooler temperature, e.g., about 20°C. With this temperature combination, the lower or interfacing surface of the photoresist layer **10** should be maintained at a temperature at least 40° lower than the temperature of the exposed surface of photoresist layer.

As a result of this differential curing step, the photoresist layer, which for purposes of this embodiment has a thickness of about 2 μ, will be partially cured with a differential state of cure, i.e., photoresist at the upper surface of layer **10** will be cured to a greater extent than the photoresist at the lower surface which interfaces with the substrate **11**.

Next, as shown in Figure 6.6c, the structure is masked with a mask **12** having transparent regions **14** and opaque regions **13**. The mask is shown in the illustration as being slightly offset from the surface merely for purposes of clarity. The exposure is to a 200 W mercury lamp for approximately 10 seconds. As a result of this exposure, the portion of photoresist layer **10** beneath the transparent region will become solubilized. The extent of solubilization will, of course, depend on the extent to which the photoresist layer has been partially cured during the previously described heating step.

Then, as shown in Figure 6.6d, the structure is subjected to a developing step utilizing a conventional alkaline developer for positive photoresist, e.g., a 2.5% by weight solids in aqueous solution of a mixture of sodium metasilicate and sodium phosphate, predominantly sodium orthophosphate having a pH at room temperature of about 12.55 for 30 seconds.

As a result, apertures **15** will be formed through photoresist layer **10** in a pattern corresponding to the transparent portions **14** in mask **12**. Each of the apertures will have a negative slope because the photoresist material at the bottom portion of the apertures will not be as extensively cured as the material in the upper portion, and thus the photoresist material in the bottom portion would be more extensively solubilized upon exposure to light in the previous step of Figure 6.6c. Thus, the photoresist defining aperture **15** will have the overhang **16** required for effective lift-off processing.

Next, as shown in Figure 6.6e, using photoresist layer **10** as a deposition mask for the substrate, a metallic film **17** is deposited at a temperature between room temperature and 100°C. Alternately, layer **17** may be an inorganic electrically insulative material such as silicon dioxide or silicon nitride. These insulative materials may be deposited in any conventional sputter deposition system. In the embodiment shown, a metallic film is used. The metal may be any metal conventionally used for integrated circuit metallization, e.g., aluminum-copper alloys, platinum, palladium, chromium, silver, tantalum, gold and titanium or combinations thereof. Metallic film **17** has a thickness of about 1 μ.

Then, using conventional lift-off techniques, photoresist masking layer **10** is completely removed by immersion into a solvent, such as N-methylpyrrolidone, a standard photoresist stripping solvent for about 10 minutes using ultrasonic agitation as shown in Figure 6.6f. This leaves thin metal film **17** in the desired, preselected configuration. The solvent selected should be one which dissolves or swells the polymeric material of layer **10** without affecting the thin metallic film. Such solvents include acetone, butyl acetate, trichloroethylene and Cellosolve acetate.

Selective Nickeling of Polyester Resin

E.A. Guditz and R.L. Burke; U.S. Patent 4,150,177; April 17, 1979; assigned to Massachusetts Institute of Technology describe a process for forming the conductor paths for electrically interconnecting a group of integrated circuit chips to form an array structure in which the chips are embedded in an insulating plastic. The plastic secures the chips mechanically and provides the base for the conductors which connect the terminal pads of the chips to the pads of other chips or to the external connection terminals of the array by means of photoformed plated conductors on the plastic. These conductors may be formed as multilayer wiring where the layers of conductors are insulated from one another by insulating layer of plastic. The conductors in the different layers are selectively connected to one another, layer to layer, by means of conduction through the insulation layers. Each chip is connected to a heat sink through thermal paths from the bottom of the chips.

This method of nickeling selected regions of a layer of polymerized polyester resin comprises etching the resin with an etching solution, covering the etched resin with a layer of liquid unpolymerized polyester resin, exposing the liquid resin to ultraviolet light through a mask which has opaque regions corresponding to the selected regions to polymerize the liquid resin to form a second polymerized layer of resin except in the regions, developing out the unexposed resin, immersing the etched layer and the second layer of resin in a palladium solution for a time sufficient to form an activating layer on the etched resin only, and immersing in an electroless boron-nickel bath for a time sufficient to provide the desired thickness of nickel on the activating layer, whereby electrically conductive regions are formed on the first layer of resin corresponding to the opaque regions on the mask.

The etching step comprises etching for 15 to 30 seconds in a room-temperature etchant consisting of the following: 10 ml distilled H_2O, add slowly 50 ml sulfuric acid (H_2SO_4), cool below 30°C, and slowly add 0.5 g chromium trioxide (CrO_3) while stirring, and rinsing. The palladium immersion step comprises immersing in a 10% palladium plating solution for 2 to 3 minutes at room temperature and rinsing. The boron bath immersing step comprises immersing in electroless boron-nickel bath for one hour at room temperature, and rinsing. All rinses are distilled water rinses.

Gold Films

L.A. D'Asaro and Y. Okinaka; U.S. Patent 4,162,337; July 24, 1979; assigned to Bell Telephone Laboratories, Incorporated describe a process for fabricating semiconductor devices with gold films made by a particular activation and electroless gold plating process. Semiconductor compounds in the III-V group which contain gallium, aluminum or indium are particularly included as, for example, GaAs, $Al_xGa_{1-x}As$, $Al_xGa_{1-x}P_yAs_{1-y}$, GaP and InP.

The process involves surface cleaning and then activation followed by electroless gold plating. Cleaning may be conventional, but a specific procedure involving plasma cleaning is often preferred. Activation involves use of an acid solution of activator species in water or weak acid. Various activator species are useful including gold-containing ions and platinum-containing ions, but palladium-containing ions are preferred because of their excellent catalyzing properties for electroless plating solutions. In addition, hydrofluoric acid is used in the activation solution. Concentrations between 0.1 and 50% by weight show good results.

Preferred activation procedures depend to some extent on the semiconductor surface being activated particularly as to the solvent used in the activation solution. Activation is carried out by wetting the surface to be electrolessly plated with activator solution. Preferred times for activation are between 1 and 5 minutes both for convenience and to insure complete activation. Activation is followed by rinsing, preferably in distilled, deionized water, and then immersion in the plating bath.

The preferred plating bath is composed of a soluble gold cyanide complex, excess-free cyanide ion, reducing agent such as borohydride or aliphatic amine borane and an alkaline agent to increase pH. Although the electroless plating may be carried out at any temperature between the freezing point and the boiling point of the plating solution, a temperature between 60° and 95°C yields good results. The process is simple, inexpensive and rapid. It is easily adapted to the technology used to produce integrated circuits and semiconductor lasers. The resulting gold films exhibit exceptionally good adherence so that devices produced in accordance with the process exhibit good temperature stability when attached by means of the gold films to heat sinks, exhibit good electrical contact to the semiconductor and produce good mechanical bonds when attached through the gold-plated surface to headers or other mechanical mounting structures.

Example: *GaAs* – The semiconductor material is first shaped and mounted, if necessary. In many applications the surface is polished, typically using a 5% by weight solution of bromine in methyl alcohol. This is followed by an acid rinse using aqueous HCl. After drying, the surface is subjected to an rf cleaning procedure using 3% oxygen and 97% helium at a power of 50 W for 5 minutes. The surface is then activated by exposure to an activation solution for at least 1 minute.

The activation solution is made by dissolving 0.3 g of palladium chloride in 9 ml of concentrated HCl and 9 ml of water. This solution is added to 864 ml of glacial acetic acid, and to this solution is added 18 ml of hydrofluoric acid (49%) which is thoroughly mixed. Variations in composition of ±50% yield equivalent results. After activation the semiconductor is thoroughly rinsed in distilled deionized water and then immersed in the plating bath.

The plating bath is maintained at approximately 70°C during plating. The sample is agitated in the plating bath so as to obtain uniform and rapid plating. It is generally preferred that the sample be moved relative to the plating solution rather than agitating the entire plating solution. The plating rate is approximately 1 μ of gold in 40 minutes under these conditions. In some cases it is useful to inspect the sample after 10 or 15 minutes and to reactivate any spots on the surface which are not plating. The plating proceeds equally well in n-type or p-type semiconductor material, as well as undoped semi-insulating crystals. This procedure yields excellent gold films with good adherence and good electrical contact. Additional gold may be put on the gold layer by gold electrodeposition procedures.

Planar Conductor Path System

By a photolithographic etching technique, conductor paths are produced on a semiconductor surface. This leads to considerable profiling of the semiconductor surface. A passivation layer applied upon the latter, and which can consist, for example, of sputtered SiO_2, frequently only insufficiently covers the edges of the etched portions and then is unable to fulfill the desired functions. Thus, for example, the passivation layer is not sufficiently flat for a second metallization layer to be applied subsequently.

G. Bell; U.S. Patent 4,098,637; July 4, 1978; assigned to Siemens Aktiengesellschaft, Germany describes a process in which the production of the conductor paths leads to as small as possible a profiling of the surface of the semiconductor arrangement. This is achieved by a process which includes forming a metal layer on a planar surface of a base, forming a masking layer on the metal layer over those regions where the conductive path is to be, and oxidizing electrolessly in aqueous solution those areas of the metal layer not covered by the masking layer.

The process will be described making reference to Figures 6.7a through 6.7e. Referring to Figure 6.7a, the base upon which the aluminum conductor paths are to be applied is referenced **1**. This base can, for example, be an insulating layer applied to a semiconductor material. For example, this can be a SiO_2 layer applied to a silicon semiconductor body. The base **1** can also be a semiconductor material itself, preferably a silicon body.

First, the base is covered on one surface with an aluminum layer **2** or a layer consisting of an aluminum alloy, for example, an aluminum-copper alloy. Next, a forming layer (Al_2O_3) **3** is applied to the whole of one surface of the aluminum or aluminum alloy layer **2**. This layer **3** is deposited on the layer **2** over its entire surface. Here the oxide layer which because of the high affinity between aluminum and oxygen is constantly deposited on the aluminum layer, is reinforced by anodic oxidation to form a layer which exhibits a good surface quality and other favorable properties. The aluminum or aluminum alloy layer **2** serves as anode.

When a suitable selection has been made for the electrolyte, pore-free thin layers can be deposited in a known manner. Here when the current is constant, the layer thickness is determined by the voltage, which is applied. The process is well known in the prior art.

Using photolithographic process steps, the structure of the conductor paths or the conductor path system is produced. For this purpose, the aluminum oxide layer **3**, as illustrated in Figure 6.7b, is first covered on the whole of its surface with a photolacquer layer **6**. Then this photolacquer layer is exposed and developed in a further process step.

Here, as illustrated in Figure 6.7c, after the removal of the remaining photolacquer layer, those parts **31** of the forming layer **3** remain, under which a conductor path or an electrically conductive aluminum circuit is to be produced.

Figure 6.7: Planar Conductive Path System

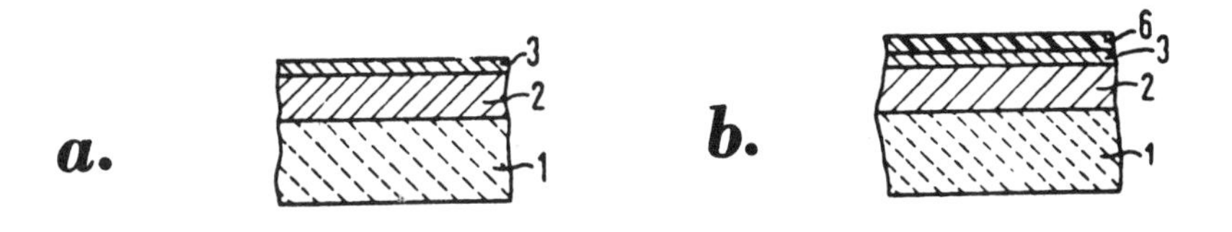

(continued)

Figure 6.7: (continued)

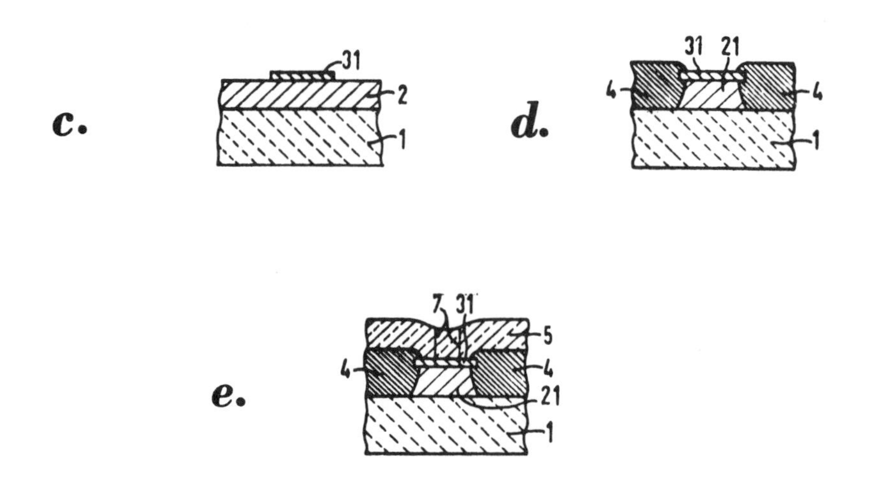

Source: U.S. Patent 4,098,637

Thereafter, as illustrated in Figure 6.7c, the open zones of the aluminum or aluminum alloy layer **2** which are not covered with aluminum oxide layers **31** are oxidized electrolessly in aqueous solutions such as described, for example, in the *Handbuch der Galvano-Technik,* Dettner-Elze, Edition III. It should be noted that this oxidation step is not carried out electrolytically and does not have the associated disadvantages of an electrolytic oxidation step.

The oxide layers **4** which are formed by the electroless step, are advantageously porous. Because of this, a high-speed growth is assured, since the aqueous solutions can always penetrate through to the remaining aluminum or to the remaining aluminum alloy through the aluminum or aluminum alloy which has already been oxidized in porous fashion. With the aid of this process step, metallization layers are completely converted in the technologically normal layer thickness in the order of microns. During the conversion, the pore-free forming oxide layer **31** serves as a mask. For example, the oxidation is carried out in an aqueous solution consisting of 1% hexamethylenetetramine in H_2O at a temperature of 90°C for approximately one hour.

As likewise illustrated in Figure 6.7e, a protective layer **5**, which, for example, is a SiO_2 layer, is subsequently applied by sputtering to the arrangement as shown in Figure 6.7d. In further process steps, connection spots are etched in a known manner for the purpose of contacting. Here openings are produced which extend through the protective layer and through the forming oxide **31**. In Figure 6.7e, such an opening is indicated by the broken lines **7**.

Solderable Nickel Layer

G. Schaal; U.S. Patent 4,132,813; January 2, 1979; assigned to Robert Bosch GmbH, Germany describes a method for production of solder-compatible metallized layers of high transverse electrical conductivity on an insulating or semiconductor substrate, particularly a method in which an aluminum layer with the desired transverse electrical conductivity is first applied to the substrate and then a comparatively thin nickel layer is applied on top, after which a tempering treatment is carried out in a nonoxidizing atmosphere, the product being held for a particular time somewhat below the temperature at which a liquid phase is formed between the substrate and the aluminum layer.

Briefly, the thickness of the nickel layer applied on top of the aluminum layer and the time and temperature of the tempering treatment are so adjusted to each other that, on the one hand, adequate contact between the substrate and the metallization is obtained and, on the other hand, the solderability of the metallization is preserved. In particular, the thickness of the nickel layer is kept within the limit of 0.6 to 1.0 μm, the tempering temperature within the limit of 400° to 480°C and the tempering time within the limit of 10 to 20 minutes. The thickness of the aluminum layer is preferably between 3 and 10 μm. In the preferred method, the tempering temperature is about 475°C and the tempering time, about 15 minutes.

The method is applicable to the manufacture of multiple semiconductor devices for providing both connection paths within the device and contact areas for the soldering on of external leads. The metallization may be on top of a protective insulating layer of silicon oxide or nitride, or such a protective layer topped by a glass layer, contact windows being provided through the protective layer for the electrodes of the various semiconductor devices previously manufactured in the substrate.

Figure 6.8a is a diagrammatic cross section of a portion of a multiple semiconductor device which includes a Darlington transistor combination prior to the application of an aluminum layer. Figure 6.8b is a cross section of the semiconductor device portion of Figure 6.8a after the application of an aluminum layer in accordance with the process. Figure 6.8c is a diagrammatic cross section of the semiconductor device portion of Figure 6.8b after the application of a nickel layer in accordance with the process. Figure 6.8d is a diagrammatic cross section of the semiconductor device portion of Figure 6.8c after an etching step to provide a desired contour of the metallization layer that precedes the tempering step.

Figure 6.8: Semiconductor Device

a.

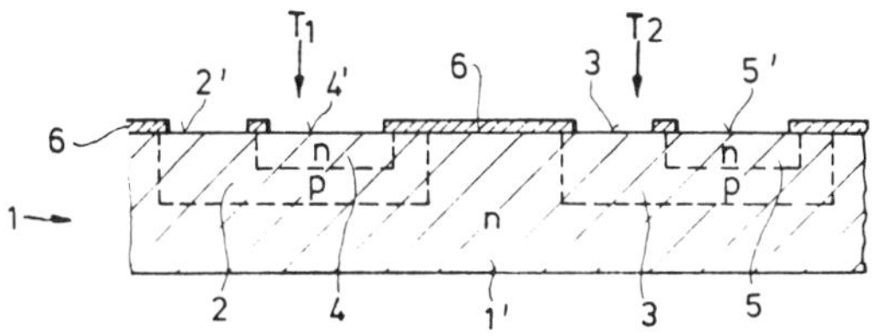

(continued)

Figure 6.8: (continued)

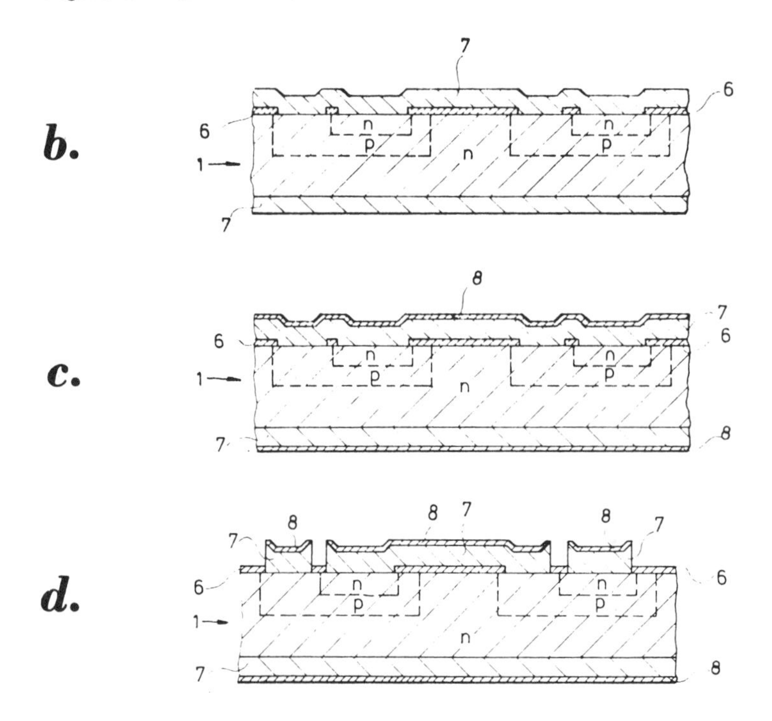

Source: U.S. Patent 4,132,813

The monolithic Darlington amplifier in the portion of a monolithic multiple semiconductor device shown in the above figure has the following known characteristics, produced by known methods of semiconductor device manufacturing, prior to completion of the device by the use of this process. In a semiconductor wafer **1** of **n**-type silicon, two base zones **2** and **3** of **p**-type conductivity have been provided by diffusion of an appropriate dopant and in both of these base zones, the emitter zones **4** and **5** of **n**-type conductivity have, respectively, been provided by diffusion of another dopant, so as to produce a double **npn** zone sequence. Thus, there have been produced in the substrate **1** two transistors with a common collector **1'**, including a first transistor T_1 formed of the zone sequence **1'**, **2** and **4** and a second transistor T_2 formed of the zone sequence **1'**, **3** and **5**.

These two transistors T_1 and T_2 are to be connected to each other in a Darlington circuit. Assuming that the first transistor T_1 is the driver transistor and the second transistor T_2 is the power transistor of the Darlington amplifier, it is necessary to provide an electrically-conducting connection between the emitter zone **4** of the first transistor T_1 and the base zone **3** of the second transistor T_2. This electrically-conducting connection can be provided by a layer of metallization that will extend from the emitter zone **4** of the first transistor T_1 over the silicon dioxide layer **6** to the base zone **3** of the second transistor T_2. Furthermore, it is necessary to provide metallization for the connection of external leads to the base zone **2** of the first transistor T_1 to the emitter zone **5** of the second transistor T_2 and to the common collector zone **1′** of the two transistors T_1 and T_2. This metallization can be applied with great advantages by the process in the manner described below.

The silicon wafer **1** already has its base zones **2** and **3** and emitter zones **4** and **5** formed therein by diffusion of the appropriate conductivity-affecting impurities. Furthermore, for the protection of the **pn** junctions of the semiconductor devices formed in the semiconductor body, the silicon wafer is covered with a silicon dioxide layer **6**. By means of a photolithographic process, contact windows **2′**, **3′**, **4′** and **5′** are etched into the silicon dioxide layer **6** by conventional processing for that purpose. After removal of the photoresist through which the etching is done, the wafer is carefully washed and dried. This brings the wafer into the condition shown in Figure 6.8a, where it constitutes the substrate on which metallization layers are to be provided.

The wafer is then placed in a high vacuum metallizing chamber and at a pressure between 10^{-6} and 10^{-5} torr and aluminum layer **7** is applied by vapor deposition and, without interruption of the vacuum, a nickel layer **8** is then applied likewise by vapor deposition, these layers being applied to both sides of the wafer. The thickness of the aluminum layer **7** on the upper side of the wafer is in the range between 7 and 10 μm and the thickness of the nickel layer **8** (Figure 6.8c) is in the range from 0.6 to 1.0 μm. Because of reduced requirements for transverse conductivity of the aluminum layer on the underside of the wafer, the aluminum layer **7** on the underside may be thinner than the corresponding layer on the upper side. In order to prevent the formation of a brittle contact, however, the thickness of the aluminum layer there should not go below 3 μm.

Figure 6.8b shows the silicon wafer **1** after the deposition from vapor of the aluminum layer **7** and Figure 6.8c shows it after the vapor-deposition of the nickel layer **8**.

The metallization consisting of the layers **7** and **8** on the upper side of the semiconductor wafer is then removed where it is not necessary, that being done by means of a well-known photolithographic process. In such a process, the surface of the wafer is covered with a photoresist from which a mask is made by a photographic process, leaving the portions of the metallization **7, 8** that are not to be removed protected by the photoresist mask, whereas the portions of metallization to be removed are exposed. (Photoresists for this purpose are commercially available.) The drawings do not show the stage in which the wafer is covered with a photoresist, nor the stage in which the photoresist has been developed to form a mask. The masked wafer is then dipped in a bath containing preponderantly phosphoric acid, with an additional content of nitric acid and water. In a typical composition the proportions of phosphoric acid, nitric acid and water are respectively 74, 11 and 15%. The treatment temperature for this step is preferably between 50° and 70°C.

In this bath the portions of the metallization **7, 8** not protected by the photoresist mask are etched away. The photoresist mask is then removed by a suitable solvent, typically acetone. The resulting state of the semiconductor wafer with the remaining portions of the metallization layers **7** and **8** is shown in Figure 6.8d.

After removal of the photoresist mask, the wafer is thoroughly washed and dried and thereafter it is tempered in a forming gas (largely N_2 and H_2) atmosphere at a temperature of 475°C for a period of 15 minutes.

The wafer can thereafter be broken up into individual semiconductor devices by known methods. Thus, the Darlington amplifier illustrated in the figure could constitute a single chip made from the original wafer, which is commonly in disk shape, and each such chip can then be completed as a separate semiconductor device.

Each of the individual chip devices can then be provided with electrode leads that are soldered onto the nickel surface of the metallization, utilizing the usual soldering techniques. The nickel surfaces produced in accordance with the above procedure are easily wetted by the usual soft solders composed of lead and/or tin.

Control of Shrinkage

L.A. Blazick and L.F. Miller; U.S. Patent 4,109,377; August 29, 1978; assigned to International Business Machines Corporation have found that the use of a metal oxide powder with the metal powder for shrinkage adjustment allows an excellent shrinkage adjustment in the formation of multilayer ceramic substrates. Commonly used ceramics, such as aluminas, mullite, beryllias, titanates and steatites are usable as the ceramic component. Metallizing compositions which are useful include molybdenum, tungsten, and noble metals that can form oxides such as silver and palladium. The metallization composition is adjusted by ratios of the metal oxide to the metal in the range of 1:1 to 1:9, depending upon the shrinkage condition of the ceramic to be cofired with the conductive composition. The particulate mixture containing the metal and metal oxide is deposited in a suitable pattern on at least a portion of the plurality of unfired ceramic or green ceramic layers or substrates which will make up the multilayer level ceramic substrate.

The patterns are dried. The plurality of layers of ceramic are then laminated by stacking together and then applying a substantial pressure of an order of greater than about 2,500 lb/in^2 thereto. The laminate is then fired at an elevated temperature and then cooled. The result is a multilayer ceramic substrate which is free of stresses, cracks and warpage. The fired metallurgy is dense and conductive.

In the preferred metal and metal oxide embodiment, the molybdenum powder and molybdenum trioxide powder are mixed dry in the ratio of 1:1 to 1:9 molybdenum to molybdenum trioxide. The average preferred particle size for molybdenum is about 1.5 to 3.5 μ and molybdenum trioxide, 2 to 5 μ. A suitable vehicle or solvent is mixed with the dry powder and then milled in a suitable mill, such as a three-roll mill, into a paste.

The vehicle chosen must be one which may be given off at or below the firing or sintering temperature of the ceramic being utilized so that only the residual metallization remains after the process is completed. The conductive paste is then screened onto the green sheet to form the desired circuit patterns by the conventional silk screening techniques. Where it is desired to have electrical connections between the layers, it is necessary to punch holes in the sheet prior to silk screening, and a second silk screening operation may be done to fill the via holes. Alternatively, one silk screening can be used to simultaneously coat the surface and force the paste into the via holes. Thereafter the paste is dried by placing the sheets in an oven and baking them at a rather low temperature, for example, 60° to 100°C for 16 to 60 minutes, or the paste may simply be air-dried.

Figure 6.9 illustrates a plurality of layers of the ceramic having a variety of conductive patterns thereon which are being stacked in the proper sequence. The stack may be carefully registered using registration pins (not shown) so that all conductive lines from layer to layer are properly registered and aligned.

Figure 6.9: Multilayer Ceramic Structure

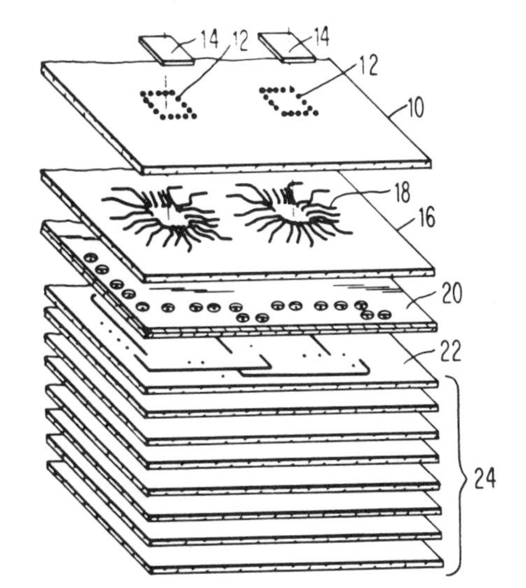

Source: U.S. Patent 4,109,377

The top or surface layer **10** is provided in the figure with two patterns **12** that are suitable for joining semiconductor chips **14** thereto. These particular chips are of the flip-chip or contacts-down variety. The next level **16** has two conductive patterns **18** which connect through conductive via holes through the layer **10** to the conductive lines. Other via holes through the layer **16** make circuit connections to the succeeding layers **20, 22** and the remaining group of layers **24** so as to provide the required circuit connections for the input and output of signals to the semiconductor chips **14**. These vias are between about five to seven thousandths of an inch punched diameter and on centers about ten to twelve thousandths of an inch.

The registered stack of green ceramic layers is placed in a laminating press. Moderate heat and pressure is applied. The preferred pressure for alumina ceramic is greater than 2,500 psi and a temperature of about 80° to 100°C. In this step, the thermoplastic binder in the green ceramic sheets softens and the layers fuse together, deforming around the metallized pattern to completely enclose the lines. The result is that the unfired stack will show no signs of individual layers. The stack of green sheets is then sawed or punched to the size of the finished module plus an allowance for shrinkage. The green module is fired in a suitable furnace where the module is raised from room temperature to a temperature greater than 1450°C at a rate of 140°C/hr and the furnace is then maintained at 1500° to 1600°C for 1 to 5 hours for the firing of green ceramic. The firing ambient is wet hydrogen. The temperature is then reduced to room temperature at a rate of about 200°C/hr.

InSb Thin Film Element

As the computing speed of an electronic computer has become higher, it has been required to render the operating speed of transistors for use in the electronic computer higher. Therefore, a MOS-IC for the electronic computer also needs to have the speed increased. Since, in general, the operating speed of a transistor is substantially proportional to the electron mobility of a constituent semiconductor material thereof, a MOS-type FET constituted of a semiconductive material of high electron mobility becomes desirable. The electron mobility of silicon (Si) having heretofore been used for transistors is several hundred square centimeters per volt-second. Therefore, insofar as Si is used, it is subject to the limitation to render the operation of a transistor highly speedy.

On the other hand, experiments have revealed that an InSb film having a thickness of, for example, about 1.5 μm exhibits as high an electron mobility as 6×10^4 cm^2/V-sec. A MOS-type FET using the InSb film is expected to realize a speed exceeding the limitation of the operating speed of the conventional Si-MOS type transistors.

With the MOS-type FET using the InSb film, however, when the thickness of the InSb film is greater than about 0.2 μm, an electric field normal to the film surface is not effectively applied by the provision of a gate electrode, and the operation as a FET is difficult. Therefore, an InSb thin film whose thickness is about 0.2 μm or below and which has good electrical characteristics, especially an excellent electron mobility, has come to be desired. However, it has been extremely difficult to acquire such InSb thin film.

N. Kotera and N. Miyamoto; U.S. Patent 4,128,681; December 5, 1978; assigned to Hitachi, Ltd., Japan describe a method for readily producing an InSb thin film element whose proper or main body is an InSb thin film being at most 0.2 μm in thickness, having an even surface and exhibiting good electrical properties. In order to accomplish the object, the method comprises the following steps:

(a) Preparing a substrate at least one surface of which is made of alumina (i.e., all aluminum oxides) or an inorganic insulating material containing at least 12 mol % of alumina, and forming an InSb thin film of a thickness of at most 0.2 μm on the surface of the substrate;

(b) Depositing on the InSb thin film a film (an overlying layer) which is made of alumina or an inorganic insulating material containing at least 12 mol % of alumina;

(c) Heating the InSb thin film above the melting point of InSb; and

(d) Cooling the InSb thin film and recrystallizing InSb.

In the example, 7059 Glass refers to 7059 Glass produced by Dow Corning Inc. which has a chemical composition consisting of 50 mol % SiO_2, 12 mol % Al_2O_3, 25 mol % BaO, 13 mol % B_2O_3, 0.05 mol % MgO, and less than 0.01 mol % Na_2O.

Example 1: Referring to Figure 6.10a, on a 7059 Glass substrate **1** having a thickness of 1 mm, an InSb film **2** is evaporated to a thickness of 0.15 μm by the known three-temperature evaporation method. By keeping the temperature of the glass substrate **1** at 350°C during the evaporation, the InSb film **2** comes to have a mirror surface and is formed to be flat. Thereafter, an alumina film **3** is formed to a thickness of 0.6 μm on the InSb film **2** at 400°C in an atmosphere of argon by the thermal decomposition of aluminum isopropoxide. The resultant substrate is zone-molten three times by the known hot-wire microzone melting method. A sectional view of the element thus fabricated is shown in Figure 6.10a. The surface of the InSb film after the zone melting is flat and the film does not rupture.

The alumina film thus formed is locally removed and the electron mobility of the InSb film at room temperature is measured by Van der Pauw's method. The electron mobility which has been 210 cm^2/V-sec in the evaporated state is then enhanced to 830 cm^2/V-sec after the zone melting.

In the three temperature evaporation method described above, the temperature of the In source is 1000°C, the temperature of the Sb source is 540°C, and the temperature of the substrate is 400°C. At the zone melting, the heating temperature is 525° to 527°C, the atmosphere is a flowing He gas and the moving speed of the molten zone is 6 to 7 μm/sec. The alumina film is locally removed by the well-known photoetching method by the use of a liquid etchant in which 10 parts of water are mixed with 1 part of concentrated hydrofluoric acid. The electron mobility is measured at a current density of 5,000 A/cm^2 and a magnetic flux density of 200 gauss by the Van der Pauw method [L.J. Van der Pauw; *Phillips Research Reports,* vol.13, No. 1, p. 1 (February 1958)].

Figure 6.10: InSb Thin Film Element

Source: U.S. Patent 4,128,681

Example 2: Referring to Figure 6.10b, a silicon substrate **5** having a thickness of 1 mm is put in an equipment for performing the well-known CVD (chemical vapor deposition) in which an O_2 gas is flowing along with an N_2 gas containing triisobutyl aluminum. Thus, an alumina film **4** (0.3 μm thick) is deposited on the silicon substrate **5** heated to 400°C. An InSb film **6** is deposited to a thickness of 0.08 μm on the alumina film **4** heated to 400°C by the well-known flash evaporation method. Thereafter, using the well-known sputtering method, an alumina film **7** is formed to a thickness of 0.3 μm on the InSb film **6**. The resultant substrate is put in a high temperature oven having a temperature gradient of 800°C/cm and the InSb film is molten. Thereafter, it is cooled and solidified. At this time, when the whole substrate is once made above 525°C and then cooled, the normal freezing takes place and the solidification is conducted from one end of the substrate. This operation of the normal freezing is repeated six times. The electron mobility of the InSb film thus obtained is 530 cm^2/V-sec.

In the deposition of the alumina film **4** by the CVD process, the flow rates of the gases were 2 ℓ/min for N_2 which was passed through a triisobutyl aluminum liquid at 10° to 40°C forming the bubbles thereof, and 1 ℓ/min for O_2.

In the evaporation of the alumina film **7** by the sputtering process, the heating of a source was made by irradiation with a CO_2 laser. Figure 6.10b is a view showing a section of the element produced in this manner. The aluminum film **7** is formed with a hole **8** for an electrode as meets the purpose of use.

Gold or Platinum Chlorosilicate Films

R. DiBugnara; U.S. Patent 4,126,713; November 21, 1978; assigned to TRW Inc. describes solutions used to form a metal-containing film on the surface of semiconductor devices and methods for making such solutions. The process will be described with reference to gold film and platinum film, as these two metals have special attributes which make them particularly desirable, but other metals are also within the scope of this process.

The gold (or platinum) chlorosilicate compound film is formed by the following steps. A first alcohol solution of a gold chloride (or platinum chloride) is formed which contains a crosslinking agent in trace amounts. This first alcohol solution is mixed with a second alcohol solution containing an organic silicate. The first and second alcohol solutions are mixed, preferably in equal amounts by volume, and allowed to age for a specific length of time. It is believed that during this aging a reaction is taking place between the chloride ions and the organosilicate. The mixture thus formed can be applied by well-known techniques including brushing, spraying or spinning and produces a film coat which quickly crosslinks at room temperature to form a gold chlorosilicate glass which is then ready to be further treated.

Once the gold chlorosilicate glass is formed on a semiconductor surface, the semiconductor can be placed in a gold diffusion furnace for diffusing the gold into the silicon wafer. Under such heat, the film makes the final conversion into atomic gold and silica with a dense enough structure to permit free movement of the gold within the film. In this manner, improvements in the diffusion of the gold atoms into the surface of the semiconductor for lifetime control are achieved. To form a platinum film, the above procedure is repeated using platinum chloride. This salt forms a platinum chlorosilicate glass which is heated in order to drive the platinum into the silicon wafer.

Example: The makeup of a typical coating solution is as follows: 2 ml of gold chloride solution which is formed by making a 100 ml solution of gold chloride dissolved in water such that 1 oz of pure gold is present, i.e., sufficient gold ions to form 1 oz of free gold; 200 ml of reagent ethanol (95 wt % ethanol, 5 wt % isopropyl alcohol); 70 ml tetraethyl orthosilicate, and 40 ml of methanol containing 1% by volume of concentrated (30 wt %) HCl.

The coating solution is formed from Part A and Part B as follows:

	Milliliters
Part A	
Gold chloride solution	2
Reagent ethanol	115
Methanol and HCl	40
Part B	
Reagent ethanol	85
Tetraethyl orthosilicate	70

Part A and Part B are mixed in equal amounts by volume.

The above example has been repeated utilizing platinum chloride and the results have been just as good. In addition, when a platinum film is being formed, 1 g of chloroplatinic acid (platinum hexachloride) has been used and the water (used to make the 2 ml of gold chloride solution) is omitted.

Etch-Resistant Mask

D. Widmann; U.S. Patent 4,108,717; August 22, 1978; assigned to Siemens AG, Germany describes a process of forming a fine structure having an order of magnitude of 1 μm without a loss of dimension relative to a mask such as a conductor pattern or electrode pattern that is particularly useful in semiconductor devices, which process does not exhibit the disadvantages of a lack of dimensional stability of the pattern formed on the base.

To accomplish these tasks, the method comprises the steps of providing a base with a surface; providing a mask on the surface of the base, the mask having edges defining openings corresponding to the fine structure of material to be applied on the surface of the base; providing an etching agent which will attack the surface of the base, but does not attack the mask; etching the exposed portions of the base until an under-etching of a predetermined width exists beneath the edges of the mask; depositing a layer of material over the entire surface controlling the amount of depositing so that the layer of material being deposited on the mask and the etched portions of the base are not in contact with each other; and subsequently removing the mask with the layer of material deposited thereon. Preferably, the material being deposited is at least one layer of a material.

In many instances, the base is a nonetchable material or cannot be allowed to be etched. In such cases, an auxiliary layer of etchable material is applied on the base prior to providing the etch-resistant mask. This auxiliary layer may be either an etchable layer on top of a base having a plurality of layers or may be a layer which has been applied onto the base. In either case, after removal of the mask, the remaining portion of the etchable layer, if desired, may be removed by etching with an etching agent which selectively attacks or etches the remaining portions without subsequently etching the layer of fine structure and the base material. In some instances, it is desired that the remaining portion be retained to provide electrodes spaced by the gaps between the remaining portion and the fine layer structure.

The process is preferably used for the production of electrode structures and/or conductor path structures on semiconductor devices and particularly highly integrated semiconductor devices, preferably of the MOS or MIS technique. The process is also utilized in producing electrode structures having a small gap such as required for charge-coupled devices known as CCDs.

A specific embodiment of the process is described in Figure 6.11. In Figure 6.11a, a substrate **5**, such as a semiconductor material, is provided with an insulating layer **6**, such as a grown-on silicon dioxide layer and is provided with an auxiliary layer **7**. In this particular embodiment, the auxiliary layer **7** is deposited onto the layer **6** and consists of an aluminum-nickel alloy with approximately 0.5% nickel content. This deposited layer has a thickness in the range of 0.1 to 0.3 μm.

Figure 6.11: Etch-Resistant Mask

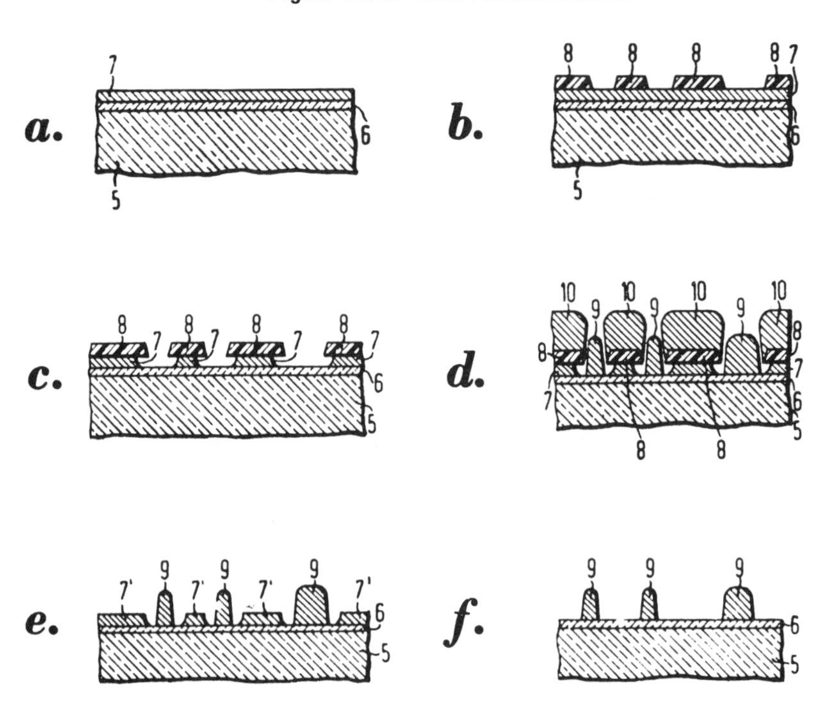

Source: U.S. Patent 4,108,717

After providing the substrate **5** with the auxiliary layer **7**, an etch-resistant mask **8** is applied on the auxiliary layer **7** (Figure 6.11b). The mask **8** is preferably a photoresist such as AZ 1350 (Shipley Company), which has been exposed and developed to provide a mask complementary to the desired conductor path structure which is to be formed on the substrate **5**.

After providing the mask **8**, the exposed portions of the auxiliary layer **7** are etched with an etching agent which does not attack the mask **8**. An example of an etching agent is a hot, concentrated phosphoric acid which is at a temperature of approximately 60°C.

During the etching step, the entire exposed portions of the auxiliary layer **7** are etched away with under-etching of the edges of the mask **8** as illustrated in Figure 6.11c. The amount of lateral under-etching of the layer **7** beneath the edges of the mask **8** corresponds approximately to the thickness of the auxiliary layer **7** and amounts accordingly to only 0.1 to 0.3 μm. Subsequent to the etching step, a pure aluminum layer is evaporated onto the structure so that the aluminum layers **9** and **10** are formed. The layers **9** and **10** will have a thickness of approximately 1 μm. After depositing the pure aluminum layers **9** and **10**, the photoresist and the overlying layers **10** are removed so that an intermediate stage represented in Figure 6.11e is formed.

After the step of removing the mask **8** and the overlying portions **10**, the remaining portion **7'** of the auxiliary layer **7** and the portion **9** are present on the layer **6**. Since the remaining portion **7'** may cause trouble in using the device, a second etching step is applied using a hot phosphoric acid at approximately 60°C. Since such an etching agent attacks the pure aluminum at approximately one-fifth the speed that it attacks the aluminum-nickel alloy, less than 0.1 μm of the pure aluminum structure **9** is removed as the portions **7'** of the auxiliary layer **7** are entirely removed. The second etching step provides the structure which is illustrated in Figure 6.11f and which comprises the substrate **5** having the insulating layer **6** with the conductor paths **9** disposed thereon.

While the abovedescribed process is more expensive than the conventional processes of the prior art, the individual process steps are relatively noncritical so that high yield can be expected with minimum loss. For example, relatively large tolerances can be permitted for the layer thickness of the auxiliary layer **7** and for the amount of under-etching without causing any difficulties with the dimension of the conductor paths **9**.

Titanium Masking Material

L. Stein; U.S. Patent 4,140,572; February 20, 1979; assigned to General Electric Company describes a method for patterning a layer of cured silicone-polyimide copolymer material disposed on the surface of the body. Preferably, it is a method used when the material is disposed on the surface of a body semiconductor material. The layer is from 1 to 10 μ in thickness. A layer of titanium metal, preferably of the order of 1000 A, is disposed on the copolymer material layer. A layer of positive photoresist material is disposed on the titanium metal material which is to act as a mask.

The layer of photoresist material is processed to form one or more windows therein. Selected surface areas of the layer of titanium metal are exposed in the windows. The exposed titanium metal is etched in fluoboric acid at room temperature for sufficient time to open windows in the titanium layer. The windows are aligned with those in the photoresist layer thereby enabling selected surface areas of the copolymer material to be exposed in the windows. The layer of photoresist is then stripped with acetone.

The copolymer material is etched in a phenol-bearing solution such as, for example, A-20 Stripping Solution (B & A Division of Allied Chemical). Chemical etching is performed until windows are opened in the layer to expose surface areas of the body thereat. The processed body is plunged into an agitated acetone bath, rinsed in isopropanol and spin-dried.

The polyimide-silicone copolymer is the reaction product of a silicon-free organic diamine, an organic tetracarboxylic dianhydride and a polysiloxanediamine which is a polymer precursor soluble in a suitable organic solvent. On curing, it yields a copolymer having recurring structural units of the formula:

$$\left[-Q-N\begin{matrix} \overset{O}{\overset{\|}{C}} \\ \\ \underset{O}{\underset{\|}{C}} \end{matrix} N-R''\begin{matrix} \overset{O}{\overset{\|}{C}} \\ \\ \underset{O}{\underset{\|}{C}} \end{matrix} N- \right]_m$$

with from 15 to 40 mol % and preferably 25 to 35 mol % intercondensed structural units of the formula:

$$\left[-N\begin{matrix} \overset{O}{\overset{\|}{C}} \\ \\ \underset{O}{\underset{\|}{C}} \end{matrix} R''\begin{matrix} \overset{O}{\overset{\|}{C}} \\ \\ \underset{O}{\underset{\|}{C}} \end{matrix} N-R\left(\begin{matrix} R' \\ | \\ Si \\ | \\ R' \end{matrix}-O\right)_x\begin{matrix} R' \\ | \\ Si \\ | \\ R' \end{matrix}-R- \right]_n$$

where R is a divalent hydrocarbon radical; R' is a monovalent hydrocarbon radical; R'' is a tetravalent organic radical; Q is a divalent silicon-free organic radical which is the residue of an organic diamine; x is an integer having a value of 1 to 4; and m and n are different integers greater than 1, from 10 to 10,000 or more.

The abovementioned block copolymers can be prepared by effecting reaction, in the proper molar proportions, of a mixture of ingredients comprising a diaminosiloxane of the general formula:

$$NH_2-R\left(\begin{matrix} R' \\ | \\ Si \\ | \\ R' \end{matrix}-O\right)_x\begin{matrix} R' \\ | \\ Si \\ | \\ R' \end{matrix}-R-NH_2$$

a silicon-free diamino compound of the formula:

$$NH_2-Q-NH_2$$

and a tetracarboxylic acid dianhydride having the formula:

$$O\begin{matrix} \overset{O}{\overset{\|}{C}} \\ \\ \underset{O}{\underset{\|}{C}} \end{matrix} R''\begin{matrix} \overset{O}{\overset{\|}{C}} \\ \\ \underset{O}{\underset{\|}{C}} \end{matrix} O$$

where R, R', R'', Q and x have the meanings given above.

Application of the block copolymers or blends of polymers in a suitable solvent (including, for example, N-methylpyrrolidone, N,N-dimethylacetamide, N,N-dimethylformamide, etc.) alone or combined with nonsolvents, to the substrate material may be by conventional means such as dipping, spraying, painting, spinning, etc. The block copolymers or blends of polymers may be dried in an initial heating step at temperatures of about 75° to 125°C for a sufficient time, frequently under vacuum, to remove the solvent. The polyamic acid is then converted to the corresponding polyimide-siloxane by heating at temperatures of about 150° to 300°C for a sufficient time to effect the desired conversion to the polyimide structure and final cure.

A preferred curing cycle for the copolymers is as follows: (a) from 15 to 30 minutes of from 135° to 150°C in dry N_2; (b) from 15 to 60 minutes at about 185±10°C in dry N_2 and (c) from 1 to 3 hours at about 225°C in vacuum. Alternately, it has been found that the coating material in other atmospheres such as, for example, air may be cured for ease of commercial application.

Multilayer Chromium Mask

In the manufacture of semiconductor integrated circuits or the like, a hard mask comprising a transparent substrate made of glass, for example, and a metal film formed thereon is used in most cases in view of its durability and high degree of resolution. A typical hard mask is a chromium mask wherein a chromium film is deposited on a glass substrate by vacuum evaporation or sputtering. However, since the reflectance of the chromium film is very high (about 40 to 60%), the resolution of a circuit pattern formed by using the chromium mask is poor due to multiple reflections of the light between a wafer and the chromium mask during exposure.

K. Sakurai; U.S. Patent 4,139,443; February 13, 1979; assigned to Konishiroku Photo Industry Co., Ltd., Japan describes a method of preparing a photomask blank comprising the steps of forming a first chromium film on a transparent substrate by sputtering, forming a second chromium film on the first chromium film by vacuum evaporation, and forming a chromium oxide film on the second chromium film by vacuum evaporation.

A preferred embodiment of this process is described as follows. A glass substrate usually used for photomasks was disposed to oppose a pure chromium target in a bell jar of a three-electrode, dc-sputtering apparatus, maintained at a vacuum of 10^{-3} torr for forming a first chromium film having a thickness of about 400 A. The substrate was then transferred into a conventional vacuum evaporation tank to deposit a second chromium film having a thickness of about 200 A while maintaining a vacuum of 5 x 10^{-6} torr in the deposition tank. Thereafter, oxygen was admitted into the deposition tank and after adjusting the vacuum to 5 x 10^{-3} torr, the vapor deposition was continued to form a chromium oxide film having a thickness of about 300 A.

The photomask blank of this process has the advantages of (a) efficient prevention of reflection, (b) resistance against chemicals, (c) high resolution of the circuit pattern, (d) substantial freedom from pinholes, and (e) the amount of side etching is small. It also eliminates all defects of the prior art photomask blanks.

Ti-Cu-Ni-Au Conduction System

N.G. Lesh, J.M. Morabito and J.H. Thomas, III; U.S. Patent 4,109,297; August 22, 1978; assigned to Bell Telephone Laboratories, Incorporated describe a process relating to thin film and hybrid integrated circuits, and in particular to a conduction system for interconnecting elements in the circuits.

The method, in accordance with two embodiments of the process, is described with reference to the flow diagram of Figure 6.12. The sequence of steps illustrated preferably begins after the deposition of the resistor and capacitor elements, usually comprising tantalum or tantalum nitride, on the insulating substrate, which is most usually alumina. The formation of the interconnection scheme began with a deposition of a layer of titanium over substantially the entire area of the substrate. The precise method used was electron gun evaporation; however, other well-known techniques such as sputtering may be used.

Figure 6.12: Ti-Cu-Ni-Au Conductive System

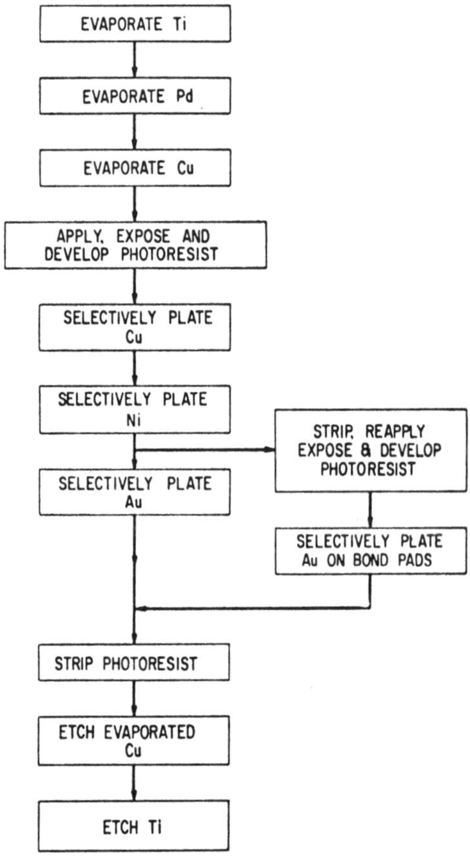

EVAPORATE Ti
EVAPORATE Pd
EVAPORATE Cu
APPLY, EXPOSE AND DEVELOP PHOTORESIST
SELECTIVELY PLATE Cu
SELECTIVELY PLATE Ni
SELECTIVELY PLATE Au
STRIP, REAPPLY EXPOSE & DEVELOP PHOTORESIST
SELECTIVELY PLATE Au ON BOND PADS
STRIP PHOTORESIST
ETCH EVAPORATED Cu
ETCH Ti

Source: U.S. Patent 4,109,297

The thickness of the titanium layer is preferably within the range 1500 to 3000 A in order to serve as an adequate glue layer and to avoid bondability problems which usually occur when the thickness is less than approximately

1500 A. A thickness of approximately 2500 A seems to be optimum. Next, a thin (approximately 500 A) layer of palladium was deposited, preferably by the same technique, on the titanium layer. This layer serves to improve adhesion between the titanium and the to-be-described copper layer. As such, the palladium layer is optional since the adhesion of Ti and Cu can be adequate if proper deposition procedures are followed. An appropriate thickness for this layer appears to be 200 to 1000 A.

Next, a thin layer of copper was deposited (also by electron gun evaporation) on the Pd layer. The copper layer serves primarily to provide a high conductivity layer for subsequent plating processes. Typically, the layer is approximately 5000 A in thickness, but a range of approximately 3000 to 7000 A would be appropriate.

Before proceeding with the photoresist step, it is desirable to coat the copper surface with a thin chrome layer to improve the adherence of the photoresist. This procedure is well-known in the art. The photolithographic process, also well-known in the art, involves essentially applying a photoresist layer to the entire copper surface, exposing desired areas through a mask, and developing the resist to remove those areas exposed to light in a pattern which will define the appropriate interconnection paths.

As shown in Figure 6.12, a layer of copper was then electroplated onto the selected portions of the evaporated copper layer not covered by the photoresist. This was accomplished by making the substrate the anode in an electrolytic cell wherein the bath comprised approximately 68 g/ℓ of $CuSO_4$ and 180 g/ℓ of H_2SO_4. The plating was carried out at a current density of approximately 20 mA/cm^2, but this parameter may be adjusted depending upon the quantity of the bath and the geometry of the circuit to be fabricated. The thickness of the copper layer is an important consideration. In this particular embodiment, the total thickness of evaporated and plated Cu was 35000 A. The subsequent Ni plating should follow the copper plating, while the copper is still wet.

In the next step, nickel was plated onto the exposed areas of the plated copper surface. The particular bath used for this plating operation is Barrett type SN (Allied-Kelite Co.) and basically comprises nickel sulfamate and boric acid. The thickness of the nickel layer was approximately 10000 A, although a useful range appears to be approximately 8000 to 20000 A for providing an adequate diffusion barrier between the plated copper and the subsequently deposited gold layer, while maintaining a proper sheet resistivity as described later. The current density was, again, approximately 20 mA/cm^2 which provides a sufficiently dense film. As before, the current density may be adjusted for particular needs. A minimum current density for the nickel film fabrication appears to be approximately 10 mA/cm^2 to produce a sufficiently dense film (of the order of 9 g/cm^3). It was also discovered that an optimum barrier layer is one which consists of essentially pure Ni without any additives.

At this point in the processing, as shown in the figure, basically two alternatives may be followed in forming the top layer of gold. In the first alternative, the photoresist layer previously formed on the substrate was utilized as the mask for electroplating the gold layer on the entire exposed area of the previously formed nickel layer. In the second alternative, this photoresist layer was stripped off and a second photoresist was applied, exposed and developed to expose only those areas of the metal which will be utilized as bonding pads for integrated circuit chips or connection to elements off the substrate. In either procedure, the electroplating processes utilized a gold cyanide bath comprising 20 g/ℓ of potassium gold cyanide, 50 g/ℓ of ammonium citrate and 50 g/ℓ ammonium sulfate at a current density of approximately 2 mA/cm^2. Again, the proportions of the bath components and the current density may be adjusted as needed. The thickness of the gold layer was approximately 20000 A, but it appears that a preferred range is 15000 to 25000 A to insure a good bonding surface.

It will be realized that the thickness of gold required in this conduction system is considerably less than that required in the previous Ti-Pd-Au system which was approximately 50000 A. Thus, a substantial cost savings is realized in accordance with the process. It will also be realized that while the embodiment involving selective plating in the bonding areas suffers from the disadvantage of requiring an extra photoresist layer, it offers an advantage in requiring less gold and permits further cost savings.

The next processing sequence involves final patterning of the interconnect scheme by etching the evaporated layers of Cu and Ti which are not covered by the plated metal layers. Thus, the photoresist utilized during electroplating was first stripped off. In both alternative embodiments the evaporated copper was removed by an ammonium persulfate solution and the titanium layer was subsequently removed by hydrofluoric acid, which solutions are known in the art. Care should be taken in removal of the evaporated copper to avoid nickel etching and undercutting of the plated copper layer. Typically, etch time for removal of 5000 A of copper in ammonium persulfate is approximately 60 seconds.

As a precautionary measure, the circuits were removed from the etchant as soon as all visible signs of copper were removed and immediately rinsed to stop further etching. It will be realized that other etchants may be used for removal of the Ti and Cu layers.

As alluded to previously, the thicknesses of the Cu, Ni and Au layers are important in providing the proper sheet resistivity for replacement of the Ti-Pd-Au system. In particular, a desired sheet resistant (R_s) can be calculated from the equation $R_s = 1/[(t/\rho)_{Au} + (t/\rho)_{Cu} + (t/\rho)_{Ni}]$ where t is the thickness and ρ is the bulk resistivity of the indicated metals. An end of life sheet resistance of approximately 0.005 ohm per square or less is desired for most applications. Thus, for a Ni layer in the range of 8000 to 20000 A and Au in the range of 15000 to 25000 A, a range of Cu thickness of 25000 to 40000 A appears to be optimum to satisfy sheet resistance requirements.

After formation of the interconnect pattern as described above, normal circuit processing such as resistor patterning, thermocompression bonding, soldering, etc., proceeds in accordance with the prior art.

Schottky Barrier Gate Field-Effect Transistors

F.S. Ozdemir and D.D. Loper; U.S. Patent 4,109,029; August 22, 1978; assigned to Hughes Aircraft Company describe an approach to the fabrication of Schottky barrier gate field-effect transistors (SBFETs) which achieves higher resolutions, higher electrode placement accuracies, smaller electrode line widths, smaller electrode spacings and higher yields than those same parameters achievable using standard multiple mask ultraviolet photolithographic processes of the prior art. At the same time, the process provides the desired separability of the two or more metallization procedures used to form, respectively, the source-drain electrodes and the gate electrode of these transistors. Additionally, the process can be combined with lower resolution mask fabrication processes in order to provide an optimum combination or partitioning of these processes for the fabrication of a particular device or integrated circuit structure.

To accomplish this purpose and in the fabrication of the SBFET, a thin, electrically-active semiconductor layer is provided at the surface of an electrically-insulating or semiinsulating semiconductor substrate and using, for example, ion implantation or epitaxial growth techniques. Standard photolighography and semiconductor processing techniques are then used to define both a semiconductor mesa (formed out of the active semiconductor layer) and a coarse or low resolution alignment mark on the surface of the substrate. The semiconductor mesa and alignment mark patterns are defined in light-sensitive polymers (photoresists) using optical masks and UV radiation, and standard semiconductor processing techniques are used to complete the semiconductor mesa and the alignment mark. The coarse alignment mark is positioned at a predetermined chosen location with respect to the semiconductor mesa structure formed from the thin, electrically-active layer.

The above structure is then coated with an electron-sensitive polymer (electron resist), and an electron beam is used to register the electrical deflection field of the beam to the semiconductor mesa. This registration is accomplished by scanning the electron beam across the abovedescribed coarse alignment mark and collecting, with a suitable detector, the backscattered or secondary electrons that are generated when the electron beam strikes the coarse alignment mark. Analysis of the data from the detector and subsequent alteration of the size and position of the electrical deflection field relative to the semiconductor mesa position brings the deflection field into registration with the chosen substrate pattern which, in this case, is the semiconductor mesa.

The electron beam is then deflected across the top surface of the abovedescribed structure in a manner to define the source and drain patterns of the SBFET which are located on the semiconductor mesa. In addition to forming these source and drain patterns, additional high resolution alignment marks are also exposed. Since these latter high resolution alignment marks are exposed during the same exposure processing step used to define the source and drain patterns, these high resolution alignment marks and the source and drain patterns are self-registered. After these above exposures are completed, the structure is developed in a suitable chemical which dissolves all the exposed electron resist. The remaining patterns are then metallized by first evaporating a selected metallization (such as Au/Ge/Ni alloys for GaAs SBFETs) over the upper surface of these mask patterns and then removing the excess metal by dissolving the unexposed electron resist on the substrate.

After the source and drain metallization patterns are completed in accordance with the above procedure, the above electron beam registration and pattern exposure process is repeated for the gate pattern of the SBFET. This time, the electron-beam-defined high resolution alignment marks are used for the gate electrode pattern registration, thereby providing the ultimate in gate, source and drain pattern registration accuracy. The gate pattern exposure consists of exposing a high resolution line pattern between the previously fabricated source and drain patterns. In addition to this line pattern, contact pad patterns which are linked to the gate pattern are also exposed. The subsequent process steps of development, metallization and dissolution are similar to those used to form the above source and drain pattern metallization, except in this case, the gate metallization is different from the source-drain metallization. In the fabrication of GaAs SBFETs, aluminum is the preferred gate metal.

Thus, separate and successive masking steps are utilized for the source-drain metallization and for the gate metallization, respectively, while simultaneously affording very high resolution metal deposition procedures which are useful in the high yield fabrication of small-geometry, high-frequency, low-noise Schottky gate semiconductor devices.

IMPATT Diodes

K. Cooper, I.S. Groves, P.A. Leigh, N. McIntyre, S. O'Hara and J.D. Speight; U.S. Patent 4,099,318; July 11, 1978; assigned to The Post Office, England describe a method of producing a semiconductor device in the form of an IMPATT diode which comprises the steps of:

(a) Selecting a slice of semiconductor material having an epitaxial layer on an n^+ substrate material;

(b) Diffusing a p^+ layer into the epitaxial layer to form an abrupt p^+-n junction;

(c) Removing the uppermost 80 to 120 A of the diffused p^+ layer by RF sputter etching under plasma conditions thereby inhibiting change in surface crystal structure;

(d) Thinning the n^+ substrate material to a desired thickness for the device;

(e) Producing a photolithographic mask on the surface of the n^+ substrate material;

(f) Depositing, by sputtering, on both the p^+ and n^+ surfaces a succession of layers of titanium, 800 to 1200 A; platinum, 1600 to 2400 A; and gold, 8000 to 12000 A; and electroplating gold layers over the sputtered gold layers; removing the photolithographic mask to expose the surface of the n^+ substrate material around a series of gold pads;

(g) By high-energy proton bombardment, passivating the zones of semiconductor material in depth through the slice to create zones of semiinsulating material to isolate the semiconductor material directly below the gold pads, the energy of protons being increased in stages at intervals of less than 0.11 MeV up to an energy in MeV approximately equal to one-tenth of the required passivated layer thickness in μm using a proton dose at each stage of from 10^{14} to 5×10^{15} cm^{-2} and a rate of from 2×10^{12} to 5×10^{12} sec^{-1} using a proton scan raster pattern ensuring uniformity of dose over the area of the device; and

(h) Mounting the processed semiconductor material on to a heat sink and connecting one or more of the gold pads to a connection block.

The photolithographic mask can be produced in two stages so that the plated gold layer has a thickness of from 8 to 12 μm and a diameter from 10 to 35% less than that of the sputtered layers. The n^+ substrate material is preferably thinned in a two-stage process comprising abrasive lapping down to 60 to 80 μm and a chemical etching down to 10 to 15 μm.

In order to obtain IMPATT diodes with the performance characteristics tabulated above, the semiconductor material selected should have less than 10^3 defects cm^{-3}.

Following p^+ diffusion, the diffuse surface should be thoroughly cleaned, preferably successively with organic solvents, concentrated acids and distilled water.

Manufacture of a Film Thermopile

A method of manufacturing a film thermopile is known whereby strips of a thermoelectric semiconductor material of one type of a conduction, e.g., a p-type (p-arms), are deposited through an aligned mask on a substrate of an insulation material. After depositing the p-type material, in order to improve its electrical and physical properties, it is common to effect heating in an inert medium or in a vacuum. The aligned mask is then changed and strips of a thermoelectric semiconductor material of an n-type (n-arms) are deposited and refined in an inert medium or in a vacuum. Usually, the chemical composition of the p-arm material differs from that of the n-arm material; therefore, the refinement conditions for the p-arm material differ from those of the n-arm material, which necessitates two separate steps of heating for the p-arms and the n-arms. Having deposited the p- and n-arms, switching buses are deposited for electrical connection of the p- and n-arms.

This method of manufacturing a film thermopile brings about the necessity of aligning masks to deposit the p- and n-arms which greatly limits the minimum width of the arms. Also, the deposited arms of one type should be protected against damage by aligned masks while depositing the arms of opposite conduction. The arms of one type being deposited, an operation of changing the masks for depositing the arms of another type is necessary, too. This operation considerably lengthens the technological process of manufacturing a film thermopile. Moreover, the changing of the masks may cause damage to the active layer of the p- or n-arms.

The known method makes it impossible to clean the substrate surface by heat treatment as the substrate heating may result in reevaporation of the p-arms already deposited. It is possible to overcome this drawback by depositing arms of both types simultaneously, e.g., evaporating the p-arms from one crucible and the n-arms from another crucible; however, films of uniform thickness and composition would not be obtained.

N.S. Lidorenko, N.V. Kolomoets, Z.M. Dashevsky, V.I. Granovsky, E.A. Zhemchuzhina, L.N. Chernousov, I.A. Shmidt, L.A. Nikolashina, D.M. Gelfgat and I.V. Sgibnev; U.S. Patent 4,098,617; July 4, 1978 describe a method of manufacturing a film thermopile including depositing p- and n-type arms in one operation and which ensures uniformity of thickness and composition of deposited semiconductor material films.

The foregoing objects are attained by providing a method of manufacturing a film thermopile by depositing a thermoelectric semiconductor material film on a substrate with further heating in an inert medium, using an n-type stoichiometric solid solution containing Bi_2Te_3 and Sb_2Te_3 as the semiconductor material. Heating is carried out in such a way that adjacent arms are at different temperatures, some being at a temperature of not above 300°C, whereas the others are at a temperature of not less than 350°C. It is advisable to use a solid solution containing 50 mol % of Bi_2Te_3 and 50 mol % of Sb_2Te_3.

Contact Electrode Bumps

High volume, fast handling and economical packaging systems are being developed for packaging microelectric components to meet a growing demand for such components and systems of such components at a reasonable cost. One of the component mounting techniques, as a part of one of these packaging methods, is the face-down bonding technique. In this technique, contact electrodes in the form of small metallic bumps are provided on selected portions of a metallization interconnection network, usually, where electrical connections to external components are desired. This metallization interconnection network is provided in the component to be mounted, typically an integrated circuit chip. The component is to both be mounted by and to make electrical connection through these bumps.

Integrated circuit chips having these bumps thereon are placed with the network side facing a surface of a mounting substrate and placed such that the bumps are in contact with conductive lead portions provided on the substrate to which the bumps are to be attached. Next, heating the bumps causes them to flow and thus come into intimate contact with these conductive lead portions of the substrate so that upon cooling the desired bond results.

L.A. Del Monte; U.S. Patent 4,113,578; September 12, 1978; assigned to Honeywell Inc. describes a process which utilizes sputtering and plating techniques to provide the contact electrode bumps. The process may involve only sputtering steps to accomplish the deposition of metals or may involve sputtering supplemented by plating steps to provide contact electrode bumps. Plating may also be used to supplement other types of initial deposition steps.

Portions of a structure, typically portions of a metallization interconnection network, are selected on which to provide contact electrode bumps. A mask, typically of metal, having a pattern of openings therein corresponding to the selected portions, is placed against the structure such that the openings are concentric with the areas selected upon which to provide the bumps.

This masked structure is to be provided in a chamber of a sputtering apparatus. The sputtering apparatus contains source electrodes, each having a type of metal which is desired to be deposited through the openings in the mask. A gas atmosphere is provided in this chamber.

The selected portions on which the contact electrode bumps are to be formed, perhaps made partially available heretofore, are first completely exposed through use of a sputter-cleaning step. Following this, each type of metal desired to be deposited through the openings in the metal mask is sputtered through these openings in the order of deposition desired. The good directivity of the ejected metal atoms resulting from use of the sputtering method leads to use of smaller amounts of each type of metal than would be the case in using an evaporation technique. In addition, forming bumps by sputtering means that each sputtered layer has a better adhesion to the material below it to give better adhering bumps because of the self-cleaning feature of the sputtering method and because of the greater likelihood of forming chemical bonds in each sputtered layer. If only metal sputtering steps are to be used, the metals sputtered must reach a sufficient height to form a contact electrode bump which is about 0.2 of a mil or more.

If the sputtering steps are to be supplemented by a plating step, the mask is removed after the metals have been sputtered through its openings to form a metal base for a plated bump with the base being of a height less than that desired for a bump. A metal flash layer, typically the same metal as the last metal sputtered, is blanket evaporated over the structure surface when electroplating techniques are to be used. A photoresist layer is then provided on this metal flash layer with openings being provided in the photoresist layer which are concentric with the selected portions. A metal is then plated onto the metal exposed in the openings in the photoresist layer, typically again the metal last sputtered. The plating metal does not plate or adhere to the photoresist layer. The photoresist layer is then removed as is the metal flash layer leaving the contact electrode bumps.

Forming a base of sputtered metal layers by use of a mask means the sputter-cleaning step is available to be used initially in the plating process so no sintering step is required to achieve relatively low electrode resistance if that is desired. The sputtered metal layers also provide a strongly adhering metal base upon which to plate the remaining metal to form contact electrode bumps. Microcracking is minimized since the sputtered layers cover but a small fraction of the structure surface.

To use electroless plating, the same steps which are performed in providing an entirely sputtered bump are repeated except the process is terminated before the last 0.2 or 0.3 mil of gold used in constructing the bump has been sputtered into place. At this point, the gold layer would be thinner by the 0.2 or 0.3 mil desired to be formed by electroless plating. The sputtering of gold is then terminated at this point and the metal mask is removed. The wafer is then immersed in a chemical plating bath and gold is thereby plated on the gold already sputtered on to provide the contact electrode bump. The gold will not chemically plate onto the silicon dioxide layer which has an unactivated surface. Thus, though the sputtering of gold provides a larger fraction of the gold in the contact electrode bump than it does in the electroplating method, the steps of providing a metal flash layer and a photoresist layer have been eliminated.

SUPERCONDUCTORS

Nb_3Sn Layer on Niobium

H. Martens; U.S. Patents 4,127,452; November 28, 1978 and 4,105,512; August 8, 1978; both assigned to Siemens AG, Germany describes a manufacturing method for forming superconductive Nb_3Sn layers on niobium surfaces for high frequency applications. In particular, uniform coating of the niobium surfaces is to be made possible without appreciable reduction of the quality factor and the critical magnetic flux density of the Nb_3Sn layers produced by the method as compared to the known methods.

To solve this problem, it is provided that, in the presence of a tin source, a tin vapor atmosphere which also contains additionally a highly volatile tin compound in the gaseous state, the saturation vapor pressure of which in a temperature range of between 200° and 1000°C, is substantially higher than that of tin at the same temperature, is developed by heating and that the surface portions to be provided with the Nb_3Sn layer are kept in this atmosphere for a predetermined period of time at a temperature of between 900° and 1500°C.

The device shown as a longitudinal cross section in Figure 6.13 contains a cup-shaped part **1** of high purity niobium, which is provided for use as a cavity resonator of the TE_{011}-field type, in the shape of a circular cylinder. Part **1** is designed for an X-band frequency of 9.5 GHz and is described, for instance, in the journal *Cryogenics*, vol. 16, pp. 17-24 (1976). The inside diameter and the inside height of this niobium part **1** are each, for instance, 41 mm. The surface portions to be coated, of the niobium part **1**, are made very smooth by repeated polishing.

Figure 6.13: Reaction Chamber for Depositing Nb_3Sn Layer on Niobium

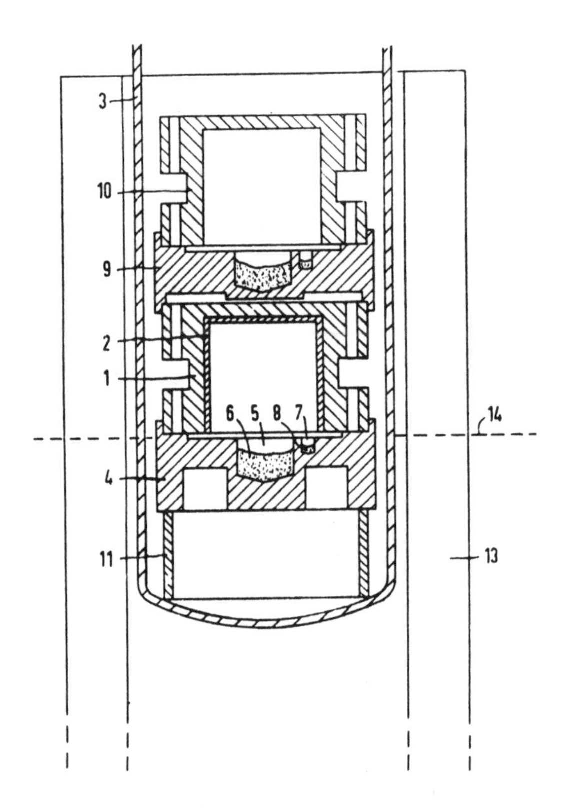

Source: U.S. Patent 4,127,452

The process serves to prepare an Nb_3Sn layer on the inside surface of the cup-shaped niobium part **1**. For this purpose, the inside surface of the niobium part can advantageously be provided, first, with a niobium pentoxide layer **2** by anodic oxidation. An appropriate method is known, for instance, from U.S. Patent 3,784,452. The thickness of the oxide layer so generated is advantageously between 0.01 and 0.3 and, preferably about 0.1 μm. Subsequently, the coated niobium part **1** is placed in a quartz tube **3**, which forms an evacuated reaction chamber. The niobium part **1** is then placed on a lower part **4** also consisting of niobium, in the center of which a depression **5** is provided, in which a tin supply **6** is placed. The purity of the tin is advantageously better than 99.96%.

The niobium part **1** and the lower niobium part **4** form a reaction zone which is delineated from the rest of the volume of the quartz tube **3** and contains the tin source **6** as well as the anodically oxidized inside surface **2** of the niobium part **1**, which is to be provided with the Nb_3Sn layer. The end face of the niobium part **1** rests on the surface of the lower niobium part **4**. Besides the depression **5** provided in the lower part **4**, a further, comparatively smaller depression **7** is provided, into which a predetermined small amount of a substrate **8** is placed. Advantageously, highly volatile tin compounds such as tin halogens, preferably tin fluoride or tin chloride, or also small amounts of other substances such as, for instance, hydrofluoric acid or hydrochloric acid which form a gaseous tin compound with tin vapor in the gaseous state can be provided as the substrate **8**.

So that another resonator can be provided with an Nb_3Sn layer in the same operation, a second lower niobium part **9**, which corresponds to the lower niobium part **4** and on which a further cup-shaped niobium part **10** stands, is placed on the niobium part **1**.

The quartz ampoule **3** with the parts **4, 1, 9** and **10**, which are arranged on top of each other in a section of quartz tube **11** in tower-like fashion, is evacuated at room temperature until a predetermined pressure prevails at its open end, not shown in the figure.

At the beginning of the coating process, the quartz ampoule **3** is introduced into a vertical, tubular resistance furnace **13**, which is only indicated in the figure and has a temperature of about 750°C. The temperature of the furnace is then increased to about 1050°C so that the niobium parts gradually assume this temperature. In this process, part of the tin from the tin source **6** evaporates, and the highly volatile substrate **8** evaporates completely. Thus, an atmosphere consisting of tin vapor and the vapor of the tin compound develops. The amount of substrate is advantageously selected so that the vapor pressure of the tin compound is at least 10 times, and preferably 50 times, higher than the tin vapor pressure.

The niobium parts are then kept at the temperature of 1050°C for a predetermined time (~3 hours). During this time, an Nb_3Sn layer of high quality and high critical magnetic flux density develops on the inside surface of the niobium part **1**. The thickness of the Nb_3Sn layer generated can be influenced by the length of this reaction time. This time is advantageously chosen so that the thickness of this layer is between 0.5 and 5 μm.

By an optionally provided temperature lead of the tin source, i.e., by heating the tin source more than the surface parts of the niobium parts **1** to be coated, the thickness of the Nb_3Sn layer on these surface parts can, likewise, be influenced distinctly. Such a process step which is accompanied by considerable expense, is not absolutely necessary, however. Also, with the temperature lead, uniform if relatively thin Nb_3Sn layers can be obtained. A temperature lead of the tin source can be achieved, for instance, by inserting the quartz tube **3** into the furnace **13** only far enough that the tin source **6** is located at the height of the upper edge of the furnace. This position of the upper edge is indicated in the figure by a broken line **14**.

If the process parameters are chosen so that excessively thick Nb_3Sn layers are obtained, this hardly has an effect on the quality factor and the critical flux density of these layers, as layers that are too thick can be worked down to their optimum thickness by oxy-polishing. After the end of the reaction time, the quartz ampoule is slowly cooled down and the Nb_3Sn layer cleaned.

The Nb_3Sn layers produced in this manner can advantageously be given a post-treatment in accordance with U.S. Patent 3,902,975 by generating on them an oxide layer by anodic oxidation and subsequently dissolving this layer again chemically.

Nb_3Ge Films

R.A. Sigsbee; U.S. Patent 4,128,121; December 5, 1978; assigned to General Electric Company describes a process whereby a superconductive film of Nb_3Ge is produced by providing within a vacuum chamber a heated substrate and sources of niobium and germanium, reducing the pressure within the chamber to a residual pressure no greater than about 10^{-5} mm Hg, introducing oxygen into the resulting evacuated chamber in controlled amounts and vaporizing the niobium and germanium to deposit a film of Nb_3Ge on the heated substrate.

Shown in Figure 6.14 is substrate **2** mounted in vacuum chamber **3**. Sources of niobium **4** and germanium **5** are placed within chamber **3** which is also provided with vacuum pump **6** and a source of oxygen **7**.

A movable protective means **8** is used to protect substrate **2** until niobium and germanium are each vaporized to the levels required for producing the impinging Nb/Ge composition. Partitioning means **9** prevents Ge flux from source **5** from striking monitor **10** and niobium flux from source **4** from striking monitor **12** so that monitor **10** reads niobium flux only and monitor **12** reads germanium flux only. By flux is meant herein moving species of niobium or germanium vapors emitted from their respective sources. Due to the low pressures of the process, niobium and germanium flux travel in straight lines and both strike all areas of the substrate surface on which the film is to be deposited.

Monitor **10** is connected to controller **11** and is used to determine the flux level of niobium within the chamber whereas monitor **12** is connected to controller **13** and is used to determine the flux level of germanium within the chamber. Controllers **11** and **13** control the heating power to sources **4** and **5**. An electron gun (not shown) is used to vaporize niobium whereas a number of heating means (not shown), including an electron gun, can be used to vaporize germanium **5**. Residual gas analyzer **14** is used to determine the partial pressure of oxygen within chamber **3**. Ionization gauge **15** was used to measure the total pressure in chamber **3**.

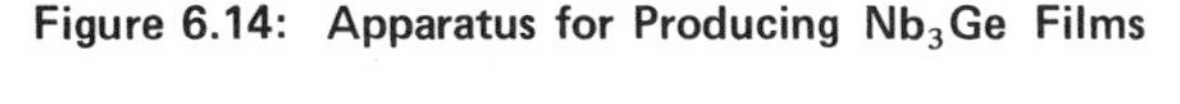

Figure 6.14: Apparatus for Producing Nb_3Ge Films

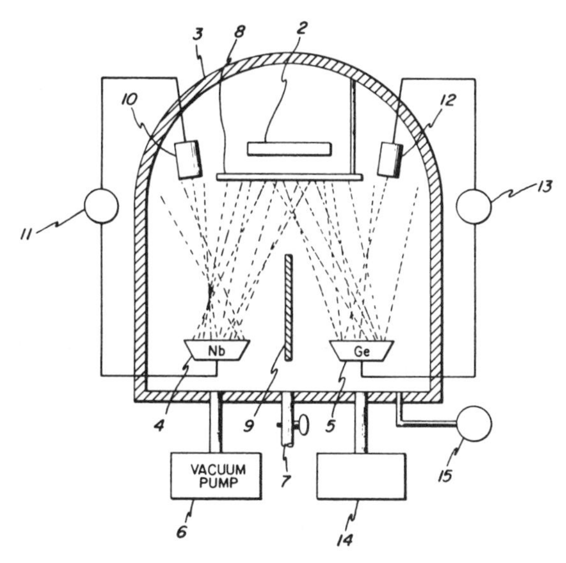

Source: U.S. Patent 4,128,121

Representative of the substrates useful in the process are sapphire, silicon or oxidized silicon, polycrystalline alumina, niobium, copper, molybdenum, stainless steels and superalloys such as the nickel and/or cobalt-based superalloys.

In the process, the vacuum chamber is provided with a source of oxygen which can be introduced into the chamber in controlled amount to provide the chamber with a particular partial pressure of oxygen. Such a source of oxygen can be provided in a number of ways, such as a variable valve or other means capable of providing a low controlled leak rate into the chamber.

In carrying out the process, protective means **8** such as a shutter which is easily movable within the chamber by remote control, is positioned to protect the substrate from impinging niobium and germanium. Vacuum chamber **3** is then evacuated by conventional means in association therewith such as a vacuum pump **6** so that it has a residual pressure no greater than about 10^{-5} mm Hg, and generally such residual pressure ranges from about 10^{-5} to 10^{-7} mm Hg or lower. The substrate is heated to a temperature ranging from 750° to 1100°C and maintained at this temperature during deposition of the film.

Niobium and germanium are then vaporized and oxygen is introduced into the chamber. The particular rates of vaporization of niobium and germanium can be controlled by controlling the temperature at which each is vaporized. Niobium and germanium should be vaporized to produce an impinging flux composition on the substrate comprised of about 1 to 3 parts of niobium for about 1 part of germanium, i.e., an impinging Nb/Ge flux ratio ranging from 3/1 to 1/1.

Oxygen is introduced into the evacuated chamber in a controlled amount to produce a partial pressure of oxygen therein ranging from about 1 x 10^{-7} mm Hg to about 1 x 10^{-5} mm Hg. The particular partial pressure of oxygen to be used during the deposition process is determinable empirically, for example, by determining its effect on the critical temperature T_c and compositions of the resulting deposited film.

R.A. Sigsbee; U.S. Patent 4,129,166; December 12, 1978; assigned to General Electric Company describes a similar process using an incremental air pressure of about 2.5 x 10^{-5} mm Hg instead of oxygen.

In another similar process, *R.A. Sigsbee; U.S. Patent 4,129,167; December 12, 1978; assigned to General Electric Company* describes the use of nitrogen instead of oxygen at an incremental pressure of about 1 x 10^{-5} mm Hg.

Superconducting Wire

G.H. Miller; U.S. Patent 4,098,920; July 4, 1978; assigned to Texaco Inc. describes a method of continuous production of a superior superconducting compound in wire form. It comprises the steps of passing a filament of niobium wire substrate through a plating chamber, and heating the wire filament in the plating chamber to a temperature of about 1200°C by generating a standing half wave of radio frequency current in a predetermined length of the wire and by simultaneously passing a dc electric current through the wire in the chamber.

It also comprises introducing gaseous germanium chloride and niobium chloride and hydrogen in relative amounts to produce optimum rate of plating of the superconductor compound niobium-germanium on the niobium substrate and introducing an inert gas to assist in temperature control, and passing the plated filament out of the chamber for freezing the compound in place on the filament.

The method which is applicable to continuous production of superconducting wire comprises at least the following steps. However, they are not necessarily carried out in the order recited. In Figure 6.15, a filament of a metallic wire substrate is passed through a plating chamber. In the apparatus illustrated, this is the filament **12** which is wound onto the reel **78**, while unreeling from the other reel **77**. The material of this substrate might be tungsten or other substrate material including niobium, zirconium, tantalum, titanium or other metals. The wire is heated in the plating chamber **11** to a high enough temperature to cause plating on the surface thereof. The particular temperature may vary, depending upon the substrate and the superconductor compound being plated. Plating gas compounds are introduced into the chamber **11** along with sufficient hydrogen to promote a chemical reaction which forms an intermetallic superconducting compound plated onto the heated substrate wire. The plate wire is passed out of the plating chamber where it rapidly cools as it is collected onto winding reel **78**.

It will be appreciated by one skilled in the art that various combinations of niobium, vanadium, titanium, gallium, tin, germanium and many others could be combined by chemical vapor plating in accordance with this process. However, it is particularly to be noted that this method makes practical the production of the superconducting material niobium-germanium which has extremely beneficial qualities of a transition temperature that is higher than the boiling point of hydrogen.

Figure 6.15: Continuous Production of Superconducting Wire

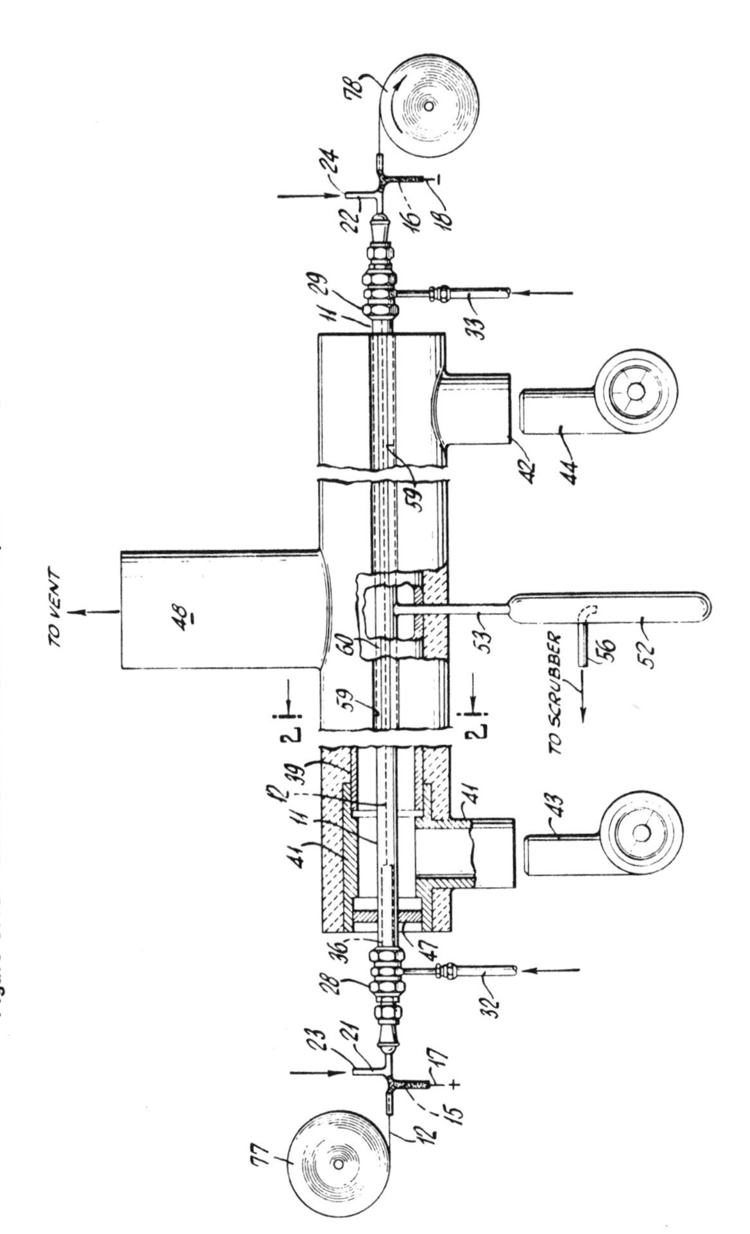

Source: U.S. Patent 4,098,920

THICK FILM CIRCUITS

Addition of Organometallic Compound

Hybrid integrated circuits are becoming increasingly important as a result of the miniaturization of electronic components, circuits and systems. Thick films are extensively used as capacitors, conductors and resistors in these hybrid circuits. They are also used in high power television circuits, microwave devices and other integrated circuit applications. Thick films have been used not only because they are reliable, but also because they are economical.

A conductor thick film is essentially a film of a composite of glass and a metal, usually a noble metal, on a ceramic substrate, having a thickness of the order of a few mils. The glass content is small (about 15 wt %) in conventional films, and its purpose is to provide good bonding between metal and substrate. Thick films are differentiated from thin films both in mode of preparation and in their thickness. Thin films are essentially of a thickness ranging from a few thousand angstroms to a few microns and are normally vacuum evaporated, sputtered or chemically deposited.

Conventionally, conducting thick films having a mixture of glass and metal particles in a suitable proportion in an organic vehicle with a suitable viscosity are printed onto a substrate. The glass, usually a low softening glass, with a composition having an expansion coefficient less than that of the substrate (for example, 96 wt % Al_2O_3) is powdered to a micron size (1 to 10 μm range) and mixed with the micron size (1 to 10 μm range) metallic powder, normally a noble metal. Usually the metal content varies from 70 to 90 wt % of the solids.

This mixture is then suspended in an organic vehicle which consists of a resin binder, solvent and additives to give a suitable viscosity. The liquid ratio is normally about 75 wt % depending on the viscosity. The liquid-solid mixture is screen-printed onto an alumina substrate, and the system is dried and fired for a specific time at a given temperature to give a conducting thick film.

V.K. Nagesh and R.M. Fulrath; U.S. Patent 4,130,671; December 19, 1978; assigned to the U.S. Department of Energy describe a method for preparing a thick film conductor which results in a microstructure of glass particles substantially uniformly coated with metal. The process comprises providing surface-active glass particles; mixing the surface-active glass particles with a thermally-decomposable organometallic compound, particularly an organo-compound of a noble metal, e.g., a silver resinate; and then decomposing the organometallic compound by heating, thereby chemically depositing metal on the glass particles.

The glass particle mixture is applied to an insulator base material either before or after the decomposition step. More particularly, glass particles, having generally a particle size in the range of about 20 to 44 μm, are treated with an acidic material to activate the surface thereof. The thus surface-activated glass particles are then suitably mixed with an organo-compound of a noble metal and the resulting mixture is suspended (after evaporating the solvent or after decomposing the organic compound) in a suitable organic vehicle for application to a suitable substrate. After the liquid-solid mixture is applied, as by screen printing or painting, the system is fired in an oxidizing atmosphere. The organo-metallic compound decomposes during the firing step, providing a microstructure of glass particles substantially uniformly coated with metal.

Conductivities of ~40 mΩ/□ with 40 wt % metal content and ~70 mΩ/□ with only 30 wt % metal have been obtained.

Example: The glass used was a lead borosilicate glass having a softening temperature of ~450°C. The composition of the glass used for the thick film system was chosen to give an expansion coefficient equal to that of alumina (~8 x 10^{-6}/°C). The composition was 70 wt % PbO, 20 wt % B_2O_3 and 10 wt % SiO_2. The glass was powdered using alumina balls in an organic lined mill with isopropyl alcohol as the liquid medium. After milling, the glass was size separated and a particle size fraction between 20 and 44 μm was used for the experiments. The surface treatment involved essentially surface reactions with dilute HCl (0.01 N and 0.25 N HCl) for 45 sec; 50 cc of acid was used for about 3 g of glass powder. The surface-active glass powder (3 to 5 g) was mixed with a silver resinate (15 wt % Ag in xylene as solvent). The metal content of the glass-metal mixture was controlled by controlling the amount of silver resinate used (10 g of solution gives ~ 0.15 g of Ag).

The mixture was heated to 80°C to evaporate the solvent and then suspended in about 1.2 g of the organic vehicle, OIL 1014 (Drakenfield Colors), to form a paste with the proper viscosity for screen printing. A stainless steel 200 mesh screen and squeegee were used to apply thick films of a definite pattern onto alumina substrates (96% alumina, 1" x 1" x 0.030"). After printing, the printed substrates were dried in air for 15 minutes and then fired in air. The temperature program consisted of raising the temperature at 16°C/min to the firing temperature, holding and allowing the substrate to furnace cool. Films were fired at different temperatures for different times.

Thick films formulated with untreated glass particles were also made following similar procedures of printing, drying and firing. The electrical sheet resistivity of the fired systems was measured using the four point probe method. The sheet resistivities of the different thick films obtained with surface-activated glass particles and with varying metal contents are given in the table.

Silver Metal Content (wt %)	Concentration of HCl for 45 sec Surface Treatment	..... Firing		Resistivity (mΩ/□)
		Temperature (°C)	Time (min)	
40	0.25 N	550	8	44
	0.25 N	575	14	32
	0.01 N	500	20	80
	0.01 N	500	30	120
30	0.01 N	600	8	85
	0.01 N	640	8	70
	0.01 N	700	8	80
	0.01 N	500	8	160
25	0.01 N	600	8	165
	0.01 N	700	8	160
20	0.01 N	500	30	230

As can be seen from the table, sheet resistivities vary from 40 to 230 mΩ/□. The thick films prepared with surface passive powders and with the same metal contents as above were essentially nonconductors (ohms per square >10^6).

Resistor Circuit

J.F. Brown; U.S. Patent 4,140,817; February 20, 1979; assigned to Bell Telephone

Laboratories Incorporated describe a fabrication technique whereby a low cost thick film conductor requiring a reducing atmosphere is made compatible with a thick film resistor requiring an oxidizing atmosphere.

Fabrication of the circuit begins with a substrate, which is usually composed of approximately 96% Al_2O_3. The surface of the substrate was cleaned in accordance with standard techniques. A copper paste was prepared without the glass frits usually utilized for adhesion. The paste consisted of approximately 75% by weight of copper metal and approximately 25% by weight of an organic binder. In this instance, the binder was 2 g of ethylcellulose dissolved in 9 ml of α-terpineol and 9 ml of butyl Carbitol acetate. It will be appreciated that other binders may be used. In addition, the concentration of metal and binder may be varied, with the metal concentration ranging from 10 to 95% and the binder from 5 to 90%.

The conductor paste was screen printed on the substrate using a 325 mesh screen and a 0.6 mil emulsion pattern to produce the dried conductor pattern with a thickness of 0.9 mil. The thickness of the deposited material is preferably in the range 0.5 to 1.5 mils. The paste was dried, at a temperature of approximately 125°C for 10 minutes, and then fired in air. The firing step establishes adhesion between the conductor and the substrate by formation of copper oxide which reacts with the alumina substrate to form copper aluminate spinel at the interface. To insure adequate adhesion, it is recommended that the firing be done in the temperature range 1120° to 1220°C for 10 minutes to 1 hour. In this instance, the paste was fired at 1125°C for 30 minutes.

In accordance with a key step in the method, the copper oxide was then reduced to copper metal at a temperature of at least 700°C. The reducing ambient can be hydrogen alone or a mixture of hydrogen and nitrogen, preferably with an amount of hydrogen of at least 2%. The reduction reaction is preferably carried out until essentially all the copper oxide is converted to copper. In this instance, reduction was accomplished by heating at 1000°C for 30 minutes. A useful range appears to be heating at 700° to 1100°C for 10 to 60 minutes.

Next, the copper was reoxidized by heating in air at a temperature of approximately 850°C for 10 minutes. Again, it is desirable to convert essentially all of the copper to copper oxide. This step, in combination with the previous reduction reaction, is designed to produce a copper oxide which is less dense than the oxide formed after the initial firing of the conductor paste as evidenced by the change in thickness of the conductor layer. That is, whereas the initial copper oxide layer typically had a thickness of 0.5 mil, the thickness of the oxide layer after this step was 0.7 mil. It is recommended that the change in density fall within the range 10 to 80%.

The reoxidation step will preferably closely match the firing requirements of the resistor material to be deposited. In general, however, the temperature can range from 250° to 950°C and the time from 5 minutes to 1 hour, so long as the resulting oxide can be reduced at a sufficiently low temperature after resistor fabrication to achieve the desired conductivity without significantly degrading the resistor material.

The resistors were then deposited onto selected areas of the circuit by screen printing resistor paste through a 200 mesh screen with an 0.8 mil thick emulsion pattern. Thickness after drying typically ranges from 0.5 to 1.5 mils. The particular material used was a resistor ink containing bismuth ruthenium oxide [Birox (DuPont)]. The process is most beneficial with all ruthenium-based resistor materials, and, in general, may be utilized with any resistor material which must be fired in an oxidizing ambient.

After deposition, the resistor material was dried in accordance with standard techniques at a temperature of 125°C for 10 minutes. Firing was effected in air at a temperature of approximately 850°C for 10 minutes, which was recommended by the manufacturer of the material to achieve a resistivity of 10^5 ohms per square. Of course, other resistor materials that are available to provide a full range of resistivities can be processed in the same way.

Subsequent to the resistor processing, the copper oxide in the conductor pattern was reduced by heating in an ambient of 10% H_2 at approximately 260°C for 30 minutes. This resulted in a final conductor resistivity of approximately 0.01 ohm per square. Due to the low density of the copper oxide formed in the previous steps, a sufficiently low temperature can be utilized to achieve the desired conductivity while keeping resistor properties within tight tolerances. With these considerations in mind, a recommended range is heating at 250° to 450°C for 5 minutes to 1 hour. Again, the reducing ambient can be hydrogen alone or a combination of hydrogen and nitrogen.

Circuits fabricated in accordance with the abovedescribed embodiment consistently resulted in the desired low resistivity for the conductors, while at the same time keeping the resistivity of the resistors within the required range. Thus, it was demonstrated that the processing of the conductors had no significant adverse effect on resistor properties and vice-versa.

Air-Firable Base Metal Powder

C.Y. Kuo; U.S. Patent 4,122,232; October 24, 1978; assigned to Engelhard Minerals & Chemicals Corporation has found that a thick film conductor may be formed on a nonconductive substrate from a paste comprising in its broadest embodiment a base metal powder which would oxidize upon firing in the presence of air to become substantially nonconductive, sufficient boron powder to prevent oxidation of the base metal powder, and a vehicle. In an alternative embodiment, the paste may also contain glass frit.

The base metal and boron particles used are generally less than 325 mesh in size to assure satisfactory screen printing. When used, the glass frit particles are preferably in the submicron range, although they may be larger, up to 325 mesh as a maximum. A number of metals which otherwise would oxidize upon firing in air may be applied as conductive films, in particular, conductor films of nickel, cobalt and copper have been made.

In the preferred embodiment, the paste comprises between 50 to 80% by dry weight of nickel powder, between 5 and 20% by dry weight of boron powder, and up to 15% by dry weight of glass frit, and sufficient vehicle to provide a suitable viscosity.

In a process for forming an air-firable base metal conductor, the paste as described above, is mixed in the desired proportions and then applied on a nonconductive substrate by screen printing, although brushing, dipping or other techniques, such as used in the graphics industry could be used. After application, the paste is dried at an elevated temperature to evaporate the volatile components of the vehicle. Thereafter, the dried paste is fired in air to fuse or sinter the constituents and to bond the film to the substrate, while the boron prevents oxidation of the metal and preserves its conductivity. The resulting film is well bonded and is electrically conductive, making it suitable for use in electronic applications.

Example: A cobalt conductor paste is formulated by mixing 75 wt % of cobalt powder ($<$325 mesh) with 17.5 wt % of boron powder (type B, 95 to 97% purity and a submicron particle size), 7.5 wt % glass frit (lead borosilicate and a submicron particle size), and sufficient vehicle to produce a paste which can be satisfactorily screen-printed. After firing (peak temperature, 600°C for 10 minutes) in a belt furnace while being exposed to air passed through the furnace, the sheet resistance of the resulting film is between 0.1 and 5 ohms per square.

Silver on Ceramic Titanate Bodies

S.M. Marcus; U.S. Patent 4,101,710; July 18, 1978; assigned to E.I. DuPont de Nemours and Company describes conductive silver compositions of finely divided inorganic particles dispersed in an inert liquid vehicle, useful for producing in a single application step (followed by firing to sinter the inorganic particles) solderable electrodes adherent to ceramic titanate bodies. The compositions are especially useful on semiconducting titanate bodies. The inorganic particles are at least sufficiently finely divided to pass through a 400 mesh screen and consist essentially of about (by weight) either:

(a) (1) 75 to 98% silver, preferably 75 to 80%, more preferably 76%;

(2) 2 to 6% boron, preferably 3 to 4%, and more preferably 3%; and

(3) 3 to 22% glass, PbF_2 or mixtures thereof, preferably 10 to 21%, more preferably 21%; or

(b) (1) 40 to 70% silver, preferably 50 to 60%, more preferably 56%;

(2) 25 to 60% $Ni_3B_{1-x}P_x$ (where x is in the approximate range 0 to 0.6), preferably 25 to 40%, more preferably 30%; and

(3) 3 to 22% glass, PbF_2 or mixtures thereof, preferably 10 to 21%, more preferably 14%.

Mixtures of (a) and (b) may also be used. Component (3) in (a) and (b) is preferably glass. Preferred compositions contain 60 to 80% inorganic particles and 20 to 40% vehicle.

The inorganic particles are dispersed in an inert liquid vehicle by mechanical mixing (e.g., on a roll mill) to form a paste-like composition. The latter is printed as thick film on conventional dielectric substrates in the conventional manner. Any of the various organic liquids with or without thickening and/or stabilizing agents and/or other common additives, may be used as the vehicle. Exemplary of the organic liquids which can be used are the aliphatic alcohols; esters of such alcohols, for example, the acetates and propionates; terpenes such as pine oil, terpineol and the like; solutions of resins such as the polymethacrylates of lower alcohols; or solutions of ethylcellulose, in solvents such as pine oil and the monobutyl ether of ethylene glycol monoacetate. The vehicle may contain or be composed of volatile liquids to promote fast setting after application to the substrate.

After drying to remove the vehicle, firing of the compositions is carried out at temperatures and for times sufficient to sinter the inorganic materials and to produce conductor patterns adherent to the dielectric substrate. Firing is conducted at a temperature and for a duration sufficient to sinter the composition into an adherent, solderable coating which is electrically and physically continuous, according to principles well-known to those skilled in the art. Firing may be conducted in a box or belt furnace, at a peak temperature in the range 550° to 625°C, preferably at about 580°C. The peak temperature is maintained for at least 2 minutes, preferably about 10 minutes. Although firing will normally be conducted in air, firing in an inert atmosphere (e.g., nitrogen, argon, etc.) is possible.

Soldering of the fired electrodes to attach leads is done conventionally, e.g., by fluxing and then dipping in molten solder. Although special advantage is obtained by firing these compositions on semiconducting ceramic substrates of substituted barium titanate, the compositions are useful for producing conductive patterns on other ceramic titanate substrates such as barium titanate itself, etc.

ELECTRODES

Valve Metal Carrier Layer

The expression valve metals has become very popular for the group of metals including titanium, tantalum, niobium, zirconium, tungsten and molybdenum. It is known that these valve metals passivate very quickly when used in aqueous solutions, due to the development of a dense cover layer of an oxidic nature, thereby becoming extremely corrosion-resistant in many electrolytes. However, the passive layers of these metals have no electron conductivity in the electric potential ranges used, so that very high field densities occur in the layers. Above a certain potential (breakthrough potential), this leads to the destruction of the passivating layers. Despite the fact that these metals have great corrosion resistance, no anode process can be carried out with these metals in the passive state.

K. Koziol, H.-C. Rathjen and K.-H. Sieberer; U.S. Patent 4,138,510; February 6, 1979; assigned to Firma C. Conradty, Germany describe a process whereby an electrode in which the active substances, counteracting passivation, are anchored in a porous carrier layer sintered onto the valve metal base. The carrier layer sintered onto the cleaned valve metal base may consist of a powder of the same metal, or a crystallographically similar metal. The pretreatment of the valve metal base may be effected by any desired method, such as pickling, steam degreasing, rinsing, grinding, or the like. The size, shape and surface of the metal powder particles vary in accordance with the material and the production method.

The application of the powder particles to the valve metal base may be accomplished by spraying, rolling, electrodepositing, brushing, and other suitable methods, prior to the sintering operation. To facilitate the application, prior to the sintering operation, binders or adhesives or both may be admixed with the powder. It is expedient to use, as powder, various valve metal powders such as titanium powder or tantalum powder, or a mixture of valve metal powders, or a valve metal alloy present in powder form.

In order to avoid oxidation of the valve metal powder during the sintering process, the latter is carried out either in a vacuum between 1 and 5×10^{-7} torr, or in a definite gas atmosphere, such as argon. The heating rate is determined either by the quality of the vacuum or by thermally disintegrating substances limiting the heating rate in order to avoid damage to the sintered layer. The sintering temperature varies between 800° and 2800°C, depending upon the metal powder and the base metal, with the heating periods ranging between several hours and one-quarter hour, depending upon the temperature.

The infusion of the active substances, counteracting passivation, can be effected by impregnating and drying or baking them in, or both, by precipitating them from the vapor phase, galvanically, or by precipitating them from the gaseous phase. Adding a wetting agent frequently results in a further improvement. The active substances may also be ingredients of the sinter mixture before sintering.

All substances of sufficient corrosion resistance during electrolysis, and possessing good electron conductivity in the potential ranges used, so that an anode process can be carried out, are suitable as active substances. These are all metals and oxides of the platinum metal group, intermediate and mixed oxides of precious and ignoble metals or both, or oxides of ignoble metals alone, which will meet the requirements set forth above. Surprisingly, it has been found that, with this construction, even conductive materials of a base or nonprecious character lead to excellent results. Thus, the widely held opinion that the active layer always must contain noble precious metal or noble precious metal compounds, in order to remain effective, is refuted.

As a result of this treatment, there is obtained, on a valve metal base, a compound material, i.e., a metal/metal or a metal/ceramic combination. This is a mechanically strong, yet porous, crystallographically identical, well adherent valve metal carrier layer containing the active substances in well anchored form. This layer, which in part displays cermet characteristics, is characterized in that the active substances are built into a carrier skeleton having the same crystalline structures as the base valve metal, thus forming one unit with the base valve metal. Therefore, the electrical conductivity through this activated carrier layer is primarily of a metallic nature.

Example: A titanium sheet, whose dimensions are 100 x 100 x 1 mm, was pickled for 30 minutes in a 20% by weight boiling hydrochloric acid, washed with water, and rinsed with propanol. A mixture of titanium powder, polyglycol 6000 and hexanol was sprayed on the thus prepared sheet by means of a compressed air operated spray gun. After a 20 minute drying period at 120°C in a drying oven, the titanium powder coating was sintered on in an induction furnace at a heating rate of 300°C per hour and an end temperature of 1100°C.

The thus produced basic body was soaked in a 1 molar ruthenium-chloride solution, to which was added some wetting agent (Erkantol). This was followed by a heat treatment at 450°C for 30 minutes. The process was repeated three times in an identical manner.

The anode thus produced, as compared to an anode coated with the same solution in a known manner, namely without a sintered coating, has a much greater active surface, resulting, under the same load, in a small true anode current density and, consequently, in a lower cell voltage. In addition, the anode embodying the process is also considerably more amalgam-proof and short-circuit-proof.

Ruthenium Coating

H.R. Heikel and J.J. Leddy; U.S. Patent 4,112,140; September 5, 1978; assigned to The Dow Chemical Company describe a method of coating an electrode with a ruthenium compound. A valve metal substrate such as lead, molybdenum, niobium, tantalum, tungsten, vanadium, zirconium and more preferably titanium, suitable for use as an electrode is suitably cleaned to remove, for example, grease or oil, from the surface of the substrate to be coated. After cleaning and, optionally, roughening the surface, a first liquid solution is applied to at least a portion of such surface by a suitable well-known means such as brushing, spraying, flow-coating (i.e., pouring the solution over the surface to be coated), or immersing that portion of the substrate to be coated in the solution.

The first solution preferably consists essentially of ruthenium in an amount of from about 5 to 25 mg/ml of solution and titanium in an amount of from about 5 to 25 mg/ml of solution. To further improve the abrasion resistance or durability of the oxide coating, the ratio of titanium to ruthenium preferably is from about 2:1 to 1:2, and more preferably from about 2:1 to 1:1. The acid concentration of the first solution is from about 0.1 to 1 N and preferably from about 0.5 to 0.7 N. The balance of the first solution includes a solvent such as isopropanol, n-butanol, propanol, ethanol and any cations associated with the ruthenium and titanium present in the solution.

The surface to which the first solution was applied is preferably dried at a temperature below the boiling temperature of the first solution to remove the volatile matter, such as the solvent, before heating to form the oxides of ruthenium and titanium. Air drying is satisfactory; however, use of slightly elevated temperature within the range of from about 25° to 70°C and, optionally, a reduced pressure will hasten completion of the drying step.

The dried coating is heated at a temperature of from about 300° to 450°C in an oxygen-containing atmosphere for a sufficient time to oxidize the ruthenium and titanium on the substrate surface and form the desired adherent oxide layer. Generally maintaining the substrate at the desired temperature for from about 3 to 10 minutes is adequate; however, longer times can be used.

After the initial heating step at from about 300° to 450°C, the coated surface is overcoated with ruthenium and titanium using a second liquid solution with a higher titanium to ruthenium weight ratio than in the first solution. The second solution preferably contains ruthenium in an amount from about 2 to 10 mg/ml of solution, and titanium in an amount from about 20 to 40 mg/ml of solution. The titanium to ruthenium weight ratio is preferably from about 10:1 to 2:1. The solvents and acid ranges for the first solution are also suitable for the second solution. The second solution is applied to the precoated portion of the substrate, optionally dried, and heated as herein described for the first solution.

To obtain a coating with good adherence to the substrate and a low loss of ruthenium during use as an electrode, the coating resulting from the first solution has a thickness of up to about 3 μ, and the overcoating has a thickness of less than about 1.5 μ.

The second and, if desired, subsequent overcoatings applied with the second solution are preferably sufficient to form individual oxide coatings with thicknesses of up to about 1.5 μ. Increased durability of the coated surfaces is achieved by providing a number of overcoatings with individual thicknesses of up to about 0.5 μ.

A sufficient number of overcoatings is applied to obtain a total thickness of ruthenium and titanium oxides of up to about 10 μ and preferably up to about 3 μ. Coatings of greater thicknesses are operable, but are not required to provide an electrode suitable for electrolytic purposes. It has been found that a titanium substrate coated with the first solution and thereafter coated at least once with the second solution, with drying and heating steps between each coating step, in the herein described manner, results in an electrode with an effective amount of ruthenium and titanium oxides in the coating suitable for use as an anode in an electrolytic cell for producing chlorine from a sodium chloride containing brine. The coating contains sufficient ruthenium and titanium oxides to permit sufficient electric current flow between the electrodes to achieve the desired electrolysis or corrosion prevention.

Example: An electrode useful as an anode in an electrolytic cell for producing chlorine and sodium hydroxide from a sodium chloride brine was coated with adherent layers of ruthenium and titanium oxides in the following manner. A first or primer coating solution with ruthenium and titanium concentrations of 6.4 mg/ml of solution was prepared by mixing together 4.40 g $RuCl_3 \cdot 3H_2O$, 2.90 g of concentrated hydrochloric acid, 200.00 g of isopropanol and 10.20 g of tetraisopropyltitanate (TPT). This solution had a density of 0.81 g/ml. The weight ratio of titanium to ruthenium in the solution was about 1:1.

A second or overcoating solution was prepared by mixing together 1.38 g of $RuCl_3 \cdot 3H_2O$, 3.20 g of concentrated hydrochloric acid, 66.50 g of isopropanol and 13.50 g of TPT. This solution contained ruthenium and titanium in amounts of 5.3 and 22.7 mg/ml of solution, respectively, and had a density of 0.84 g/ml. The ratio of titanium to ruthenium in the second solution was about 4.32:1.

A 3" x 5" x 1/16" piece of titanium sheet meeting the requirements of ASTM standard B-265-72 was cleaned by grit-blasting with 46 mesh (U.S. Standard Sieve Series) alumina (Al_2O_3) grit using apparatus with a 7/16" diameter grit orifice and a 3/16" diameter air orifice. The grit orifice was maintained at a distance of 4" from the titanium sheet; air pressure was 70 lb/in^2 at the entrance to the blasting apparatus and the blasting rate was 15 to 20 in^2 of titanium surface per minute. The grit-blasted surfaces were determined, from photomicrographs, to have depressions therein averaging about 2 μ in depth. The depth of such depressions is not critical.

A sufficient amount of the first coating solution was poured over the cleaned titanium surfaces to wet such surfaces. Excess solution was drained from the wetted surfaces before drying such surfaces at room temperature (~21°C) for

15 minutes. The ruthenium and titanium in dried coating were oxidized by heating the dried titanium sheet in air in a muffle furnace for 10 minutes at 400°C. After cooling, the coated surface was determined to contain about 20 μg of ruthenium per square centimeter of coating.

A sufficient amount of the second solution was poured over the oxide coated surfaces to wet such surfaces. The wetted surfaces were sequentially drained of excess solution, air dried at room temperature for 15 minutes and oxidized by heating in air at 400°C for 10 minutes in a muffle furnace. A total of six overcoatings were applied to the titanium substrate using the second solution and the abovedescribed procedure. The ruthenium content of the final coating was determined by standard x-ray fluorescence techniques to be 175 μg of ruthenium per square centimeter.

The titanium electrode with an adherent coating of the oxides of ruthenium and titanium was tested as an anode in a laboratory electrolytic cell with a glass body to produce gaseous chlorine from an acidic, aqueous solution containing about 300 g/ℓ sodium chloride. The anode, with an area of about 12½ in^2, was suitably spaced apart from a steel screen cathode by a diaphragm drawn from an asbestos slurry. The cell was operated for 170 days at an anode current density of 0.5 amp/in^2 and a voltage of 2.79. The sodium hydroxide concentration in the catholyte was about 100 g/ℓ. After operating for the 170 day period, it was determined that 40 μg of ruthenium per square centimeter of anode surface had been consumed. This ruthenium loss is equivalent to 0.084 g of ruthenium per ton of chlorine produced.

Silver Iodo Solid Electrolyte Films

M.R. Arora; U.S. Patent 4,105,807; August 8, 1978; assigned to Unican Electrochemical Products Ltd., Canada, describes a process relating to the production of thin films of complex silver iodo solid electrolytes on a substrate, and is particularly concerned with the deposition on a substrate such as a ceramic substrate or a metallic substrate, e.g., forming an electrode of a coulometer or a battery, of a thin stable film of oxy-anion-substituted silver iodide solid electrolyte, such as $Ag_{19}I_{15}P_2O_7$, such film having good ionic conductivity at ambient temperature and particularly containing iodine in stoichiometric proportion, substantially without any excess or deficiency thereof; and to the procedure for producing such films and for incorporating such films into devices such as batteries, coulometers, memory devices, and the like.

It was found that stoichiometric and stable films of $Ag_{19}I_{15}P_2O_7$ which show good ionic conductivity at ambient temperatures can be deposited using the following methods and parameters of operation.

Flash Evaporation
- Evaporant–stoichiometric electrolyte ($Ag_{19}I_{15}P_2O_7$)
- Source temperature–1400° to 1600°C
- Substrate temperature–140° to 150°C
- Initial pressure–10^{-6} torr
- Total film thickness–25000 to 50000 A
- Rate of film deposition–15 to 60 A/sec

Radio Frequency Sputtering
- Evaporant–stoichiometric electrolyte ($Ag_{19}I_{15}P_2O_7$)
- Potential across electrodes–700 to 900 V

Argon gas pressure–4 x 10^{-3} torr
Source temperature–100° to 175°C
Substrate temperature–140° to 150°C
Spacing between electrodes–4 cm
Radio frequency–11 to 13 MHz
Total film thickness–25000 to 50000 A
Rate of film deposition–15 to 60 A/sec

On the other hand, stoichiometric and stable films of $RbAg_4I_5$ which have good ionic conductivity at ambient temperature can be deposited by using the following parameters of operation.

Flash Evaporation of $RbAg_4I_5$
Evaporant–stoichiometric electrolyte ($RbAg_4I_5$)
Source temperature–900 to 1100°C
Substrate temperature–60° to 80°C
Initial pressure–10^{-6} torr
Total film thickness–25000 to 50000 A
Rate of film deposition–15 to 60 A/sec

The following examples illustrate the process.

Example 1: $Ag_{19}I_{15}P_2O_7$ was prepared by vacuum fusion. Its bulk ionic conductivity was found to be 0.09 $(ohm\text{-}cm)^{-1}$. About 5 g of this material was placed in a baffle-type evaporation source formed of tantalum. At an initial vacuum of 10^{-6} torr, the evaporation source was heated using a current of 20 A to a temperature of 510°C for 10 minutes to melt and degas the solid electrolyte, and then the current was increased to 30 A and a temperature of 750°C to deposit the solid electrolyte film on ceramic substrates over a period of 9 minutes such that the film thickness so obtained was 21000 A. The substrate temperature during the film deposition was kept below 60°C. The ionic conductivity of these films at room temperature was found to be 4.5 x 10^{-5} $(ohm\text{-}cm)^{-1}$, which is too low to be of practical usefulness. Increasing the evaporation source temperature to 1150°C by using currents of 70 A results in similar films of low conductivity.

Example 2: About 10 g of the solid $Ag_{19}I_{15}P_2O_7$ electrolyte, as prepared in Example 1, was placed in a vibratory feeder. At an initial vacuum of 10^{-6} torr, the evaporation source was heated to about 800° to 1000°C using a current of about 60 to 80 A and the electrolyte particles were then fed from the vibratory feeder via a funnel to the evaporation source. Solid electrolyte films of thickness between 2000 and 25000 A were deposited at substrate temperatures of 60° to 80°C and their ionic conductivities measured, which were found to be about 2 x 10^{-5} $(ohm\text{-}cm)^{-1}$, which are again too low.

Example 3: About 40 g of the electrolyte, as prepared in Example 1, but of particle size smaller than No. 200 mesh Tyler screen, were used as the target in an rf sputtering deposition using argon gas. The potential across the electrodes was varied between 600 and 1,500 V, the substrate temperature maintained at below 75°C, the electrode spacing kept at 4 cm and the Ar^+ gas pressure at 4 x 10^{-3} torr. The films were deposited to a thickness of 10000 to 40000 A and their ionic conductivities were found to vary between 1 and 3 x 10^{-5} $(ohm\text{-}cm)^{-1}$, which are again too low.

Example 4: About 10 g of the electrolyte, as prepared in Example 1, were placed in a vibratory feeder. At an initial vacuum of 10^{-6} torr, a box-type evaporation source of tantalum material was heated to a temperature of 1400° to 1600°C, and the electrolyte particles of size between No. 40 and 70 mesh screens were then fed from the vibratory feeder directly into the heated source. These particles, on arriving at the source, instantaneously vaporized and the solid electrolyte films were deposited to thicknesses between 8000 and 50000 A with the substrates maintained at a temperature between 140° and 150°C. The ionic conductivity of these films at room temperature was found to vary between 2×10^{-2} and 3.5×10^{-2} $(ohm\text{-}cm)^{-1}$, whereas the ionic conductivity obtained by using the methods of Examples 1 and 2 was about 2 to 5×10^{-5} $(ohm\text{-}cm)^{-1}$. Therefore, the method outlined in this example yields solid electrolyte films of high ionic conductivity and shows the marked improvement of the process over the conventional methods.

When a mixture of powders of AgI and $Ag_4P_2O_7$, in the molar stoichiometric ratio of 15:1 ($15\ AgI + 1\ Ag_4P_2O_7 = Ag_{19}I_{15}P_2O_7$) was used in place of $Ag_{19}I_{15}P_2O_7$, essentially similar results were obtained.

Germanium Disulfide Coating

A. Nidola and P.M. Spaziante; U.S. Patent 4,132,620; January 2, 1979; assigned to Diamond Shamrock Technologies SA, Switzerland describe an electrode particularly useful for electrochemical reactions involving oxygen evolution and oxygen ionization which comprises an uncorrodible substrate or base having a coating containing an electrocatalytically effective amount of germanium disulfide. Preferably the coating contains 10 to 50, more preferably 20 to 30 g/m² of germanium disulfide.

The substrate is preferably electrically conductive and may be of any suitable material which is resistant to the corrosion in the specific embodiment, e.g., wherein oxygen is either evolved or ionized at the electrode surface. Examples of suitable base materials for most applications are valve metals such as titanium, tantalum, niobium, tungsten, hafnium, zirconium, silicon, aluminum and alloys comprising at least one of these metals; valve metal carbides and borides such as titanium carbide, tantalum carbide, tantalum boride; and vitreous carbon as described in U.S. Patent 3,927,181.

The electrode base or substrate may have any desired configuration such as a plate, mesh or rod, but is preferably in the form of a porous body of sintered valve metals, sintered valve metal carbides or vitreous carbon strands, preferably with a porosity of 30 to 70% in order to increase the effective surface area and to improve the adherence of the germanium disulfide coating.

The coating containing the electrocatalytic germanium disulfide may also contain additional electrocatalytic agents such as platinum group metal oxides, for example, ruthenium dioxide or iridium dioxide as described in U.S. Patent 3,711,385; mixed crystal materials of a platinum group metal oxide and a valve metal oxide, for example, ruthenium dioxide-titanium dioxide or iridium dioxide-tantalum pentoxide as described in U.S. Patent 3,632,498; or nonnoble metal oxides such as β-manganese dioxide or α/β-lead dioxide. The base may also be provided with an intermediate coating to improve adherence or other characteristics, for example, a coating of a valve metal oxide or of a platinum group metal.

The electrodes may be prepared in various manners. For example, the germanium disulfide may be preformed and then secured to the base in any suitable fashion such as by plasma jet or by deposition with a binding agent such as an organic polymer resistant to the electrolysis conditions or with amorphous valve metal oxide. Another method is to coat the base with a solution of a thermally decomposible salt of germanium, heat the coated base at about 300° to 500°C in oxygen to evaporate the solvent and form germanium oxide and subject the resulting base to cathodic treatment in the presence of hydrogen sulfide to convert the germanium oxide to germanium disulfide.

In a preferred process to produce the electrode, an electrically conductive porous base made of a corrosion-resistant material is impregnated with a slightly acidic solution of a soluble germanate salt, hydrogen sulfide is added to the solution to convert the germanate salt into the corresponding thiogermanate salt and the pH is then reduced (to 2 or less) to precipitate insoluble germanium disulfide in the pores of the base. The electrode can then be removed from the solution and be dried.

Preferably, the solution of germanate salt is an aqueous solution and the salt is an alkali metal or ammonium germanate. The hydrogen sulfide is preferably added by bubbling it through the solution until an excess of the theoretical amount thereof has been added. The acidification may be effected with any mineral or organic acid such as hydrochloric acid, hydrobromic acid, sulfuric acid, nitric acid, phosphoric acid, acetic acid or propionic acid. The electrode can be dried by heating to 75° to 100°C, preferably 80° to 90°C.

The process may be used for the electrolysis of many electrolytes in which oxygen is evolved at the anode at low oxygen evolution potential, such as electrowinning of copper, zinc, nickel, cobalt and other metals from aqueous sulfuric acid solutions. The electrodes have the advantages of being resistant to acid conditions and to oxygen evolution, having a low oxygen potential and of not adding impurities to the electrolyte that may cause problems in metal deposition at the cathode.

The electrodes of the process also show a high catalytic activity for electrochemical oxidation and are useful in primary and secondary cells, for example, fuel cells wherein the electrodes provide the catalytic side for the conversion of molecular oxygen to an ionic species which then combines with the ionized fuel, for instance with H^+ in a hydrogen-oxygen cell, to form water over the active surface of the electrode with a net transfer of electrons from the electrode to the reacting species.

Ion-Sensitive Electrode

Ion-sensitive electrodes are known to the prior art for use in measuring the activity of a specific ion, or ions, in a test solution. In the case where the test solution comprises body fluids, the ion activities typically measured are those of the hydrogen, sodium, potassium, and calcium cations (respectively H^+, Na^+, K^+, Ca^{2+}). Typically, the ion-sensitive electrode and a reference electrode are immersed in the test solution. The ion-sensitive electrode may, in one instance, be constructed with an ion-exchanging membrane so that the potential difference between the ion-exchanging membrane and the test solution is a function of the activity of a particular ion in the test solution. The reference electrode is constructed so that the potential difference between the reference electrode and the test solution is a constant independent of the composition of the test solution. By measuring the voltage across the ion-sensitive electrode and the reference electrode, the activity, and therefore the concentration, of a particular ion in the test solution may be determined.

M.A. Afromowitz and S.S. Yee; U.S. Patent 4,133,735; January 9, 1979; assigned to The Board of Regents of the University of Washington describe an ion-sensitive electrode in which a conductor is bonded to the substrate, the conductor having first and second regions, with at least the first region being formed as a conducting layer on the substrate surface. An ion-sensitive membrane is bonded to the substrate and to a portion of the conductor, the membrane including at least a continuous membrane layer covering the first region of the conductor and portions of the substrate surface which are contiguous to the first region. An output means is electrically interconnected with the second region of the conductor for connecting the electrode to a utilization device.

Finally, a fluid-tight sealing means is bonded to the substrate and to the conductor and covers at least the second region of the conductor and portions of the substrate surface and output means adjacent the second region of the conductor.

An ion-sensitive electrode of the type described may be fabricated by fabricating a wafer from a substrate material, the wafer having a substantially planar surface; forming a continuous conducting layer having a desired configuration on the substantially planar wafer surface; forming a continuous ion-sensitive membrane layer on a first region of the continuous conducting layer and contiguous portions of the substantially planar wafer surface; connecting at least one lead to a second region of the continuous conducting layer; and, forming a fluid-tight seal over the second region of the continuous layer, contiguous portions of the substantially planar wafer surface and a portion of the lead adjacent the second region of the continuous conducting layer.

The conducting layer itself may be formed by either a thin-film vapor deposition process, or by a thick-film screening process. In the former case, the substantially planar wafer surface is polished and cleaned. The wafer is then placed into an evacuated chamber along with a quantity of at least one metal and the metal is heated to a temperature sufficient to vaporize the metal so that the metal uniformly deposits throughout the chamber and in a continuous conducting layer on at least the substantially planar wafer surface. The wafer is removed from the evacuated chamber. The continuous conducting layer is then photoetched to the desired configuration.

In the latter case, a first wire mesh screen is prepared, the first wire mesh screen having a predetermined mesh and a thickness approximating that of the desired conducting layer. The first wire mesh screen also has an open region therethrough corresponding in configuration to that of the desired conducting layer.

A paste is prepared by mixing, with an organic vehicle including an organic solvent and an organic binder, a conducting material in particle form, and having an average particle size less than the predetermined mesh of the first wire mesh screen. The first wire mesh screen is brought into contact with the substantially planar wafer surface and the paste is spread on the wire mesh screen so as to cover at least the open region therethrough.

The paste is then forced through the open region in the first wire mesh screen and into contact with the substantially planar wafer surface. The wire mesh screen is removed, whereupon the paste adheres to the wafer surface with the configuration of the desired conducting layer, and the wafer is heated to a first temperature for a time sufficient to drive off the organic solvent and then to at least a second temperature for a time sufficient to drive off the organic binder and to fuse the conducting material of the paste into a continuous conducting layer, and allowed to cool.

The continuous ion-sensitive membrane layer is preferably formed by a thick-film screening process. In this process, ion-sensitive material such as ion-sensitive glass, is reduced to a fine powder and mixed with an appropriate organic vehicle including an organic solvent and an organic binder to form a glass paste.

A second wire mesh screen is prepared which has a predetermined mesh and a thickness approximating that of the desired membrane layer. The second wire mesh screen also has an open region therethrough corresponding in configuration to that of the desired membrane layer.

The second wire mesh screen is then brought into contact with the substantially planar wafer surface so that the open region therethrough is in registration with the continuous conducting layer, and the glass paste is applied to the second wire mesh screen so as to cover at least the open region therethrough.

The glass paste is then forced through the open region in the second wire mesh screen and into contact with the first region of the continuous conducting layer and contiguous portions of the substantially planar wafer surface.

The second wire mesh screen is removed, whereupon the glass paste adheres to the wafer in the configuration of the desired membrane layer, and the wafer is heated to a first temperature for a period of time sufficient to drive off the organic solvent and then to at least a second temperature for a period of time sufficient to drive off the organic binder and to fuse the glass into a continuous membrane layer. Thereafter, the wafer is quickly quenched to substantially room temperature.

The structure and processes permit an active device chip, such as that including a field effect transistor, to be bonded to the wafer, preferably to the substantially planar wafer surface, and to be interconnected with the second region of the continuous conducting layer and an output lead or leads for connection of the electrode to a utilization device.

In this case, the fluid-tight seal is formed additionally over the active device chip and all exposed interconnections. If desired, the conducting layer may also be configured in the form of one or more pads which are isolated from the second region of the conducting layer on the substantially planar wafer surface, and the active device chip may be bonded to the substantially planar wafer surface and interconnected with the second region of the conducting layer and one or more of the pads by appropriate interconnecting leads, with an output lead or leads being bonded, such as by soldering, to one or more of the conducting pads.

Sodium Silicate Dust-Fixating Coating

In the storage battery industry various operational steps in the manufacture of electrode plates, particularly after intermediate drying, are accompanied by an undesirable evolution of lead dust. Experience has shown that the major quantity of lead dust is produced from formed positive plates during their subsequent processing steps up until installation in the battery.

M. Jung, E. Voss and T. Chobanov; U.S. Patent 4,135,041; January 16, 1979; assigned to Varta Batterie AG, Germany describe a lead electrode plate for lead storage batteries whose surface is provided with a dust-fixating coating wherein the coating includes sodium silicate as a film-forming component.

Electrode plates according to this process are preferably positive grid plates which are immersed in either dry or moist condition for a few seconds in dilute solutions of, for example, sodium silicate, Na_2SiO_3 (water glass) and are subsequently dried for about 1 hr at 60°C in an oven in ambient air. If desired the solution may also be applied by spraying. Obviously, positive as well as negative electrode plates may be treated in accordance with this process, and this may be done after pasting as well as after forming, preferably the treatment takes place after the pasting.

Under the general designation water glass there is commercially available a broad spectrum of sodium silicate solutions. The chemical composition and the physical properties of these materials fall approximately within the following ranges: 22 to 37% SiO_2, 6 to 18% Na_2O, 28 to 55% solid content, density 1.26 to 1.75 g/cc and viscosity (20°C) 20 to 2,000 cp. For application in accordance with the process a water glass has proven particularly desirable the analysis of which shows values which coincide generally with the lower limits of the abovementioned ranges.

The electrical behavior of the electrode plate is essentially unaffected by the coating. An additional advantage is achieved by mixing the water glass solution with a highly diluted aqueous dispersion of an organic polymer compound. Due to the organic compound the coating formed in this manner is characterized by enhanced elasticity.

The organic additives which are all water insoluble may be simple polymers such as polyvinyl acetate, polystyrol, polyacrylate, polyvinyl chloride or copolymers whose components may consist, for example, of styrol and acrylic acid ester, styrol and butadiene, or vinyl acetate and maleic acid ester. They belong to the group of thermoplastic polymerization synthetics and otherwise are used as raw materials for varnishes in the varnish industry. For example, they are available as Mowilith from Farbwerke Hoechst AG, Germany.

For utilization in accordance with this process the commercially available, about 50% aqueous dispersion of the synthetic polymer is diluted 1:10 and this dilute solution is mixed with the commercially available water glass solution in the ratio 1:1. The treatment liquid then has an overall solid content of 15 to 27 wt %. Application of the water glass solution with the synthetic additive to the electrode plate and its drying is then performed in the same manner as previously described.

There results a coating which reduces in optimal fashion the dust formation to a residue of about 2.5% of the dust emission from an untreated plate.

Gold Barrier Schottky Diodes

G. Damene; U.S. Patent 4,136,348; January 23, 1979; assigned to Societe Lignes Telegraphiques et Telephoniques, France describes a process for the manufacture of Schottky diodes with a guard ring and a gold-silicon barrier in which a film of reducing metal having a thickness between several thousand and 200 Å is deposited on the silicon prior to the deposition of the barrier metal film, the structure being subjected to a heat treatment which ensures diffusion of the barrier metal through the reducing film at a temperature below the formation temperature of the gold-silicon eutectic.

In accordance with an essential feature of this process, the solid state diffusion of gold through the metal film is effected by heat treatment at a temperature above 300°C and below the temperature at which the gold-silicon eutectic is formed (377°C).

There are shown in Figure 6.16a a substrate **20** and the epitaxial layer **21**, it being understood that the relative thicknesses cannot be shown on the same scale. The wafer provided with its epitaxial layer is then subjected to a thermal oxidation to a thickness of about 0.5 μm on both faces of the wafer, to provide masking for later manufacturing steps. These layers are respectively denoted by **22** and **22'** in Figure 6.16b.

The guard ring is formed by the usual photo-masking technique, this being followed by etching of the layer **22** to define an annular window in which a p-type impurity is diffused in the vapor phase, and then the structure illustrated in Figure 6.16c is obtained which consists of the substrate **20**, the epitaxial layer **21** on which there lies the oxide layer **22** having the window **23**, level with which there is situated the diffused annular zone **24** constituting the guard ring of the diode. The reoxidation of the window **23** resulting in the formation of a new continuous oxide layer **25** as illustrated in Figure 6.16d occurs in the course of the diffusion.

There then takes place the preparation of the cathode contact by attack of the oxide layer **22'** and doping, for example by ionic implantation, of the lower face of the substrate **20** for the purpose of degenerating the silicon in order to promote the establishment of the cathode contact, as illustrated in Figure 6.16d.

The preparation of the anode is carried out by known photoetching processes, the aperture of the anodic window **25'** being defined by the mean circumference of the guard ring **24**. A layer of reducing metal, for example chromium, is evaporated in vacuo on the previously cleaned and heated substrate. In the same enclosure, a gold layer is deposited by evaporation in vacuo.

There is shown in Figure 6.16f the diode thus treated, the two metallic layers being denoted by **26** and **27** respectively. The structure thus formed is subjected to a heat treatment at a temperature between 350° and 377°C for 1 hr, which is intended to ensure the diffusion of the gold **27** through the chromium layer **26** until it comes into contact with the epitaxial silicon layer **21**. This heat treatment may take place either directly in the vacuum enclosure used for the deposition of the metal films, or in in an independent furnace but it is carried out in air at atmospheric pressure. There is then obtained the structure illustrated in Figure 6.16g. The cathode contact is thereafter formed, in the well-known manner, preferably in the same way as the anode, but without masking (layer **28** being chromium and layer **29** gold), the cathode contact being applied to the whole of the lower face of the substrate **20**.

In the case of the cathode, it is preferable to keep the gold layer on the outer surface in order to facilitate the bonding of the cathode connection. The diagram of the completed diode is shown in Figure 6.16h. The layers **28** and **29** correspond respectively to the chromium film and the gold film constituting the cathode contact.

Figure 6.16: Diode in Various Stages of Manufacture

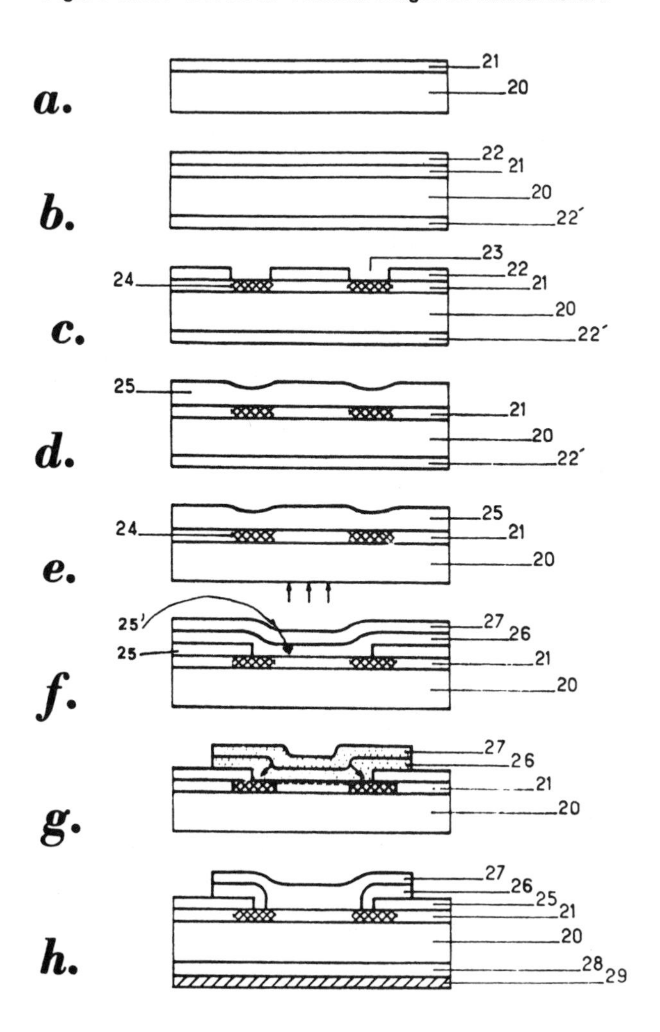

Source: U.S. Patent 4,136,348

Discharge Loop Electrode

M. Saito, K. Watanabe and K. Fukuyama; U.S. Patent 4,136,227; January 23, 1979; assigned to Mitsubishi Denki Kabushiki Kaisha, Japan describe an electrode of a discharge lamp which comprises an electron emission material containing 5 to 40 wt % beryllium oxide and 3 to 35 wt % yttrium oxide with one or more of barium, calcium and strontium components which is coated on a substrate of the electrode wherein the electron emission material comprises less than 60 wt % of total content of beryllium oxide and yttrium oxide.

When both beryllium oxide and yttrium oxide are incorporated as the heat-resistant oxide, the adhesion of the electron emission material on the substrate of the electrode is remarkably improved whereby the life of the discharge lamp is remarkably prolonged.

The electron emission material comprising the alkaline earth metal oxide and yttrium oxide and beryllium oxide is mixed with nitrocellulose and butyl acetate to form the suspensions and the suspension is coated on the substrate of the electrode and the electron emission material is adhered on the surface of the substrate of the electrode by heating it at a high temperature to prepare the electrodes.

Application to Electroacoustic Transducer Elements

F. Massa; U.S. Patent 4,123,567; October 31, 1978 describes a process for applying electrodes to the surfaces of thin ceramic discs which permits the coating of the entire disc with a conducting film and then automatically removes the conducting film from the peripheral edge of the disc as it is rolling down an inclined plane while being held within a guide slot. The bottom of the guide slot is provided with a moving abrasive belt which automatically removes the conducting film from the peripheral edge of the disc as it passes over the belt. The process will permit the advantageous use of plated electrodes such as electroless nickel without entailing the costly expense of masking the edges of the discs before plating.

Referring more particularly to Figure 6.17, the reference character **1** illustrates a table-top surface containing a longitudinal slot **2** along its length which provides clearance for the thickness of the ceramic discs **3**. Beneath the table-top is mounted a conventional endless abrasive belt **4** which is driven over the rollers **11** and **12** by an electric motor **5** in the conventional manner, as is well known in the art. A chute **6** is mounted over the slot **2** at one end of the assembly and provides a guide for permitting the ceramic discs to be fed into the longitudinal slot **2**, as illustrated.

A hopper with conventional sorting means for feeding the ceramics into the chute **6** may be provided above the chute **6**, if desired, to contain a quantity of ceramic discs which are to be processed for the automatic removal of the edge coating of conducting film. The hopper, which is not shown in the drawings, may be any one of the numerous conventionally available structures which are suitable for feeding the ceramic discs into the guide slot **2** in a continuous stream. At the opposite end of the table is a container **7** which collects the ceramics at the end of the process as they roll off the support plate **8**. The surface of the abrasive belt **4** is inclined so that the highest point is at the end nearest the chute **6**, as illustrated in Figure 6.17b. This permits the disc to roll under gravity along the surface of the abrasive belt **4** and eventually drop into the container **7**.

Figure 6.17: Application of Electrodes to Electroacoustic Transducer Elements

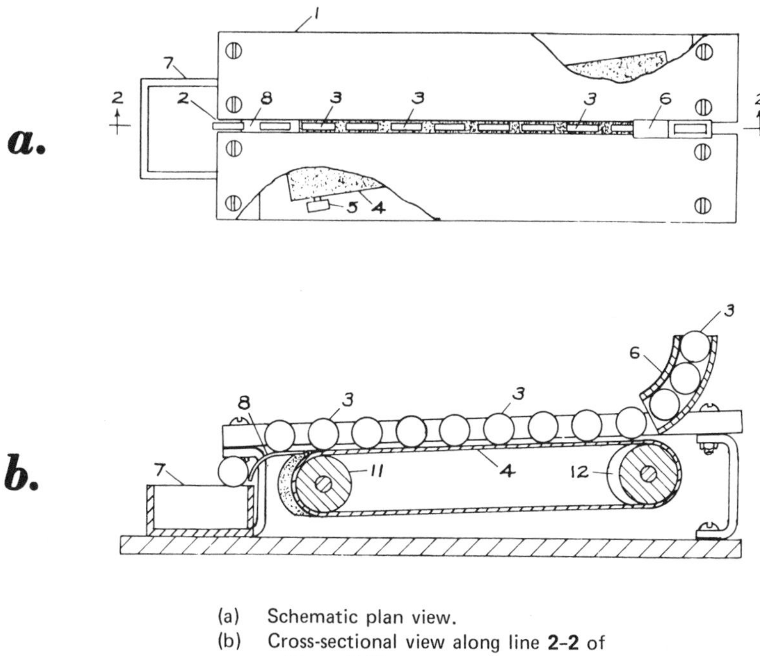

(a) Schematic plan view.
(b) Cross-sectional view along line **2-2** of Figure 6.17a.

Source: U.S. Patent 4,123,567

Transition Metal Oxides

D.L. Lewis, L.A. Schenke, M.R. Suchanski and C.R. Franks; U.S. Patent 4,125,449; November 14, 1978; assigned to Diamond Shamrock Corporation have found that an improved electrode for use in an electrolytic cell can be made of a valve metal substrate selected from the group of aluminum, molybdenum, niobium, tantalum, titanium, tungsten, zirconium, and alloys thereof; on the surface of the base substrate a semiconductive intermediate coating of tin and antimony compounds applied and converted to their respective oxides; and on the surface of the semiconductive intermediate coating a topcoating of oxides of transition metals selected from the group of chromium, manganese, iron, cobalt, nickel, molybdenum or tungsten and converted to their respective oxide forms.

The semiconductive intermediate coating of tin and antimony oxides is a tin dioxide coating that has been modified by adding portions of a suitable inorganic material, commonly referred to as a dopant. The dopant of the process is an antimony compound such as $SbCl_3$ which forms an oxide when baked in an oxidizing atmosphere. Although the exact form of the antimony in the coating is not certain, it is assumed to be present as Sb_2O_3 for purposes of weight calculations. The compositions are mixtures of tin dioxide and a minor amount of antimony trioxide, the latter being present in an amount of between 0.1 and 30 wt %, calculated on the basis of total weight percent of SnO_2 and Sb_2O_3. The preferred amount of the antimony trioxide is between 3 and 15 wt %.

There are a number of methods for applying the semiconductive intermediate coating of tin and antimony oxides on the surface of the valve metal substrate. Typically such coatings may be formed by first physically and/or chemically cleaning the substrate, such as by degreasing and etching the surface in a suitable acid (such as oxalic or hydrochloric acid) or by sandblasting; then applying a solution of appropriate thermally decomposable compounds; drying; and heating in an oxidizing atmosphere.

The compounds that may be employed include any inorganic or organic salt or ester of tin and the antimony dopant which are thermally decomposable to their respective oxide forms, including their alkoxides, alkoxy halides, amines, and chlorides. Typical salts include: antimony pentachloride, antimony trichloride, dibutyl tin dichloride, stannic chloride and tin tetraethoxide. Suitable solvents include: amyl alcohol, benzene, butyl alcohol, ethyl alcohol, pentyl alcohol, propyl alcohol, toluene and other organic solvents as well as some inorganic solvents such as water.

The solution of thermally decomposable compounds, containing salts of tin and antimony in the desired proportion, may be applied to the cleaned surface of the valve metal substrate by brushing, dipping, rolling, spraying, or other suitable mechanical or chemical methods. The coating is then dried by heating at about 100° to 200°C to evaporate the solvent. This coating is then baked at a higher temperature such as 250° to 800°C in an oxidizing atmosphere to convert the tin and antimony compounds to their respective oxides. This procedure is repeated as many times as necessary to achieve a desired coating thickness or weight appropriate for the particular electrode to be manufactured.

The topcoating of the electrode, of a transition metal oxide, selected from the group of chromium, manganese, iron, cobalt, nickel, molybdenum or tungsten, can be applied by several methods such as dipping, electroplating, spraying or other suitable methods. The topcoating can be layered in the same fashion as the intermediate coating to build up a thickness or weight per unit area as desired for the particular electrode.

Major uses of this type of electrode are expected to be in the following: electrodeposition of metals from aqueous solutions of metal salts, such as electrowinning of antimony, cadmium, chromium, cobalt, copper, gallium, indium, manganese, nickel, thallium, tin or zinc; production of hypochlorite; and in chlor-alkali cells for the production of chlorine and caustic. Other possible uses include: cathodic protection of marine equipment, electrochemical generation of electrical power, electrolysis of water and other aqueous solutions, electrolytic cleaning, electrolytic production of metal powders, electroorganic synthesis, and electroplating. Additional specific uses might be for the production of chlorine or hypochlorite.

Cathode Heaters

P.D. Williams; U.S. Patent 4,126,489; November 21, 1978; assigned to Varian Associates, Inc. describe a method of making an improved cathode heater for use in an electron tube comprising the steps of oxidizing the surface of a tungsten or molybdenum wire and reducing the oxide to its metallic state. By this method the surface of the heater wire is roughened and darkened thereby improving its heat radiating character. The heater may be further roughened and darkened by coating the oxidized wire with a soluble salt of a refractory metal. Upon subsequent heating the refractory salt is also reduced.

In more detail, the surface of a tungsten or molybdenum wire is first oxidized by heating the wire in an air atmosphere for 5 min at 725°C. The wire is then coated with alumina by cataphoresis or spraying techniques. The coated wire is next fired in a wet hydrogen furnace for 5 min at 1650°C. This heat treatment causes the alumina to sinter onto the heater at the same time that the tungsten oxide reduces to its metallic form. The resulting alumina-coated wire has a slight gray color whereas in the prior art the surface of the tungsten under the alumina has been bright. One must thus conclude that the foregoing steps increase the spectral emissivity of the tungsten, and thereby improve the transfer of heat from the heater wire to its electrically insulating coating.

It is possible now to further improve heater radiation by darkening the alumina coating itself. This is done by dipping the wire in an ammonium tungstate solution at room temperature. If desired, a spraying technique could be alternatively used. The ammonium tungstate solution may be purchased or prepared by adding concentrated ammonium hydroxide to a 10% suspension of tungsten trioxide in distilled water. If capped to prevent loss of ammonia, the solution will not crystallize or precipitate.

After dipping, the heater wire is shaken slightly to remove drops, and dried in an upright position by inserting the heater legs in holes drilled in a plastic block. The heater is dried uniformly without the presence of drops under a heat lamp or at 110°C in an air oven. Following this step the heated wire is placed in a molybdenum boat and fired for 10 min in a wet hydrogen furnace at 1000°C. Upon subsequent cooling the heater is ready for assembly into tube assemblies or cathode subassemblies.

Coating of Co_3O_4

D.L. Caldwell and M.J. Hazelrigg, Jr.; U.S. Patent 4,142,005; February 27, 1979; assigned to The Dow Chemical Company have found that highly efficient, insoluble electrodes are prepared by depositing on a suitable electroconductive substrate a coating of a single-metal spinel, Co_3O_4 formed by thermal decomposition of inorganic cobalt compounds.

Preferably, the electroconductive substrate is one of the film-forming metals which are found to form a thin protective oxide layer when subjected directly and anodically to the oxidizing environment of an electrolytic cell. Electroconductive substrates which are not film-forming metals are also operable, but generally are not preferred due to the possibility of chemical attack of the substrate if it contacts the electrolyte or corrosive substances.

Preferably, the electroconductive substrate is one of the film-forming metals selected from the group consisting of titanium, tantalum, tungsten, zirconium, molybdenum, niobium, hafnium, and vanadium. Most preferably the electroconductive substrate is titanium, tantalum, or tungsten. Titanium is especially preferred.

Furthermore, it has been found to be advantageous to codeposit a modifier oxide along with the single-metal spinel. The modifier oxides may be, preferably, one or more of the oxides of, e.g., titanium, tantalum, tungsten, zirconium, vanadium, lead, niobium, cerium, or molybdenum.

The amount of modifier oxide metal or metals may be in the range of zero to about 50 mol %, most preferably 5 to 20 mol % of the total metal of the coating deposited on the electroconductive substrate. Percentages, as expressed, represent mol percent of metal, as metal, in the total metal content of the coating. The modifier oxide is conveniently prepared along with the Co_3O_4 from thermally decomposable metal compounds, which may be inorganic metal compounds or organic metal compounds.

The Co_3O_4 coatings of the process (with or without the modifier oxides) are conveniently prepared by repeated applications of inorganic cobalt salt compounds, each application (or layer) being thermally decomposed to yield the metal oxide. The coating step is repeated as necessary until the desired thickness (preferably about 0.01 to 0.08 mm) is reached. The inorganic cobalt compound may be any inorganic cobalt compound which, when thermally decomposed, gives the spinel structure, Co_3O_4. For example, the inorganic cobalt compound employed as the precursor of the Co_3O_4 may be cobalt carbonate, cobalt chlorate, cobalt chloride, cobalt fluoride, cobalt hydroxide, cobalt nitrate or mixtures of two or more of these compounds.

Preferably the cobalt compound is at least one compound selected from the group consisting of cobalt carbonate, cobalt chloride, cobalt hydroxide and cobalt nitrate. Most preferably, cobalt nitrate is employed.

The application of inorganic cobalt compounds to the substrate may be done by any of the methods commonly known to practitioners of the relevant arts, such as brushing, spraying, dipping, pouring, or painting.

A preferred method of preparing the single-metal spinel coatings is as follows.

(a) Prepare the substrate by chemically or abrasively removing oxides and/or surface contaminants.

(b) Coat the substrate with the desired thermally-decomposable inorganic cobalt compound (e.g., one or more cobalt salts of inorganic acids).

(c) Heat the coated substrate to a temperature high enough, and for a time sufficient, to decompose the coating and yield the Co_3O_4 coated substrate. For most compounds, e.g., the cobalt nitrate, temperatures in the range of 200° to 600°C and baking times of 2 to 60 min are generally preferred.

Example: An ASTM Grade 1 titanium sheet approximately 3" x 3" x 0.080" was immersed in HF-HNO_3 etching solution for 15 sec, washed with deionized water and dried. It was then blasted with Al_2O_3 grit to a uniform rough surface and dipped in coating solution prepared by dissolving reagent grade $Co(NO_3)_2 \cdot 6H_2O$ in sufficient technical grade acetone to give a cobalt concentration of 2.5 mols/ℓ of solution. After dipping, the anode was brushed, baked in a 400°C oven about 10 min, removed, and cooled in air about 10 min. A second coat was then brushed on, baked at 400°C for 10 min, removed, and cooled in air 10 min. Twenty-one additional coats were applied in a similar manner. A twenty-fourth coat was applied and baked 60 min at 400°C. The anode was placed in a test cell and operated continuously. Initial voltage was 2.942 V; voltage after 275 days continuous operation was 2.934 V.

RESISTORS

Zero TCR Bi-Film Resistor

W.K. Jones; U.S. Patent 4,104,607; August 1, 1978; assigned to U.S. Secretary of the Navy describes a thin-film, low TCR (temperature coefficient of resistance), bi-film resistor system in which the materials used for the bi-film resistors are commercially available resistor materials having nominal properties which, when combined, can provide a system having a TCR no greater than ±50 ppm/°C.

Figure 6.18: Plan View of Bi-Film System

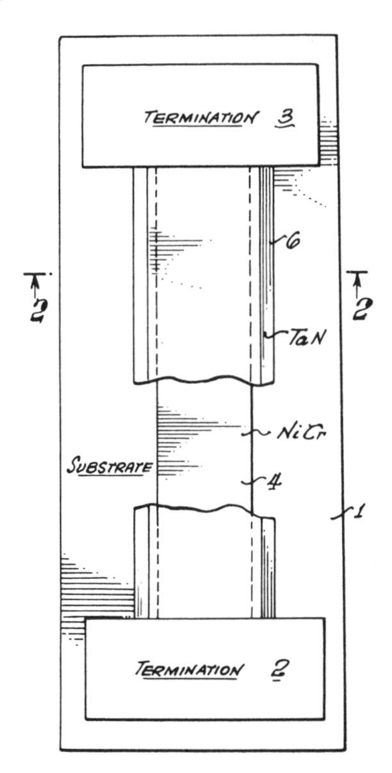

Source: U.S. Patent 4,104,607

Referring to Figure 6.18, this system includes a conventional substrate member **1** formed of a ceramic material or the like, such as alumina. Mounted on the substrate are end terminal means **2** and **3** in the form of thin conductive films deposited in well-known manners to couple the resistor circuit to external components which may be components of a microcircuit. The parallel resistor circuit itself is, as shown, a bi-film arrangement including, preferably, a NiCr film **4** adhered directly and intimately to the substrate and a tantalum-based film **6** overlying film **4** in intimate electrical contact with it. The procedures for applying the bi-film members to the substrate are well-known and should need no special explanation.

The resistor materials used in the circuit preferably are TaN and NiCr although materials, such as TaAlN also can be used.

Physically considered, the two resistor systems are disposed on the substrate with the TaN film completely covering the NiCr so as to provide a protective passivation layer for the NiCr. For the purpose of zero TCR, the TaN layer or film, which can be called R_1, obviously must be wider than the NiCr or R_2 film. For example, in a zero TCR illustration, the relationship is one in which the width of R_2 equals 0.6 that of R_1. The electrical as well as physical lengths, however, of R_1 and R_2 are equal, both terminating at spaced end terminals **2** and **3**. A further constraint of particular importance is that the unit resistance value of the material used for R_1 and R_2 must be closely matched or, in other words, substantially equal.

For these purposes this resistance value is expressed by the conventionally used term of ohms/square ($\Omega/\square$) which, as is known, is the ohmic value for a unit square of the material or a square area having a unit length and width. In this regard, TaN is a trimmable resistor which easily can be trimmed by anodization or by thermal methods to produce the desired electrical properties. Also, of considerable significance is the fact that the trimming of the TaN inherently passivates it by producing a Ta_2O_5 layer. Consequently, the use of a trimmed TaN resistor over the reactive NiCr film assures a passivation layer for the NiCr. The use of the TaN layer provides a trimmable resistor which protects the underlying NiCr and also achieves the zero TCR capability.

Two-Layer Contact Plating

B. Seebacher; U.S. Patent 4,131,692; December 26, 1978; assigned to Siemens AG, Germany describes a ceramic electric resistor having a semiconductor body and being contacted with two layers formed of palladium or in association with palladium chloride and nickel or nickel-phosphorus or nickel-boron alloys.

The process relates to a ceramic electric resistor with a positive temperature coefficient of resistance which is referred to herein as a cold conductor. The cold conductor body is stoved in an oxidizing atmosphere and consists of a material which is of a perovskite structure and is semiconducting (n-conducting) as a result of doping with alien-lattice ions. This body is provided on its surface with contact platings which are free of blocking layers and which consist of two layers of different materials to which an outer contact plating is in each case soldered.

Ferroelectric materials which possess a perovskite structure are, e.g., the titanates and/or stannates and/or zirconates of the alkaline earth metals with and without lead additives. Antimony, niobium, bismuth or rare earths are used for example as doping substances for the production of the semiconductor.

The advantages of this process are that no sensitization of the surface is effected, and that at the same time the deposition speed of the second layer is substantially increased, whereby the reproducibility of the blocking-layer-free contacting is improved. The usual processes suitable for the application of metal preparations, such as painting, rolling, spraying or silk-screen printing can be used to apply the activation substance ($PdCl_2$), and it is possible to form desired insulation zones. Preferably a 0.01 to 0.2% $PdCl_2$ solution is used which is stoved in at temperatures of 300° and 550°C in 10 to 60 min.

This produces a layer thickness of the first layer of approximately 0.2 to 50 nm in dependence upon the concentration used. In dependence upon the stoving-in temperature, a layer of palladium alone or of palladium in association with non-decomposed palladium chloride (decomposition temperature of $PdCl_2$ approximately 500°C) will form. At stoving-in temperatures less than 300°C the palladium is not held sufficiently firmly on the surface of the cold conductor so that it passes into the nickel bath which it can cause to decompose. At higher temperatures than 550°C, the hot resistance can be reduced, i.e., the resistance ratio at the leap temperature of the cold conductor is undesirably reduced.

The quoted concentration values of the palladium chloride bath and the quoted stoving temperatures proved favorable in the case of cold conductors consisting of barium titanate and barium-lead-titanate. These values could also be used, however, for cold conductors consisting of other materials.

Nickel baths known from the metallization of synthetics can be used for the production of the second layer, and in dependence upon the composition of the bath either nickel-phosphorus or nickel-boron alloys will be obtained.

In an exemplary embodiment, cold conductors consisting of barium titanate and barium-lead-titanate were coated with 0.01 to 0.2% $PdCl_2$ solutions and stoved at temperatures of 300° to 550°C. The subsequent nickel-plating was carried out in a bath of similar composition to that described in the U.S. Patent 3,586,534. The bath contained the following: 3 wt % $NiCl_2 \cdot 6H_2O$; 1 wt % $NaH_2PO_2 \cdot H_2O$; 3 to 7 wt % $Na_3C_6H_5O_7 \cdot 2H_2O$ and 89 to 93 wt % water.

Bath temperatures of between 65 and 95°C were used. The deposition times amounted to between 1 and 60 min. The deposition time depended upon the surface of the cold conductor, the palladium concentration, the bath temperature and the citrate content of the bath. The lower the bath temperature, the more resistant is the bath but the longer are the periods of dwell in the bath.

Subsequently, the samples were well flushed and tempered at temperatures of between 200° and 300°C for 2 to 15 hr. Tempering carried out at higher temperatures produced layers which were difficult to tinplate.

Figure 6.19 illustrates a cold conductor of the type described above. Arranged on the ceramic body **1** are the contact platings **2** which consist of a first layer **3** of palladium and palladium chloride and of a second layer **4** of a nickel-phosphorus or nickel-boron alloy. The outer contact elements **5** are soldered to the contact plating **2** by means of a solder **6**. For clarity the proportions have been exaggerated in Figure 6.19.

Figure 6.19: Ceramic Electric Resistor

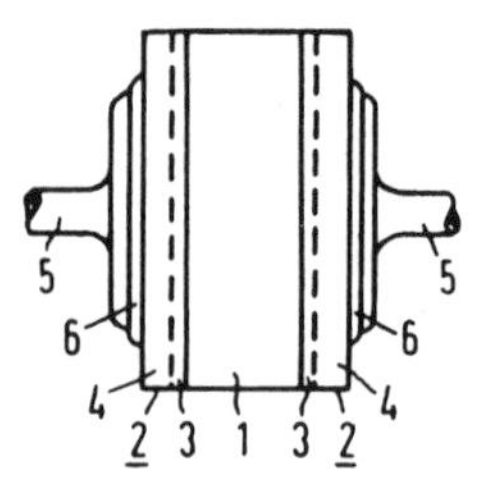

Source: U.S. Patent 4,131,692

Addition of Silica or MnO_2

J.P. Maher and T.W. Johnson; U.S. Patent 4,104,421; August 1, 1978; assigned to Sprague Electric Company describe a process whereby a resistor is formed on an insulating substrate having metal film terminations. A resistor layer of the conventional cermet type contains glass and conductive material that may be metal or metal oxide or mixtures thereof. The resistor glass is preferably a low temperature glass containing an oxide selected from Bi_2O_3, CdO, PbO or mixtures thereof.

These oxides are among known ingredients of glasses that depress the melting temperature from that of glasses containing only the so-called glass formers such as silica and boron oxide. The glass containing resistor layer of this process has distal portions lying in contact with the metal film terminations. These terminations contain particles of silica or of manganese dioxide to reduce the residual stresses normally remaining after firing. The terminations preferably contain more than 20 wt % of the silica or more than 17% of the MnO_2 to improve the resistor-to-termination junction.

This improvement is realized whether the resistor is formed on a glazed or an unglazed substrate, and is independent of the metals used in the terminations or cermet resistor layer. It is further independent of the glass that is employed in the resistor layer, leaving the choice of the cermet resistor layer composition entirely open for determining and adjusting the resistor performance as may be desired. For the first time it becomes practical to manufacture a wide variety of cermet resistors having submicron film terminations which results in substantial cost savings.

The resistor is made by forming on a substrate submicron metal film terminations containing particles selected from silica and manganese dioxide, and depositing thereover a glass-containing resistor layer. Both the film terminations and resistor layer are preferably formed by sequentially screen printing noble metal resinate-containing pastes and firing, the termination paste containing the silica or manganese dioxide particles, and the resistor paste containing a low temperature glass frit.

Resistance Thermometer with SiO Insulation Layer

The most commonly used resistance thermometer includes a sensor constructed of a wire coil, the resistance of which changes in a predetermined known manner as a function of temperature. These wire coils have been made of nickel, platinum, tungsten, Nichrome (an alloy of nickel and chromium) and other materials having a suitably high temperature coefficient of resistance (TCR). To achieve the requisite high degree of accuracy with such a wire sensor, the material must have high electrical resistance which, in turn, necessitates the use of a relatively long length of wire of small diameter.

The reason for this is that a high resistance sensor has a high change of resistance for a change of temperature, and, therefore, is more easily calibrated than a low resistance sensor would be. In addition, a wire coil can only be loosely supported on an insulating substrate and must be annealed in order to obtain a predictable and repeatable resistance. All of these requirements result in the wire coil being relatively fragile and susceptible to breakage from vibrations, shock, and as well, contamination from external materials.

On the other hand, thin-film temperature sensors can be constructed having very high resistance and at the same time be exceptionally rugged and not readily damaged by normally occurring external circumstances. In addition, thin-film temperature sensors may be deposited on very small substrates providing an improved advantage with respect to size, weight and response time over coil sensors. Still further, shocks and vibrations do not affect deposited film resistors since the substrate is relatively rigid and the resistor may be coated, making it substantially immune to contamination from the outside.

T.S. Kirsch; U.S. Patent 4,139,833; February 13, 1979; assigned to Gould Inc. describes a process whereby a polished ceramic substrate is provided with an insulation layer of silicon monoxide (SiO). A nickel metal thin-film is then laid down onto the insulation layer in a helical or serpentine pattern taking up a desirably small area, but at the same time giving a high electrical resistance. Finally, a cover or protective layer of silicon monoxide is then deposited over the resistor serving to protect it from the possibility of outside contamination.

With reference to Figures 6.20a and 6.20b, the temperature sensor is enumerated generally as at **10**, and is seen to include a base or substrate **11** on a surface on which there is arranged a serpentine resistor **12**, at the ends of which connector pads **13** and **14** are interconnected with external apparatus, via leads **15** and **16**. More particularly, and as best shown in Figure 6.20b, the substrate **11** has one surface formed into a flat surface **17** onto which an insulation layer **18** is deposited with the resistor **12** deposited thereover. Finally, the connector pads **13** and **14** and leads **15** and **16** are laid down and the entire conductive film portions covered with an insulating and protective layer **18** (e.g., SiO).

As to detailed aspects, the substrate **11** is preferably constructed of high density alumina (Al_2O_3) and in a practical embodiment was finished to 0.140 x 0.140 x 0.0015", although other geometries may be used such as circular (Figure 6.20a). A major surface is ground and polished to form the flat, smooth surface **17** and thoroughly cleaned. The substrate **11** is then loaded onto a suitable deposition fixture which, in turn, is placed on a rotating substrate carrier and entered into a vacuum evaporation system. The vacuum system includes four different deposition stations for depositing, respectively, insulation layer **18**, resistor **12**, cover insulating layer **19** and connector pads **13** and **14**.

Figure 6.20: Thin-Film Resistor Temperature Sensor

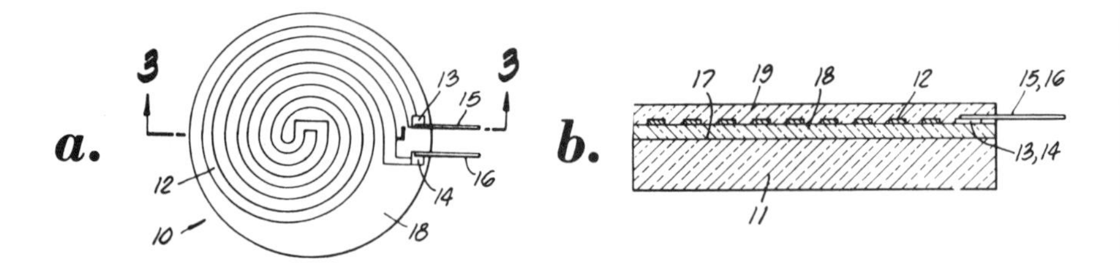

(a) Plan view.
(b) Elevational sectional view taken along line **3-3** of Figure 6.20a.

Source: U.S. Patent 4,139,833

In the process, the first step is the vapor deposition of silicon monoxide (SiO) onto the flat, polished substrate surface to form the insulation precoat **18**. Then, the precoated substrate is moved to the next station where metallic nickel is vapor-deposited via a suitable mask to provide a spiral-shaped or serpentine resistor **12** on the insulating layer **18**. Next, the partially completed unit is moved to a further station where the connector pads **13** and **14** of gold or nickel alloy are deposited.

At this stage the partially completed sensors are removed from the vacuum system and vacuum annealed at 805°F to effect both stabilization of grain structure and resistance value. In a practical construction the final annealed resistance of **12** was 1,000 ohms at 70°F.

Gold leads **15** and **16** are then secured to the connector pads **13** and **14** (e.g., by resistance welding), after which the assembly is once more placed in the vacuum deposition chamber where it is overcoated with silicon monoxide to form the protective cover **19**.

As a final matter, the completed temperature sensor is removed from the vacuum deposition chamber, cemented to a metal end cap, after which it is subjected to 400°F for 48 hr to stabilize the resistor **12** further, and, as well, cure the cement used to secure the substrate and metal cap together.

As alternatives, the substrate may be constructed of beryllium oxide and the temperature sensitive resistor **12** of platinum.

Resistance Thermometer with Differential TCR

W. Diehl and W. Koehler; U.S. Patent 4,103,275; July 25, 1978; assigned to Deutsche Gold- und Silber-Scheideanstalt vormals Roessler, Germany describe resistance elements for resistance thermometers which have a short response time, are also producible in small dimensions without special expense and, above all, have a TCR between 0° and 100°C of at least 3.85 x 10^{-3}/degree.

The process comprises application of resistance elements consisting of an insulating member as support and a thin platinum layer as resistance material wherein as the support for the platinum layer there must be used a material which has a greater thermal coefficient of expansion between 0° and 1000°C than platinum.

Especially approved as support is magnesium oxide whose thermal coefficient of expansion is 12 x 10^{-6}/degree while platinum has a corresponding value of 9.3 x 10^{-6}/degree. Besides magnesium oxide there can be used as supports, for example, various heat resistant nickel alloys, such as Inconel, with an insulating coating. As thin insulating coating there can be used, for example, magnesium oxide, aluminum oxide or a silicate glass, e.g., a soda-lime silicate glass.

By sputtering (cathode sputtering) or vacuum vaporization there is placed a platinum layer having a thickness of 1 to 10 μ on the insulating support. For the production of meander designs the platinum film is then coated, for example, with a photosensitive lacquer and the desired structure produced on this by partial covering, exposure to light and development. The desired conductor path then can be produced by ionic etching or other processes. In this way, there are producible conductor paths up to a width of about 2.5 μ. The adjustment of these conductor paths to a fixed R_0 value (R_0 = 100±0.1 Ω), is known from microelectronics and, preferably, takes place by means of a laser beam.

There are produced especially high temperature coefficients of the electrical resistance if the thin platinum layer is produced by sputtering in an oxygen-containing atmosphere. There has been found particularly valuable an argon-oxygen mixture in which the oxygen content is preferably 5 to 60 volume %. However, there are also usable other noble gas-oxygen mixtures. Among other suitable noble gases are helium and neon. The layer applied by sputtering or vaporization must be subsequently tempered at temperatures above 800°C, preferably in the range of 1000° to 1200°C, to reach a maximum grain growth which again is a prerequisite for a high TCR.

Example 1: A commercial sputtering apparatus with an argon-oxygen mixture, containing 17% oxygen under an operating pressure of 6 x 10^{-3}, was used to sputter onto flat magnesium oxide plates of 20 mm x 20 mm a platinum layer of 4.2 μ. The high frequency output was 1,100 W, the applied voltage 2,600 V and the backlash voltage (bias) 100 V. The platinum layer was subsequently tempered for 3 hr at 1000°C in air; meanders were produced by photoresist technique: the platinum film is coated with a photosensitive lacquer, and the desired structure on this lacquer is produced by partially covering it with a mask, exposure to light through this mask and development. The desired conductor path in the platinum layer then is produced by ion etching (sputter etching), the parts of unremoved photosensitive lacquer preventing the platinum covered by them from being etched off. The measured temperature coefficient of the electrical resistance was (3.86±0.01) x 10^{-3}/degree.

Example 2: Using the apparatus and conditions of Example 1 there was applied by sputtering to an Inconel sheet (80 Ni, 14 Cr, 6 Fe) measuring 20 mm x 20 mm and previously coated with about 10 μ magnesium oxide, a platinum layer having a thickness of 6.3 μ in an argon-oxygen mixture containing 50 volume % of oxygen and an operating pressure of 8 x 10^{-3} torr. After the tempering (2 hr, 1050°C) and production of the meanders, there was measured a TCR of (3.89±0.01) x 10^{-3}/degree.

CAPACITORS

Base Metal Glass-Ceramic Capacitor

I. Burn; U.S. Patent 4,101,952; July 18, 1978; assigned to Sprague Electric Company describes a bme (base metal electrode) monolithic ceramic capacitor having a reduction resistant body of high dielectric constant capable of sintering and fully densifying at temperatures below 1080°C.

The monolithic ceramic capacitor of this process comprises a reduction resistant glass-ceramic body and at least one buried base-metal electrode. Generally, at least another metal electrode is in contact with the body and is in capacitive relationship with the buried electrode. This process recognizes the principle that an alkaline earth aluminoborate glass and a high firing barium titanate ceramic may be combined to form a very low firing reduction resistant glass-ceramic body that has a high dielectric constant.

The reduction resistant glass component consists of an alkaline earth borate glass. No more than trace amounts of the easily reduced oxides of cadmium, lead and bismuth should be included. Partial replacement of boron oxide by silica retards densification at sintering, lowers the dielectric constant and should be avoided. Small quantities of zinc oxide are believed to be permissible and may even be beneficial up to 3 wt %. Partial replacement in these glasses of the alkaline earth metal with alkali metals such as Li, Na and K may also have suitable properties. Alumina is not essential in these glasses but generally improves chemical durability and inhibits devitrification.

The glass content in the glass-ceramic body is restricted to no more than 15 wt %, again to provide a body having a high dielectric constant. Lower amounts tend to provide even a larger dielectric constant but also tend to raise the minimum firing temperature at which the body fully densifies. At least 5 wt % of the glass is needed to provide a characteristic sintering and full densification temperature at least as low as 1080°C.

The ceramic component is a crystalline ceramic phase, preferably but not necessarily a single phase, having a characteristic sintering and densification temperature of greater than 1200°C. It is further characterized as a barium titanate wherein at least 50 mol % of the barium may be replaced by others of the alkaline earth metals, strontium, calcium and magnesium. Also, up to 20 mol % of the titanate may be replaced by a zirconate. These formulation limitations of the ceramic phase of the glass-ceramic body are necessary to provide a dielectric constant of the glass-ceramic body greater than 1,000.

Example: A ceramic powder of formulation $Ba_{0.65}Sr_{0.35}TiO_3$ was prepared by blending in water 987.2 g of $BaCO_3$ with 397.8 g of $SrCO_3$ and 615.0 g of TiO_2, which was then dried, granulated and calcined at 1230°C. No acceptor (or donor) dopants were added to the ceramic formulation as is typically necessary in bme capacitors of the prior art. After being jet pulverized, 100 g of the powder was mixed with 7.3 g of glass powder of composition $4BaO \cdot Al_2O_3 \cdot 2B_2O_3$ prepared as described in U.S. Patent 3,902,102. The mixture was milled with 30 g of organic binder for 12 hr in a 200 cc capacity porcelain mill. The slip was cast on a glass plate after milling, using the doctor-blade technique, and when dry was cut into small squares approximately 10 x 10 x 0.5 mm. Copper paste, made by mixing 17.0 g of ethyl cellulose binder with 28.5 g of copper powder being of 99.9% purity and from 1 to 5 μ particle size was then painted on both sides of the squares, and dried before firing.

The squares were fired for 2 hr at 1050°C in a CO_2-CO mixture that produced an oxygen partial pressure of approximately 5×10^{-10} atmospheres of oxygen as indicated by a zirconia oxygen monitor extending into the hot-zone of the furnace. The capacitors of this example had a dielectric constant (K) of approximately 3,500 and an insulation resistance (IR) of 4,000 ohm-farads (ΩF) at 25°C and 350 ΩF at 125°C.

MnO_2 Solid Electrolyte Layer

A. Nishino, H. Hayakawa, A. Yoshida and J. Umeda; U.S. Patent 4,105,513; August 8, 1978; assigned to Matsushita Electric Industrial Co., Limited, Japan describe a solid electrolyte capacitor which has an anode body of a valve metal and an anodized dielectric oxide film on the surface of the anode body as in conventional solid electrolyte capacitors. The solid electrolyte capacitor of this process is characterized by comprising a manganese dioxide solid electrolyte layer which is formed on and in intimate contact with the dielectric oxide film by pyrolytic decomposition of an aqueous solution of manganese nitrate and is formed of densely accumulated manganese dioxide particles within the range of the particle size between 0.1 and 50 μm and a cathode collector layer formed by spraying of a metal powder on and in intimate contact with the manganese dioxide layer.

Metals useful as the material of the cathode collector layer are copper, tin, lead, nickel, silver and aluminum, including alloys thereof, and the cathode collector layer is preferably 5 to 60 μm in thickness.

It is preferable that the solid electrolyte layer is entirely formed of the above-described manganese dioxide particles. However, it is permissible that only a surface region of the solid electrolyte layer is formed of such particles while the remaining region is formed of more coarse manganese dioxide particles.

A method of producing the solid electrolyte capacitor according to this process comprises the following steps:

(a) Anodizing the surface of a valve metal anode body in a known manner to form a dielectric oxide film,

(b) Wetting the dielectric oxide film with an aqueous solution of manganese nitrate,

(c) Heating the wet anode body to cause pyrolytic decomposition of manganese nitrate to manganese dioxide in a heating chamber of a radiant furnace thereby to form a solid electrolyte layer on the dielectric oxide film, which heating chamber is semiclosed and in communication with the atmosphere exclusively through at least one vent formed in the wall of the heating chamber with such a total vent area that the pyrolytic decomposition proceeds under a positive pressure of a small magnitude attributable only to a natural restriction by the vent to the outflow from the heating chamber of gaseous substances generated from the solution during the pyrolysis, and

(d) Spraying a metal powder, preferably by a plasma spraying technique, onto the surface of the solid electrolyte layer to form a metallic cathode collector layer.

The sequential steps (b) and (c) may be repeated several times to attain desirable denseness and thickness of the manganese dioxide electrolyte layer. It is permissible that the sequential steps (b) and (c) are preceded by at least one cycle of conventional pyrolysis procedure wherein the anode body, which is wetted with a manganese nitrate solution, is heated in a heating chamber other than the semi-closed chamber, for example, a heating chamber of a hot-air circulating type, on condition that the formation of the solid electrolyte layer is ended by the above defined step (c).

These capacitors are superior to conventional manganese dioxide capacitors particularly in the loss, leakage current, frequency-capacitance characteristic and impedance characteristic.

Metal Termination on Wound Capacitor

W.W. Schroeder, N.C. Sears and J.C. Boni; U.S. Patent 4,115,600; September 19, 1978; assigned to Sprague Electric Company describe a method for forming a metal termination at the end of a wound capacitor section which comprises rotating the section about an axis that is essentially parallel to the axis of the section, placing a shield plate adjacent to and spaced from an end of the rotating section and spraying a molten metal onto the section end. The shield plate is offset with respect to the axis of rotation to expose each region of the section end to the sprayed metal for substantially less than the full period of a revolution of the section.

This method is particularly advantageous for depositing a metal termination to the end of a capacitor section which section has two metal sheet electrodes that are spaced by and may be supported by thermoplastic dielectric layers. Some thermoplastic dielectric materials typically employed in wound capacitors are polypropylene, polyethyleneterephthalate, polycarbonate, polystyrene, and certain fluorocarbons.

The alternate deposition and cooling that is effected in this method is capable of preventing substantial distortion of the thermoplastic material at the section end being sprayed. Such distortion and melting typically results in poor and inefficient electrical connection between the sprayed termination and the electrode and often results in a short between the adjacent electrodes. At the same time, a thick strong termination layer may be deposited by this method within a matter of seconds with a minimum of handling. The result is an improved quality termination having a lower cost of manufacturing.

This process recognizes the principle that during metal spray deposition, the more frequently the spray deposition is interrupted by a cooling period, the lower the average temperature of the object being sprayed becomes.

Also accompanying this trend is a lower value of peak temperature that is reached by the object being sprayed. The number of alternate spray deposit and cooling cycles employed to deposit a given thickness of metal coating in a given time may readily be increased by increasing the rate of rotation at no additional cost.

Figure 6.21: Formation of Metal Termination of Wound Capacitor Section

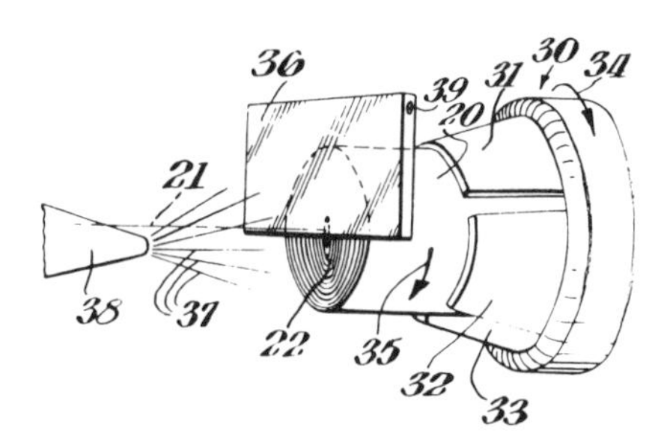

Source: U.S. Patent 4,115,600

In Figure 6.21 a wound capacitor section **20** is shown mounted and firmly held by a rotating chuck **30**. Chuck jaws **31, 32** and **33** grip the capacitor **20** such that rotation of the chuck causes the capacitor section to rotate about section axis **21** in a direction as shown by arrows **34** and **35**. A shield plate **36** is mounted adjacent to but spaced from capacitor section end **22** and is offset with respect to the axis of rotation so as to expose only a portion of capacitor section end **22** to the sprayed metal **37** being discharged from the metal spray nozzle **38**. Any of the various standard metal spray equipments will be suitable including those employing a plasma arc, flammable gas, compressed air and combinations thereof.

The abovedescribed method of holding and rotating the capacitor section is particularly suitable for sections having large diameters. For example, a capacitor section of diameter 2.0" and length of 4" was held in a chuck and rotated at 200 rpm. The spray nozzle was adjusted to produce a 1½" diameter spot at 5" distance from the capacitor section. A shield plate was employed to cover approximately half of the section end. The metal shield was effective in shielding about half of the section end at any instant of time from sprayed metal and from radiant heat emanating from the spray gun. Thus, each incremental region in the section end was alternately subject to metal spray deposition and to cooling. The rotation of the section in the ambient air enhances the cooling.

Base Metal Pigments

J.J. Hertz; U.S. Patent 4,130,854; December 19, 1978; assigned to Erie Technological Products, Inc. describes a pigment for metallizing ceramics comprising borate-treated nickel with or without additions of tin, zinc, or mixtures of tin and zinc.

Referring to Figure 6.22, the disc capacitor **1** is an exemplar of devices in which metallizations may be used. The term metallization as used herein refers to a powder of pigment sized particles which may be dispersed in an inert liquid vehicle to form a metallizing composition. The latter is useful to print desired electrode patterns on dielectric substrates, which upon firing produce conductors. It comprises a dielectric **2** of barium titanate or the like, electrodes **3, 4** fired on the dielectric and in capacity relation to each other through the dielectric and leads **5, 6** soldered or otherwise mechanically and electrically bonded to the electrodes as indicated at **7, 8**. The leads **5, 6** must withstand a pull test.

Figure 6.22: Metallization of Disc Capacitor

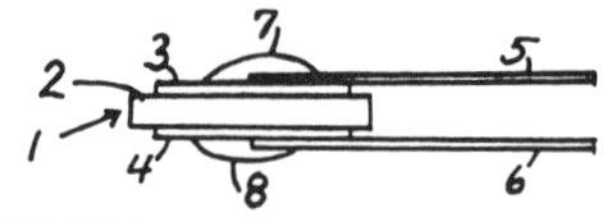

Source: U.S. Patent: 4,130,854

After the dielectric has been fired, the electrodes **3, 4** are applied as patterns of paint of pigment size particles of tin (MP about 232°C) and zinc (MP 419.4°C) dispersed or suspended in a vehicle of inert or indifferent material which does not react with the pigment particles or the ceramic.

The size of the pigment particles of tin and zinc should be small so that the particles make conductive contact with each other. There is no advantage in particles of diameter smaller than 1 μ. Particles of diameter greater than 40 μ may not make the desired electrical contact with each other.

In a tin-zinc pigment system, preferred proportions by weight are 65 to 95% tin and 35 to 5% zinc. Preferably the tin particles which are the major component of the paint range from 1 to 40 μ in diameter and the zinc particles range from 1 to 20 μ in diameter. The zinc and tin particles are ball milled together in any suitable inert or indifferent vehicle previously used for silver paint and the paint is applied by silk screening or by any of the other methods previously used for the silver paint. After the application of the electrode patterns, the paint is fired on in air at a temperature of 1000° to 1450°F (539° to 788°C). At temperatures below 1000°F, the particles of tin and zinc have poor contact with the ceramic and high losses and low capacitance are obtained.

At temperatures above 1450°F, the conductivity of the electrode is excessively reduced by oxidation of the particles. Within the 1000° to 1450°F range the tin and zinc particles are sintered together forming a new compound of good conductivity and adherence to the ceramic. Prior to the sintering, the characteristic x-ray diffraction spectral lines of tin and zinc can be observed. After the sintering, new spectral lines appear. Also, and unexpectedly, the tin and zinc particles without any frit are bonded to the ceramic as tightly as the prior art frit containing silver paint. This takes place without any adverse reaction between the pigment particles and the ceramic.

The bonding of the tin-zinc pigment to the ceramic is surprising. Tin or zinc particles alone will not bond to the ceramic. Tin-lead solders will not bond to ceramic, nor will noble metals stick to the ceramic. The combination of tin and zinc apparently forms an alloy which either wets the ceramic or has a surface reaction with the ceramic which forms a bond.

The tin-zinc electrodes are completely interchangeable with the standard silver electrodes. The variations in properties which appear are within the normal limits of variations for dielectrics with the standard silver electrodes.

The leads may be attached to the electrodes by several methods. First, the electrodes may be copper sprayed and then soldered by solder dip methods. Second, the electrodes may be solder sprayed and the leads attached either by solder reflow or by solder dip. The leads may also be attached by electrically conductive epoxy cement.

Another base metal pigment for metallization which replaces silver is nickel (MP 1452°C). As obtained, for example, from nickel carbonyl, the nickel particles have a surface oxide coating which must be removed and replaced by a coating which prevents oxidation. Oxide removal may be accomplished by treatment in acid solutions. A specific acid treatment consists of washing the powder in a water-nitric acid solution from 5 to 30% by volume nitric acid for a period of 1 to 5 min followed by a distilled water rinse followed by a dilute 1 to 2% by volume sulfuric acid wash followed by a 30 sec rinse in distilled water followed by a hot water rinse.

The powder is submerged immediately in a solution of 10 g sodium borate plus 10 g ammonium borate in 100 g of water, the solution decanted and the powder oven dried. The nickel powder has an extremely thin coating of nickel borate, possibly monomolecular, an oxidation-preventing coating. The powder may be used as the sole pigment of a metallization paint or paste or the powder may be mixed with other base metal pigments such as tin or zinc. The range of proportions is 99 to 50% nickel powder to 50 to 1% base metal.

SOLAR CELLS

Four-Layer Metal Contacts

J. Lindmayer; U.S. Patent 4,124,455; November 7, 1978 describes a solar energy cell in which a multimetal contact comprises a layer of titanium group metal adhered to the cell; a layer of a mixture of a titanium group element and a platinum group element overlies that first layer. Still a third layer is formed by the platinum group element alone. Then, a body of silver or other contact metal, which has substantially greater mass than any of the other layers, is firmly adhered to the platinum group layer, preferably by plating or electroplating.

The titanium layer serves to provide firm adherence to the silicon body. The platinum layer serves as a metal on which the silver is readily platable. The problem of adherence between the titanium and platinum was solved by interfacing the titanium and platinum layers with a layer comprising a mixture of both platinum and titanium. It was found that by providing such an interface layer, the normal problems of adherence between titanium and platinum is overcome, and that there is firm and lasting adherence between the titanium group and the platinum group metals. With such firm adherence, there is no difficulty in securing the platinum group metal indirectly to the body of the solar cell, and with the platinum group metal in position, there is no problem in applying a relatively large mass of silver directly on the platinum group layer by an economical means.

In the preferred embodiment of the process, a titanium source, i.e., titanium wire, is placed in one of the filaments and palladium wire within the other filament. A vacuum is drawn, and deposition takes place at a pressure of about 10^{-5} torr. When the desired pressure has been reached, the power is turned on to the filament containing the titanium and evaporation of the titanium is commenced. The titanium will be deposited on a plurality of solar cells that have been placed within the vacuum chamber.

After a layer of titanium of 500 to 600 Å in thickness had been deposited, the filaments containing the palladium were also activated and deposition continued with both sets of filaments activated so that simultaneously titanium and palladium in substantially equal atomic amounts were deposited. When a further layer of 700 to 800 Å of the mixture of titanium and palladium had been deposited, the set of filaments containing the titanium were shut off and an additional layer of palladium alone was deposited in a thickness of about 500 Å. Then the entire vacuum system was deactivated. The solar cells were removed from the Veeco high vacuum deposition system and, where photolithography had been used, a fine pattern was defined having an exposed palladium layer.

An electroplating bath was prepared utilizing a potassium silver cyanide formulation (Silver Sol-U-Salt by Sel-Rex). The material was stated to contain 54% silver. An electroplating bath was prepared using the Sel-Rex potassium silver cyanide composition and the solar cells with their metal grids consisting of three layers, the outer layer being the palladium, were placed in the bath.

It has been found that on exposure to air the palladium had acquired a light coating of palladium oxide. Consequently, the cells were immersed in the bath for a period of about 10 sec, during which time the palladium oxide decomposed. Then the current was turned on and silver was plated from the Sel-Rex solution to a depth of 6 to 10 μ. Such plating took place only at the palladium-surfaced grid and not at the remainder of the front surface of the solar cells. Since the depth of the silver was 6 to 10 μ, the mass of the silver was considerable compared to the layers of the other metals that form the electrically conductive grid.

Such method of depositing silver has definite economic advantages over other methods of deposition, e.g., vapor deposition. Further, the body of silver that formed the main portion of the contacts was firmly adhered to the body of semiconductor material, since the titanium adhered well to the silicon, the palladium was firmly affixed to the titanium through the intermediation of an interface layer of titanium and palladium, and the palladium formed a good substrate on which the silver could be plated. The resulting contact was highly corrosion-resistant.

CdS-Cu_2S Photovoltaic Cell

J.F. Jordan and C. Lampkin; U.S. Patent 4,104,420; August 1, 1978; assigned to Photon Power, Inc. describe a method of fabricating a photovoltaic cell, including applying a layer of uniformly oriented cadmium sulfide microcrystals on an electrically conductive surface of a substrate, forming a layer of Cu_2S on the cadmium sulfide so applied as to form a heterojunction, applying a layer of $CuSO_4$ on the Cu_2S layer and mutually isolated electrodes of Cu and Zn on the $CuSO_4$ which is heated to form a Cu-CuO_2 rectifying junction, while the Zn diffuses down through the layers underlying and thereby renders them conductive. The CdS may be or may not be impregnated with zinc. The Cu-$CuSO_4$ junction provides oxygen for the rectifying junction, which reduces or largely prevents shorting of the Cu_2S-CdS layer, when the Cu_2S layer or the CdS layer is defective due to the presence of holes in the layers. The CdS and the Cu_2S layers are deposited by successively spraying respectively a cadmium salt-thiourea solution and a copper salt N,N-dimethyl thiourea solution while the glass is floating in molten metal baths of suitable temperatures, allowing CdS microcrystals and a heterojunction with Cu_2S to develop only on contact of each complex with a suitable heated surface.

Figure 6.23: CdS-Cu_2S Photovoltaic Cell

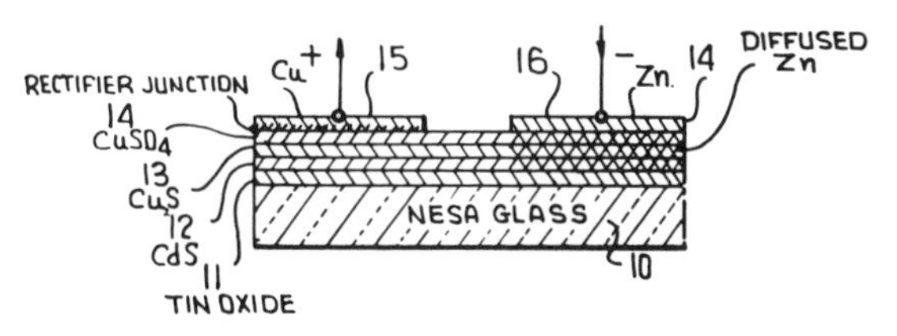

Source: U.S. Patent 4,104,420

In Figure 6.23, **10** is a plate of Nesa glass, i.e., nonconductive glass having on one of its surfaces a thin layer **11** of tin oxide, which is conductive. Overlying the layer **11** of tin oxide is a layer **12** of polycrystalline CdS. Overlying the layer **12** of CdS is a further layer **13** of Cu_2S. Overlying the CdS is a thin layer **14** of $CuSO_4$, on which are deposited positive and negative electrodes **15, 16,** of Cu and Zn respectively.

The CdS and Cu_2S layers, at their interface, form a voltage generating heterojunction, the Cu_2S being positive and CdS being negative, when the CdS is illuminated by light of the proper wavelength. Specifically, the cell is responsive to sunlight.

The voltage generated at the heterojunction between the microcrystalline CdS and the Cu_2S is communicated via the $CuSO_4$ layer to the Cu electrode **15.** A reaction occurs between the $CuSO_4$ and the Cu electrode when the latter is heated to 500°F for about 12 min, forming a rectifying like junction R of Cu-CuO_2, which is conductive of current out of the cell, so that there is no interference with operation of the cell.

The glass plate which forms a substrate in the process must be hot, about 700°F, while being sprayed, and the spraying must be sufficiently slow to permit uniform growth rates for the CdS microcrystals of the layer. It has been found that any nonuniformities of temperature of the glass plate, producing temperature gradients along the surface of the plate, result in imperfect crystal growth, and therefore, a defective cell. To avoid this contingency, the glass plate **10** is sprayed while the plate **10** is floating in a bath of molten metal, specifically tin. The glass plate is not wet by the tin, so that when the glass plate is removed from the molten tin bath, after it is sprayed, the underside of the plate is clean, or easily cleaned. The spray is provided via an oscillating nozzle which repeatedly retraces a planar path designed uniformly to cover the plate with spray.

The spray is a true water solution of cadmium chloride and thiourea. As the fine droplets of the spray contact the hot surface of the glass plate **10**, the water is heated to vaporization and the dissolved material is deposited on the plates forming CdS, plus volatile materials, and the CdS, if it has nucleating areas available, grows as small crystals. The nucleating areas are provided by the tin oxide, and if the spray is sufficiently uniform and sufficiently slow, and if the temperature of the glass surface is adequately high and uniform, crystal growth is uniform and all the crystals have nearly the same spatial inclinations, so that a uniform layer of nearly identical microcrystals is produced.

It has been found that irradiating the crystals, as they grow, with high intensity UV light assists in the crystal growing process and produces a higher yield of near perfect layers than is otherwise the case.

The method of forming a CdS layer and a Cu_2S layer is summarized as follows. A plate of Nesa glass is floated in a tin bath heated to 800°F, to provide 700°F at the upper surface of the glass plate. $CaCl_2 \cdot 2\frac{1}{2}H_2O$ of 0.01 M solution is employed, and an excess of thiourea, in deionized water, for the reaction desired. The desired thickness of the CdS microcrystal layer is about 1 to 2 μ.

The Cu_2S is developed by floating the glass plate previously coated with polycrystalline CdS, in a bath of molten metal at about 200° to 300°F and spraying with a water solution of 0.0018 M copper acetate and 0.001 M of N,N-dimethyl thiourea, to a thickness of about 1000 Å. The $CuSO_4$ is sprayed over the Cu_2S layer to a thickness of about 250 to 1000 Å, and the Cu and Zn are deposited as interdigitated electrodes. The entire cell is then heated to 500°F for about 12 min causing a Cu-CuO_2 layer to form at the copper, and causing the Zn to diffuse.

The copper and zinc electrodes may be radiation heated via separate masks to provide optimum heating in each case for the chemical and/or physical effects desired.

CdSe-SnSe Photovoltaic Cell

M.P. Selders; U.S. Patent 4,101,341; July 18, 1978; assigned to Battelle Development Corporation describes a photovoltaic cell which can be fabricated with low cost deposition techniques because of the good crystallographic match between semiconductive material layers. The good match also allows thin polycrystalline layers of less than about 3 μ to be used. The cell contains a photo-sensitive semiconductor junction, a conductive substrate for supporting the semiconductor layers and acting as one electrode, and a second electrode on the semiconductor junction face opposite the substrate. The second electrode is either transparent or is a grating or grid pattern layer which allows light to strike the upper semiconductive layer of the junction.

The improved semiconductor junction comprises regions of n-type cadmium selenide and p-type tin selenide. The layers are preferably polycrystalline, but at least one may be a single crystal. The substrate and the second electrode may advantageously be silver, indium, cadmium, zinc or gold.

Referring to Figure 6.24, a photovoltaic cell according to this process is shown and consists of polycrystalline semiconductor layers of n-type cadmium selenide

1 and p-type tin selenide **2** on a conductive substrate **3**. The second electrode **5**, in a grating pattern, is disposed on the tin selenide layer. Incident radiation is represented by arrows **4** striking the upper surface of the tin selenide layer and the second electrode. Leads **6** are connected to the substrate and the second electrode for tapping electricity from the cell.

To produce the photovoltaic cell, the two selenide layers are advantageously vapor-deposited in the desired sequence on the heated substrate at a pressure of about 10^{-5} torr. Preferably, the first deposited layer is about 3 μm in thickness, and the second layer is about 0.3 to 1 μm in thickness. The electrode on the side which is towards the light is thereafter also vapor-deposited.

In another embodiment of the production method, a thin cadmium selenide layer is vapor-deposited on the substrate and then partially converted into a tin selenide layer by immersion in a tin ion-containing solution. Optionally, the tin selenide may be first deposited, followed by cadmium for tin ion-exchange at the exposed surface.

In a third production method, the cadmium selenide and the tin selenide layers of the cell can also be produced in either order by spraying and thermally decomposing cadmium and selenium-containing or tin and selenium-containing solutions, respectively, on a substrate which is heated to 100° to 300°C.

Carrier concentrations (n-type) necessary for successful photoelectric behavior were obtained in CdSe by nonstoichiometry or by doping with, for example, indium or cadmium. The p-type carrier concentrations were obtained in SnSe by nonstoichiometry, but could be obtained by known dopants.

Example 1: An n-type cadmium selenide layer about 3 μm in thickness may be vapor deposited at a pressure of about 10^{-5} torr on a metallized glass or plastic substrate having a silver coating which is about 1 μm in thickness. During the deposition process, the substrate temperature is set to between 150° to 250°C, preferably about 200°C, in order to achieve the best possible crystalline structure of the layer and resulting good photoelectric properties (high mobility and operating life of the electrical charge carriers produced by sunlight). A carrier concentration of about 10^{17} cm^{-3} is obtained by nonstoichiometry.

Subsequently, at a substrate temperature of between 250° to 350°C, preferably about 300°C, and also under a pressure of about 10^{-5} torr, a p-type tin selenide layer of about 0.3 to 1.0 μm in thickness may be vapor deposited on the cadmium selenide layer. A carrier concentration of about 10^{17} cm^{-3} is again obtained by nonstoichiometry.

A front electrode **5** (see Figure 6.24) may also be vapor deposited on the tin selenide layer under a high vacuum. The electrode **5** comprises gold in the form of a grating with narrow bars, so that the light can reach the upper semiconductor layer. However, it is important, because of the small degree of contact by the electrode, that the electrical resistance in the material layer under the grating is sufficiently low so that there is not a substantial voltage drop in the cell, which would thus reduce the level of efficiency. Instead of a vapor-deposited metal grating, it is also possible to apply a ready-made net of, for example, gold-plated copper, by pressing onto the tin selenide layer.

Following the deposition processing the cell is preferably annealed for several hours at 150° to 170°C in inert gas to increase carrier mobility and lifetime.

Figure 6.24: CdSe-SnSe Photovoltaic Cell

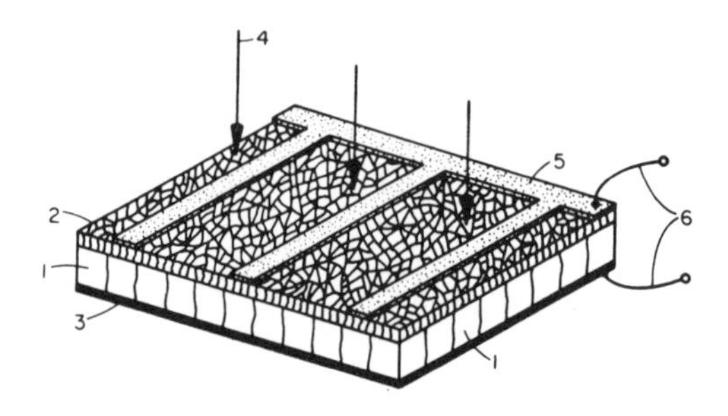

Source: U.S. Patent 4,101,341

Example 2: An n-type cadmium selenide layer about 4 μm in thickness was first vapor deposited on a heated molybdenum substrate. The layer was subsequently converted in a surface layer, to tin selenide by dipping in a solution of $SnCl_2$ acidified with HCl to a pH of about 2, whereupon the ion exchange of Sn for Cd was effected at about 90°C. The cadmium chloride product is water soluble and was merely washed away. A front electrode of gold was then applied by vapor deposition on the surface of the tin selenide region.

Example 3: The n-type cadmium selenide layer may be produced on a substrate by a chemical transport reaction of cadmium-bearing solutions (for example, the chloride, propionate, acetate, formate or nitrate of cadmium) and selenium-bearing solutions (such as selenourea; N,N-dimethyl selenourea; allyl selenocyanate, etc.). In this method the cadmium-bearing and the selenium-bearing solutions are sprayed together onto a heated substrate and then the formation of the desired polycrystalline cadmium selenide layer is effected by thermal decomposition at a temperature of about 300°C.

Subsequently, the tin selenide layer is preferably formed by dipping the cadmium selenide layer into a $SnCl_2$ bath, in a similar manner to the production method of Example 2, or thermal decomposition of tin and selenium-bearing solutions sprayed onto the cadmium selenide layer.

The succession of the semiconductor layers applied by such transport reactions can also be reversed, for adaptation to the desired structure of the cell to be produced. According to the literature SnSe can be chemically grown from metal organic solutions such as tetramethyl or tetraethyl tin.

Method of Silicon Deposition

J. Meuleman and J.-P. Besselere; U.S. Patent 4,124,411; November 7, 1978; assigned to U.S. Philips Corporation describe a process relating to the manufacture of slices and layers of a semiconductor material, especially silicon, of large area to manufacture semiconductor devices, in particular solar cells.

According to this process, a method of providing a layer of solid material on a flat substrate surface, in which a liquid material from which the solid material is formed is provided dropwise on the substrate, while the substrate is given a rotating movement is characterized in that the rotating movement of the substrate is combined with at least one other periodic movement which produces a change of the position of the plane of the substrate surface, such that the points of the surface which are situated outside the point of passage of the axis of the rotating movement change in height with respect to the point of passage.

A homogeneous spreading is further promoted when the frequencies of the rotating movement and the other periodic movement are chosen to be different. The frequency of the other periodic movement is preferably chosen to be larger than the frequency of the first-mentioned movement. In that case the rotating movement may be restricted so as to prevent liquid from hurling away off the substrate, while by the combination with the more rapid other periodic movement a rapid spread of the dropped liquid material can be obtained.

In principle the axis of rotation of the rotating movement can remain vertical, whereas the other periodic movement varies the position of the plane relative to the axis of rotation, for example, by using a tilting movement.

By way of example, the use of the device shown in Figure 6.25a, but with the movement device shown in Figure 6.25b, for providing a layer of silicon on a flat substrate will be described.

The substrate **7** of graphite is circular and has a diameter of 100 mm and a thickness of 100 μm. It is heated on the turntable **8** to a temperature between 1390° and 1410°C; the heating element **11** is arranged at a height of **5** cm with respect to the center of the substrate surface.

The height of the outlet of the nozzle **4** may be adjusted between 8 and 12 cm, measured from the center of the substrate surface. The heating element **3** heats and maintains the silicon in the crucible **1** at a temperature which is higher than the melting temperature, for example 1430°C. A constant flow of argon is sent through the vessel **12**. According to a modified embodiment a reduced gas pressure in the order of 16 mm Hg, for example, also argon or another protective gas, may alternatively be used in the vessel.

The orientation of the substrate surface is varied by means of a rotation of the axis **CC** about the vertical axis **AA** with a frequency of 5 rps (300 rpm). These two axes which intersect each other in the point **G** in the center of the surface of the substrate **7** to be covered enclose an angle of 3° with each other. Since the axis **CC** is also at right angles to the substrate surface, the substrate surface has an inclined position with an angle of 3° deviating from a horizontal position. Since the axis **AA** is vertical, the substrate surface in its various positions during the periodic movement, will always deviate by an angle of 3° from the horizontal position. The movement of the axis **CC** about the vertical axis **AA** is obtained by rotating the sleeve **28** at the speed of 300 rpm.

In order to vary the height of the points of the surface of the substrate situated outside the point **G**, the spindle **23** with the gear wheel **24** should not be rotated at the same speed of rotation (in the same direction) as the sleeve **29**.

Figure 6.25: Method of Silicon Deposition

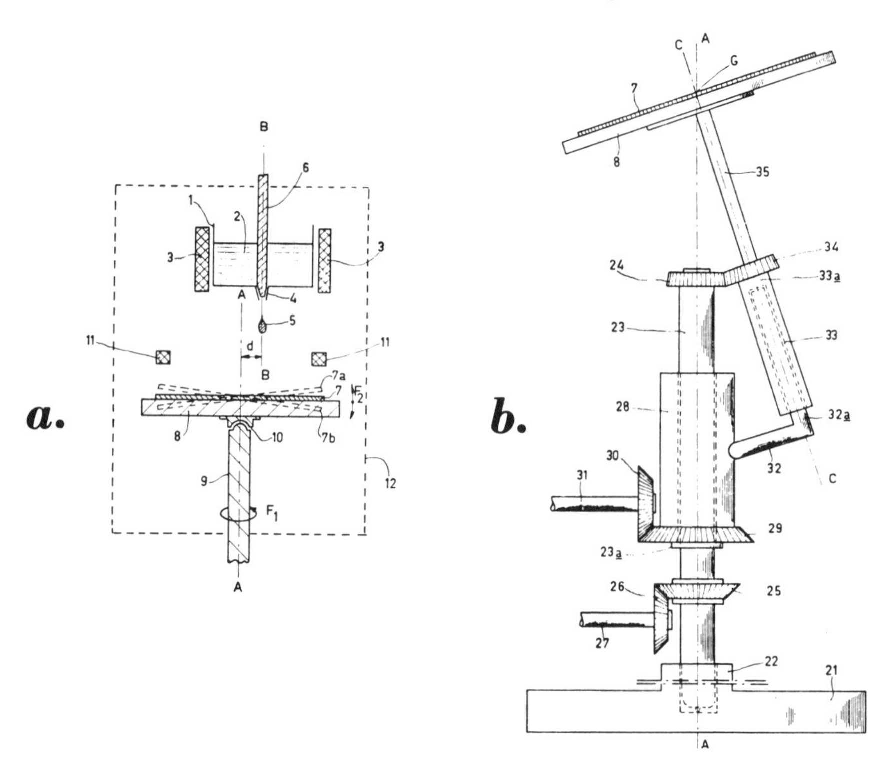

(a) Device for providing a layer of solid material on a substrate.

(b) Device for combination of rotating movement and another periodic movement of a flat substrate.

Source: U.S. Patent 4,124,411

Furthermore, it is desired that the horizontal speed of movement of the place on the substrate surface on which a drop of liquid material falls, is not too large so as to prevent hurling away of the drop. In this example a rotation about the axis **CC** is added in such manner that the points of the substrate surface situated beyond **G** perform rotations about the axis **AA** with a frequency of only 20 rpm. For this purpose the spindle **23** is rotated in the same direction as the sleeve **28** but with a frequency of 580 rpm (twice the frequency of the sleeve **28** reduced by the desired frequency of the revolutions of the points on the substrate surface).

The distance **d** (see Figure 6.25a) between the track of the molten drops and the vertical axis **AA** is 20 mm. The average volume of a drop **5** is 20 mm^3. 20 drops of molten silicon are allowed to fall on the substrate at a frequency of 3 drops per second.

In this manner a layer of polycrystalline silicon is provided on the whole upper surface of the plate **7**, the thickness of which is uniform and equal to 50 μm and the crystallization of which is completed between 6 and 10 sec after the last drop of the abovementioned 20 drops has fallen. This layer adheres very intimately to the underlying graphite and has a strongly shining appearance which is an indication of a good homogeneity and purity.

The graphite plate **7** which is covered with the layer of polycrystalline silicon may be used for the manufacture of a solar cell dependent upon the extent to which the silicon has suitably been doped. For this purpose, a suitable quantity of dopant may previously be added to the quantity of liquid material **2**. According to an embodiment, the dopant used is boron. The crystalline silicon of the resulting layer then has a p-type conductivity. Over a thickness of 0.3 μm, taken from the surface of the polycrystalline layer, a zone of n-conductivity is formed by diffusion of phosphorus, the zone forming a p-n junction with the underlying p-type silicon. A contact with the n-type zone is provided on the surface of the layer, while the other contact of the cell is provided on the uncovered surface of the graphite plate **7**.

Solar Collector System

J.C. Evans, Jr.; U.S. Patent 4,122,214; October 24, 1978; assigned to U.S. Administrator of the National Aeronautics and Space Administration describes an improved collector system which can be applied to the surfaces of known photovoltaic devices. The collector system provides an electrically conductive, protective transparent coating for the underlying photovoltaic substrate and at the same time provides for a very effective metallic grid system which conducts the current generated by incident photons to the external circuitry of the device.

The method of fashioning a transparent electrically conductive coating for photovoltaic substrates having metallic conductive channels therein can be more clearly understood by reference to Figure 6.26. A heat sink **1** or heat conductor is fashioned from a metal block of a metal such as copper, silver, gold, or platinum. The block **1** is provided with channels **3** for the flow of a coolant such as water through the heat sink, and is further characterized by raised portions **5** which protrude from the block **1** in such a fashion as to define the pattern of the metallic grid system desired for the collector system coated on photovoltaic substrate **7**. The heat sink is of a size sufficient to define a grid system pattern on a photovoltaic substrate.

Figure 6.26: Solar Collector System

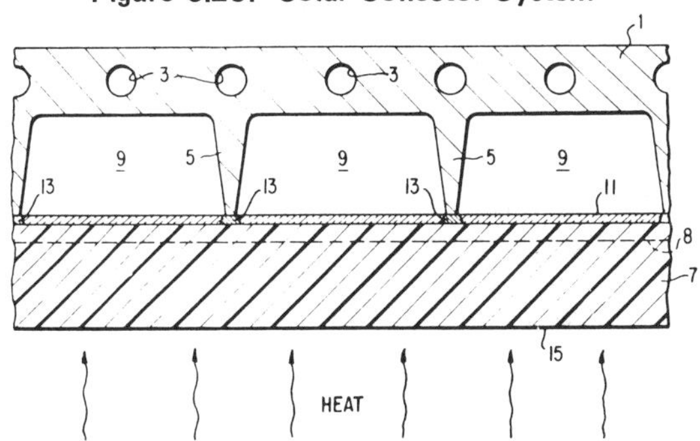

Source: U.S. Patent 4,122,214

The obscured area is reduced to only 3 to 5% while at the same time the collection efficiency of the cell is increased as a result of reduced recombination losses. Protruding portions **5** of the block **1** also define the pattern of recessed regions **9** of the block.

A photovoltaic substrate or solar cell such as a film of cadmium sulfide or a silicon substrate whose surface has been diffused with a counter dopant in a conventional manner is coated with a conductive layer of indium-tin, indium-antimony or indium-arsenic by any conventional technique such as vacuum evaporation or vacuum sputtering to a thickness such that the thickness of the mixed metal oxide conductive layer formed ranges from 500 Å to 10 μm. Thicker perimeter conductors may be added by known methods such as solder dipping, plating, and the like.

A factor in determining the thickness of the layer **11** is that it must be thick enough so that the total collector grid system has sufficient conductivity. Other suitable photovoltaic substrates to which the collector grid system of this process can be applied include gallium arsenide and numerous other types of solar cells. Other suitable substrates include Schottky barrier and heterojunction devices.

The p and n dopants used in the surface layers of these devices must be accounted for in the design and doping of the transparent conductive grid system. However, it has been established that properly formed transparent conductive layers are compatible with both conductivity types. Line **8** defines a surface diffused region or junction region if the photovoltaic substrate employed is an n/p or p/n photovoltaic diode or solar cell.

The block **1** is positioned and secured against the solar cell **7** such that the protruding portions **5** of the block **1** are in contact with the conductive layer **11** of the solar cell. The solar cell **7** is then subjected to heat on the bottom side **15** of the cell. The cell can be heated simply by placing the substrate in open air on a hot plate or heating block controlled by a thermocouple. Also, a flow of hot air from any suitable device can be used. However, the use of an electrically heated block probably provides better heat control.

At the same time a gaseous oxidant such as moist air or oxygen is forcefully passed through or allowed into the recessed regions **9** of the block **1** which causes the oxidation of all portions of the conductive layer **11** exposed to the oxidizing atmosphere to a transparent, mixed metal oxide layer. The heat sink **1** is positioned on the mixed metal layer **11** of the solar cell substrate such that protruding portions **5** are in contact with the metal layer **11**.

In the design of heat sink **1**, it is only necessary that recessed regions **9** be of sufficient relief or clearance above layer **11** to allow the free flow of oxidizing gas molecules in the regions as they strike and rebound from the surface and combine with the metal elements in the layer. The gas molecules will have rather large mean free paths which are dictated by the temperature of the substrate, the extent of access to more of an atmosphere and the ambient pressure of the gas used. The depth of the recessed regions **9** can be determined empirically. Furthermore, the relative closeness of the conductive metal channels **13** in the oxide layer **11** will determine the spacings between protruding portions **5**. These factors are not likely to be critical over a great range.

OXYGEN SENSORS

Magnesium or Calcium Oxide Protective Bonding Layer

J.D. Bode and S.K. Rhee; U.S. Patent 4,107,018; August 15, 1978; assigned to Bendix Autolite Corporation describe a solid electrolyte sensor for sensing oxygen in exhaust gases having a solid electrolyte body of a thimble-like design with an inner conductor thereon and a nonreactive but compatible bonding layer of magnesium or calcium oxide over the outer surface of the electrolyte body and a conductive catalyst layer superimposed on the bonding layer.

Referring to Figure 6.27, there is illustrated an oxygen sensor element **S** where a generally hollow tube or thimble-like solid electrolyte body **1** for transferring oxygen ions is provided composed of known oxygen-ion transferring material of the art. Zirconium dioxide is a preferred solid electrolyte for the body **1**, while the same may incorporate various known stabilizing materials such as yttrium oxide, thorium dioxide, calcium oxide or the like, with the body, as is known, being open at one end for entrance of reference gas, such as the atmosphere, while the other end is closed and inserted into the exhaust gases to be monitored. The inner surface of the solid electrolyte body **1** has a conductive means **3** thereon which may be a strip of conductive material or a layer or film of conductive material, such as platinum, and which is applied to the inner surface of the body **1** by known methods.

A nonreactive but compatible bonding layer **5** is provided on the outer surface of the electrolyte body **1**, this layer **5** providing for good adherence of the outer conductive catalyst **7** while spacing the latter from the solid electrolyte body **1**. The porous, protective bonding layer **5** is preferably of magnesium oxide.

Another alkaline earth metal, calcium, can also be used in its oxide form in the protective bonding layer, or mixtures of magnesium and calcium oxides used.

The alkaline earth metal oxide may be applied directly as a layer to the solid electrolyte body such as applying an aqueous slurry thereof by dipping, spraying or painting. Other forms of the alkaline earth metal compounds may be applied to the body which upon firing will form the oxide, for example, magnesium or calcium carbonates, hydroxide, nitrate or other decomposable inorganic compounds as well as organic compounds such as magnesium or calcium acetate, propionate, oxalate, or the like.

An alternate means of applying the alkaline earth metal oxide is by sputtering the oxide directly onto the solid electrolyte or by sputtering the oxide directly onto the solid electrolyte or by reactive sputtering of the metal in an oxygen atmosphere using known techniques of sputtering.

The outer conductive catalyst layer **7** is then superimposed over the protective bonding layer **5**. The conductive catalyst layer **7** is platinum or a platinum family catalyst which is applied over the layer **5**, after firing thereof to a temperature of 2800° to 3000°F, to provide a conductive and catalytic film. This layer **7** may be applied by known methods, such as vapor deposition, sputtering, spraying, painting or the like. In one embodiment, platinum may be applied as a paste to the protective bonding layer before the latter is fired at 2800° to 3000°F so that particles of platinum are physically adhered to the protective layer with subsequent deposition of the platinum layer on this formation to provide direct platinum to platinum bonding points for additional bonding of the conductive catalyst layer superimposed thereover.

Figure 6.27: Sensor Electrode Element

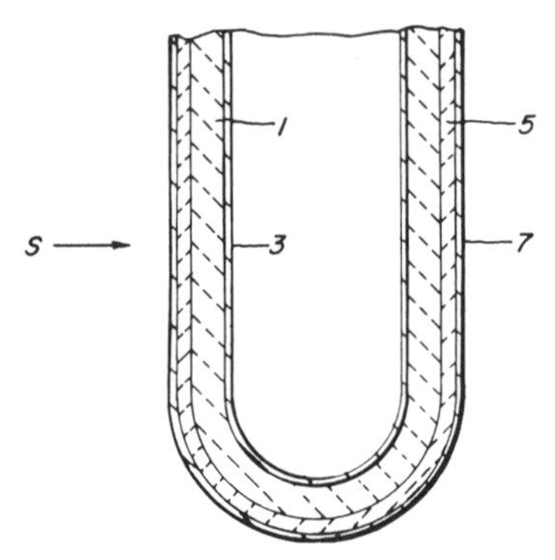

Source: U.S. Patent 4,107,018

The resultant sensor element comprises a solid electrolyte body with a protective bonding layer over the outer surface and a conductive catalyst layer superimposed. The protective bonding layer is applied as an integral part of the solid electrolyte and the conductive catalyst layer improves sensor performance. The construction can prolong the active life of the sensor by preventing deleterious chemical reactions that would otherwise occur between the outer platinum electrode and the solid electrolyte. It can prolong the sensor life by providing a much greater surface area to enhance and extend the desired catalytic activity of the external electrode, the conductive catalyst layer, and with platinum particles embedded in the surface of the protective bonding layer, the bonding layer can act as an anchor to improve adherence of the platinum electrode to the sensor.

Application of Chloroplatinic Acid

M.S. Shum, R.E. Svacha, K.R. Janowski and A.V. Fraioli; U.S. Patent 4,121,989; October 24, 1978; assigned to UOP Inc. describe an industrial air/fuel sensing device which incorporates features common to most of the prior art sensors in that it is a solid electrolytic oxygen ion conducting cell with noble metal electrodes. The electrolyte is in the form of a disc and shares the advantages of other industrial and automotive disc sensors when compared to the tubular varieties of these sensors. It is much simpler in design, however, than other sensors and most importantly, the features of this design have been incorporated so that it can be used in industrial air/fuel control systems without the disadvantages of other designs.

The electrolyte disc is preferably ZrO_2-8 mol % Y_2O_3. This disc is sealed into the recessed end of a fabricated forsterite tube, using glass, such as Corning 1415, as a sealant. The forsterite tube is machined prior to firing to produce a retaining shoulder, a recess into which the electrolyte disc is sealed, and two grooves on the opposite end of the tube. The recess is machined to a depth greater than necessary to accept the electrolyte disc. The reason for this extra recess will be discussed below.

After sealing the disc into the tube, a commercially available fluxed platinum paste, such as Plessey 4276, is deposited on the two exposed surfaces of the disc. In addition, one platinum paste stripe is painted along the inside diameter of the forsterite tube, over the end and into the bottom recessed groove on the outside diameter of the tube. A second stripe is painted from the top surface of the disc along the tube outside diameter to the upper groove. While the platinum paste electrodes are still tacky, felted ceramic fiber discs are pressed into the paste on both electrodes.

The felted ceramic fiber material can be obtained as No. 300 ceramic paper (Cotronics Corp., New York) and serves a number of functions. During operation of the sensing device, it acts as a combination filter to remove particulate matter and gaseous poisons from the stack gas being measured and a barrier to protect the electrode from any gaseous erosion that might occur. Another important function of the ceramic fiber disc is its role as a wick during application of chloroplatinic acid (CPA) to be discussed below. The open fiber structure of the ceramic fiber disc allows uniform deposition of the CPA on the paste substrate. The importance of this deposition will also be discussed below. A fiber disc thickness of about 0.040" has proved to be quite satisfactory.

The openness of the felted material is important in the operation of the device for industrial applications due to the relatively slow gas flow rate often inherent in such applications. Experiments conducted substituting a more dense material such as gamma-alumina for the felted ceramic disc were not successful due to the inability of the gas to penetrate through the layer and reach the electrode material where oxygen ionization occurs.

This application technique involves pressing the felted fiber disc into the tacky platinum paste and then firing in a furnace at 1750°F in air for 30 min to cure the platinum paste and firmly bond the two ceramic fiber discs to the platinum paste. So bonded they will remain in place during the operation of the sensor in the industrial application.

Chloroplatinic acid is then applied through the ceramic fiber discs directly onto the surface of the platinum paste substrate. Sufficient CPA is applied to insure adequate coverage of the platinum paste substrate. This requires about 0.5 mg of platinum or about 0.01 ml of CPA per sensor where the electrolyte disc has a diameter of about 0.388" and a thickness of about 0.060". The amount required is low due to the small area of the electrolyte and the efficient wicking action of the felted disc. Ionization of the oxygen in the stack gas environment takes place at the surface of the electrolyte, in the presence of both the platinum deposited as the paste substrate and high surface platinum from the CPA.

If sufficient platinum surface is not present on the faces of the electrolyte disc, as would be the case if only platinum paste were present, ionization of residual oxygen would not be complete, and nonrepreducible sensor performance would result. It is important that both surfaces of the electrolyte disc have sufficient platinum surface area present to complete the ionization-deionization reactions which occur on these surfaces. Therefore, sufficient CPA must be added to both the gas and reference sides of the electrolyte.

The purpose in bringing both the front and back conducting stripes to separate grooves fabricated in the forsterite tube is to insure the elimination of the dissimilar material junctions.

Such junctions have been shown to introduce irreproducible and unwanted signals onto the true, theoretically predictable cell output. While not important in stoichiometric sensors, this feature is critical in importance for nonstoichiometric industrial control. By bringing all of these junctions out of the cell, these unwanted voltages can be eliminated and sensor reproducibility and stability insured if the other sensor features discussed above are included.

The purpose of the extra deep recess in the electrolyte end of the forsterite tube is to assist in eliminating the erosion effects of a flowing gas stream. In addition, this recess promotes turbulence in the stream at the sensor electrolyte location and allows continuous interaction of the stack gas with the electrolyte, assuring gaseous measurements are valid representations of the gas stream.

The final assembly operations include painting a gasket material such as graphite paste onto the forsterite tube at the locating shoulder or flange and compressing this gasket material between the metal body and the pressure nut. Alternatively, the forsterite tube can be produced without the locating ring and a compression gland can be used for sealing. This would simplify the design even further, thereby reducing costs.

The advantages of this sensor are the following: (a) it is simple; (b) uses very few parts; (c) the amount of platinum used is very small; (d) it is accurate and reproducible so that it can be used in industrial applications to indicate true oxygen content; (e) all dissimilar material junctions have been eliminated; and (f) the sensor has demonstrated long term operation at 1300°F lower than the 1500°F operating temperature of other industrial sensors. This relatively low operating temperature in an industrial environment is a marked improvement over previously available commercial sensors. In addition, it has been shown that the installation of this sensor directly into stack gas systems can be done very simply compared to other available sensors.

MISCELLANEOUS PROCESSES

Electrically Conductive Contact Layer

Gas discharge display devices for indicating numbers and symbols are known in flat constructions. They comprise two surface carriers made from insulating material which lie opposite one another at a small spacing, of which at least one carrier comprises glass. On the inner surfaces of these substrates facing one another are arranged electrodes which comprise electrically conductive coatings, particularly coatings made of tin oxide. The oxide (stannous or stannic oxide) coatings are used at least on the substrates because of their transparency which makes optical visibility possible.

The two substrates are kept at the desired spacing by means of spacing members and are connected together in an airtight manner at the periphery. Inside there is a gas atmosphere. When applying a voltage to opposite electrodes a blue glow discharge, used for indication, occurs at the cathode. The cathode electrodes on one substrate are arranged as necessary to enable numerical, alpha numeric, cross or point raster representation. In order to be able to apply the required voltage to the electrodes, airtight ducts are necessary. These ducts have already been proposed in great numbers. Ducts are even known in the form of electrically conductive coatings.

H. Keiner, E. Schmid and W. Richly; U.S. Patent 4,113,896; September 12, 1978; assigned to Licentia Patent-Verwaltungs GmbH, Germany describe a method for manufacturing contact layers which have a good airtight adherence (adhesion) both to electrically conductive coatings as well as to substrates made of glass or ceramics.

According to this process, there is provided a method of manufacturing an electrically conductive contact layer comprising producing a pastelike mixture of 60 to 98 wt % of silver or silver oxide, 2 to 40 wt % of glass powder and a vaporizable liquid agent, applying the paste to a surface on which the layer is to be formed, drying the paste and baking the paste in air at 350° to 550°C.

The paste type preparation takes place adviseably by mixing a bonding or thinning agent, such as anisaldehyde, methylglycolacetate, ethyl- and/or butylacetate. Preferably a fluid is used which is of organic nature and which volatilizes at temperatures below 400°C. In a preferred example, a paste was used which contains 50 g commercial silver or silver oxide powder and 15 g glass powder. The paste type preparation was carried out with methylglycolacetate after the solid substances had been carefully ground finely in a mortar. A glass which contains 5 to 25% ZnO, 65 to 85% PbO and 5 to 25% B_2O_3 is preferably used as a glass powder. In a preferred example, a glass powder was used which comprised 13% ZnO, 75% PbO and 12% B_2O_3. All percentages are by weight. The grain size both of the glass powder and of the silver powder or silver oxide powder should preferably be 5 to 50 μm.

The previously described paste was painted onto the surface to be contacted and dried by means of air or hot air and then baked in at temperatures of approximately 400° to 550°C. The baking in took place in a preferred time period of 10 to 20 min. Such a layer manufactured in accordance with this process had an electrical resistance which was smaller than 50 Ω/cm (particularly smaller than 30 Ω/cm) and was used for contacting during gas discharge display.

Aluminum Sulfide Coating on Sodium/Sulfur Cell

The sulfur reactant container in sodium/sulfur cells is subject to attack in contact with molten sodium polysulfide and sulfur reactants. Aluminum has been identified as a nondegrading material in this environment, because it forms a continuous layer of aluminum sulfide over its exposed surfaces. Although this layer is protective, it is electrically insulating and prevents the use of untreated aluminum as an electrode material in sodium/sulfur cells.

Other metals, such as nickel, iron and their alloys, are inadequate as electrode and container materials in their untreated form. They form porous metal sulfide scales which lead to extensive physical degradation and contamination of the catholyte melt under cycling. This interferes with efficient cell operation and causes discharge capacity losses, cell resistance increases and degradation of the electrolyte.

A preferred sulfur reactant container or other electrically conducting component subject to corrosive attack by the battery reactant would be a low cost material or composite material, which exhibits the chemical stability of aluminum and the strength and electrical properties of iron or nickel or their alloys.

D. Chatterji and R.R. Dubin; U.S. Patent 4,123,566; October 31, 1978; assigned to Electric Power Research Institute, Inc. describe a method to convert an iron or nickel based container or component surface, such that on exposure to the cell reactants, particularly the sulfur-sodium polysulfide melt, a doped, electrically conductive aluminum sulfide scale will form, instead of the pure electrically insulating aluminum sulfide scale found on pure aluminum under the same environment.

Sodium/sulfur batteries have three basic components: sodium, sulfur and a solid β-alumina electrolyte. The battery is run at elevated temperatures, frequently 300°C or higher, which in conjunction with the high chemical reactivity of the cell reactants creates serious materials problems. There is the further consideration, that sulfur is nonconductive. Therefore, means must be provided for introducing and removing electricity from the sulfur in the sulfur reactant cell. Conveniently, a metal container may be employed for the sulfur cell which is electrically conducting and graphite or carbon fibers distributed through the sulfur. Means are then provided for electrical connection between the container and the carbon fibers.

Conventional structural metals are susceptible to corrosive attack by sulfur and sodium. In accordance with this process, structural metals such as iron, nickel and their alloys, particularly with cobalt, are aluminided on the surface subject to corrosive attack.

The metal component, which can be unshaped or shaped, can be coated with the metal aluminide by a variety of conventional processes. Such chemical or physical processes known in the art include pack aluminiding, plasma spraying, hot dipping, metalliding, vacuum deposition, sputtering, and the like. Where necessary, the coated product may be heated to a temperature of about 900°C for a time sufficient to cause reaction between the aluminum coating and the base metal or alloy. The temperatures and times will vary widely, depending upon the base alloy.

The coating will generally be at least about 50 Å thick and may be up to 1 mil or greater. The metal aluminide is then subjected to molten sulfur or sulfides for a sufficient time to form a coating of aluminum sulfide. This can occur in the sodium-sulfur cell itself.

In order to exemplify the process, pack aluminiding was carried out. Alloy samples (53.7% Fe, 29% Ni, 17% Co, and 0.3% Mn; all percents are by weight) were aluminided by the pack processing technique. The process involved aluminiding for 3 hr at 1060°C employing argon using an Inconel retort. The composition of the pack was 1% aluminum, 0.25% ammonium fluoride and the balance aluminum oxide (percents by weight). After processing, the sample cross sections revealed formation of a uniform metal aluminide skin on the sample surface. The degree of aluminiding to form this skin can be varied by controlling the aluminum activity in the pack, the process temperature, and the aluminiding time. The surface was then exposed to molten sodium polysulfide at a temperature of about 300°C.

The resulting aluminum sulfide surface was found to be electrically conductive and inert to sulfur. Samples prepared by the above method were cycled in molten sodium polysulfide for up to 775 hr at 30 mA/cm^2 and showed no serious signs of chemical attack.

Composite Body

B.S. Dunn; U.S. Patent 4,131,694; December 26, 1978; assigned to General Electric Company describes a method of forming a composite body which has a body of solid ion-conductive electrolyte material, the body having a casing portion with one open end and a header integral with the casing adjacent its open end, and a surface portion of an ion-insulating material on only the exterior surfaces of the header portion of the body.

In Figure 6.28, there is shown generally at **10** a composite body made in accordance with this process. Composite body **10** has a body **11** of solid ion-conductive electrolyte material of sodium β-alumina. Body **11** has a casing portion **12** with one open end **13** and a header **14** integral with casing **12** adjacent its open end **13** and surface portion **15** of an ion-insulating material on only the exterior surfaces of header portion **14** of body **11**. Ion-insulating material **12** is a strontium substituted ion β-alumina.

Figure 6.28: Composite Body

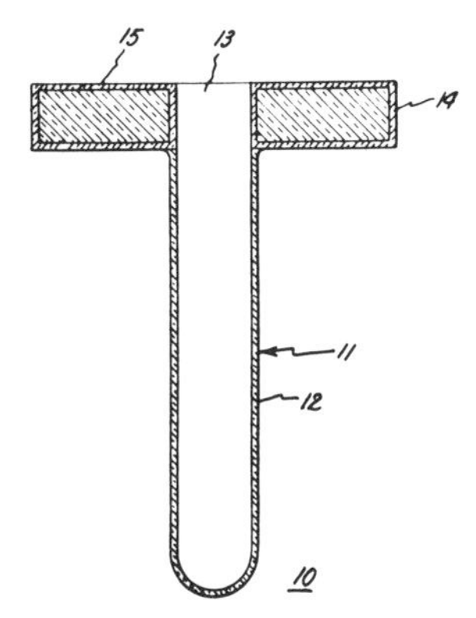

Source: U.S. Patent 4,131,694

The method comprises providing a solid ion-conductive electrolyte material, such as, sodium β-alumina, as a body having a casing portion with one open end and a header portion integral with the casing adjacent its open end. Only the header portion of the body is submerged in a molten salt bath of a salt containing a divalent substitution ion, such as, a strontium(II) ion, and the surface portion of only the header portion is converted to an ion-insulating material, such as, a strontium substituted ion β-alumina thereby providing a composite body.

Example: A solid ion-conductive electrolyte material body of sodium β-alumina is provided which body includes a casing portion with one open end and a header portion integral with the casing portion adjacent its open end. A bath composition of $3Sr(NO_3)_2 \cdot NaNO_3$ is in molten condition at a temperature of 600°C. The molten bath contains a divalent substitution ion of Sr^{2+} which ion is to be exchanged into the surface portion of the header portion of the body. The header portion is submerged into the molten bath for a period of 1 hr thereby converting the surface portion or exterior surfaces of the header portion to an ion-insulating material of strontium substituted β-alumina. The penetration depth of the strontium divalent ion is calculated to be 70 x 10^{-4} cm.

Thin Film Strain Gauge

L.J. Boudreaux and J.H. Foster; U.S. Patent 4,104,605; August 1, 1978; assigned to General Electric Company describe a thin film strain gauge and the method of fabrication thereof.

Dynamic strain sensors are known to be useful for testing of mechanical components, especially when in use in their intended environment. Under certain conditions, such as in low-cycle-fatigue and flutter dynamic strain testing, associated with development of blades for aircraft engine compressors and steam turbines, strain gauges are located on critical stages of the apparatus and, being exposed to conditions of high oxidation, erosion and vibratory stress, experience unacceptable fatigue damage (to the strain sensor element and its associated conductive leads) and premature failure. Strain gauges having extended operating times, when utilized in hostile environments, will eliminate test delays and repeated sensor replacement, to result in potentially large cost savings on test programs.

Referring to Figure 6.29, thin film strain gauge **10** is fabricated upon a surface **11** of a part or substrate, such as jet-engine compressor blade **12**, to be tested. The strain gauge includes a pair of conductive lead means **14a** and **14b** to facilitate suitable electrical connection to be made between gauge **10** and conductor means, such as low resistance leads **15** and **16**, to allow reading of the value of strain gauge resistance. External equipment is coupled to conductor means **15** and **16** at a point on article **12** removed from the location of strain gauge **10** and from the relatively high stresses, strains and vibratory motion of the working environment of the strain gauge.

Strain gauge **10** is fabricated upon substrate surface **11** by initially depositing a first film **20** (Figure 6.29b) of a high temperature insulative material, such as alumina (Al_2O_3) or forsterite ($2MgO{\cdot}SiO_2$) and the like, to a first thickness $\mathbf{T_1}$ of between 2 and 6 μ. Preferably, the high temperature insulating film is deposited by rf sputtering, especially when fabricating a strain gauge on a part, such as a jet-engine compressor blade, which is fabricated of a material (such as titanium) which cannot be subjected to a sufficiently high temperature to allow chemical vapor deposition of the film to be feasible. Also, rf sputtering is preferable as the alumina or forsterite film thus deposited is apparently harder and somewhat less porous than plasma- or flame-sprayed films of the same material, whereby porosity (leading to leakage of electric current through layer **20** to the generally conductive substrate **12**) is avoided.

At least one strain-sensing member **21** is fabricated by depositing a thin film of a resistive material, such as Nichrome, platinum, MoSi, CrSi and the like, upon the top surface **20a** of the high temperature insulative film. The resistive material may be deposited by rf sputtering or vacuum evaporation and may be shaped by masking techniques or deposited as a continuous film and then formed to the desired resistor pattern utilizing process such as photoetching, laser-machining and the like.

The resistive film is deposited to a thickness $\mathbf{T_2}$ having a range from 200 Å to greater than 1 μ, dependent upon the desired magnitude of resistance to be achieved and the physical dimensions of the resistor pattern, as constrained by the area of the substrate available for strain gauge deposition.

Figure 6.29: Thin Film Strain Gauge

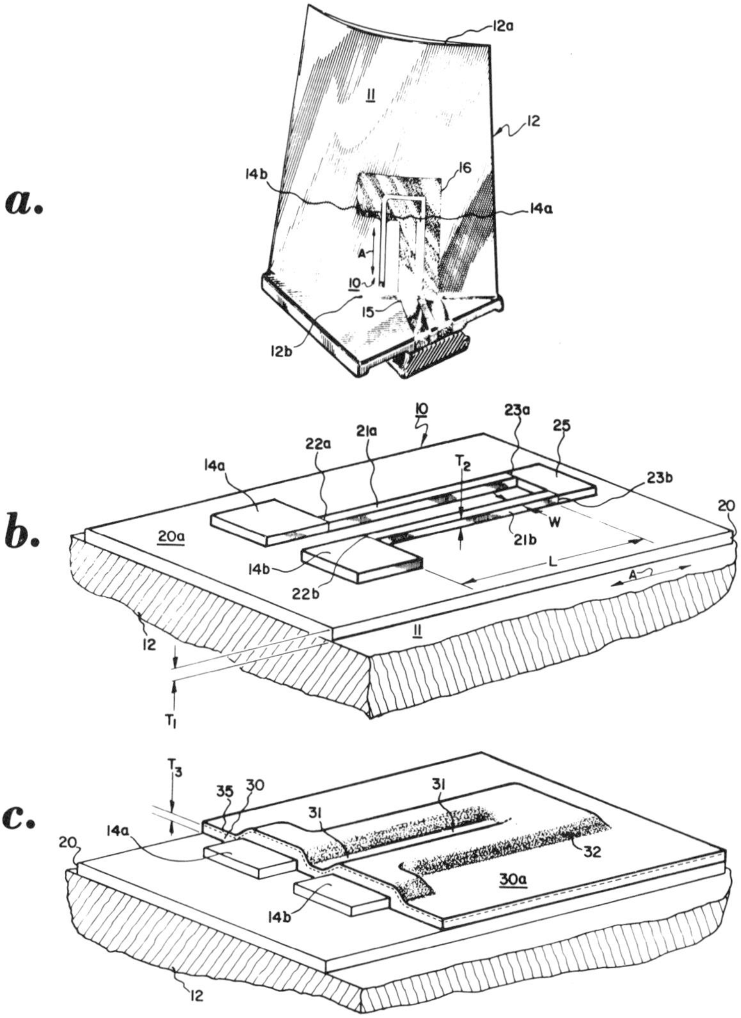

Source: U.S. Patent 4,104,605

After deposition of the resistive member **21**, a layer of conductive lead-forming material, such as nickel, gold and the like, is deposited on the insulating layer surface **20a** either through a mask or as a continuous film with subsequent pattern formation via the aforementioned photoetching, laser-machining and like processes, to form at least a pair of pads **14a** and **14b**, each in electrical connection with one of the ends **22a** or **22b** of the resistive pattern.

Dependent upon the particular shape, or shapes, selected for the one or more resistive members **21**, a series-connective conductive pattern **25** may be required to couple the remaining ends of a plurality of resistance elements in the desired electrical pattern.

After deposition of the required resistive pattern, its conductive leads and any interconnections, upon top surface **20a** of the insulative base layer **20** (Figure 6.29b) a second film **30** (facilitating corrosion-erosion protection) also preferably formed by rf sputtering of alumina or forsterite, is deposited upon, and cooperates with, a portion of film top surface **20a** to completely enclose all of the resistive members **21**, any interconnecting conductive portions **25** utilized therewith, and at least a portion of each conductive lead-connection pad **14** in the region adjacent the connection of each pad to the resistive pattern.

Protective film **30** may be deposited with a thickness T_3 of as little as about 2 μ, which relatively thin protective layer tends to have a surface **30a** relatively conformally coating the underlying strain gauge portions and, being somewhat undulent, may disturb air flow over the surface **11**. Smoother protective layer surfaces may be achieved by depositing relatively greater thicknesses of film **30**, whereby the deposited molecules tend to fill in such features as the channel **31** between the parallel resistive members as well as to smooth the slopes **32** of the boundaries of the underlying pattern.

It was found that the rf sputtered films of alumina and forsterite possess an apparent hardness and relative smoothness preventing erosion or chipping away of surface **30a** during operation in a hostile environment. In extremely hostile environments, the use of a top protective film **35**, of a metallic material such as nickel, chrome, nickel-platinum and the like, may be deposited upon the protective film surface **30a**, with thicknesses in the range of 2 to 5 μ to provide additional protection against corrosion and erosion.

Frequency Adjustment of Piezoelectric Resonators

G.R. Buynak and F.L. Sauerland; U.S. Patent 4,112,134; September 5, 1978; assigned to Transat Corp. describe a vacuum deposition method for adjusting the resonant frequency of piezoelectric resonators wherein the speed and accuracy of the process are enhanced by abrupt termination of the deposition through injection of a gas into the vacuum chamber at the end of the adjustment cycle.

In the vacuum deposition process, a conductive or nonconductive material is evaporated in vacuum and deposited on the surface of a piezoelectric resonator, thereby lowering its resonance frequency (FR) toward a desired target frequency (FT). A basic conventional system may comprise a vacuum chamber containing an electrically heated filament wire for evaporating the deposition material, a mask for establishing a deposition pattern, and a resonator plus socket and electrical connections for monitoring the resonance frequency. The adjustment can be done by turning off the filament current when FR equals FT. However, due to the thermal inertia of filament and deposition material, the deposition cannot turn off immediately and causes FR to overshoot FT.

The resulting adjustment error increases with the deposition rate at the time of filament turnoff. As a consequence, high operating speed and accuracy, two of the most important performance criteria, are mutually exclusive in this basic system.

There are several known ways to improve the performance. One approach is to control the deposition rate from an initial high rate to a low rate as the resonator frequency approaches the target frequency. Another is to interject a shutter between the filament and resonator at the end of the adjustment cycle. Some systems combine both of these approaches.

The shutter has several drawbacks, among them: (a) movable parts partially exposed to material deposition and, therefore, in need of periodic cleaning and repair; (b) complexity and bulk inside the vacuum chamber, which for reasons of fast and economical pumpdown should be small; (c) added feedthroughs, either mechanical or electrical, through the vacuum chamber wall; (d) an adjustment error due to mechanical inertia of the shutter; and (e) an adjustment error due to the deposition of the vapor that at the time of shutter closing is trapped between the shutter and the resonator and moving toward the latter. These drawbacks are overcome by this process.

This process is based on abruptly terminating deposition when FR equals FT by releasing the vacuum and injecting a gas, preferably inert, into the vacuum chamber. The method can be implemented by a modification of conventional frequency adjustment methods. Its advantages are a combination of high speed, high accuracy performance with simplicity of operation.

For a better understanding of this process, reference is made to Figure 6.30 which shows a schematic diagram of a frequency adjustment system. In this diagram there are shown a source for vapor deposition **1**, a mask **2**, a resonator **3**, and a network **4** with electrical connections to the resonator assembled in a vacuum chamber **5**. The source **1** may comprise the deposition material, heated by an electrically heated filament. The network **4** may in its simplest form comprise two wires directly connecting the resonator to the electrical conductors **6** which interconnect network **4** with the control circuit **7**.

The control circuit serves to compare the resonator frequency to the target frequency and to control the current supply **8** which supplies current to the source **1** and thereby controls the deposition rate of source **1**. A vacuum pump **9** is connected to the vacuum chamber **5** via the pipe **10** and valve **11**. Also a vent valve **12** is connected to the vacuum chamber **5** via pipe **13** on one side and to pipe **14** on the other side. Pipe **14** is connected to a vessel **15** which usually is open to the surrounding air.

The system described so far is conventional; a typical operating cycle comprises the following steps: (a) insert a resonator; (b) close the vacuum chamber, close valve **12**, open valve **11**, and pump down; (c) initiate deposition by turning on the filament current; (d) turn off the filament current when |FR – FT| reaches a given value, usually zero; (e) after deposition has stopped, close valve **11** and open valve **12** to vent the chamber; and (f) take out the finished resonator.

In the conventional method, a small amount of deposition continues after the filament has been turned off. This is due to the thermal inertia of the deposition material and filament and causes adjustment errors.

To implement this method, the timing in operating the valves **11** and **12** is changed such that valve **12** is opened when |FR – FT| reaches a given value, usually zero, and that valve **11** is closed before that. Also vessel **14** preferably contains an inert gas. Further, an extension **16** may be added to pipe **13**, thereby extending the venting orifice into the area between deposition source **1** and resonator **3**.

Figure 6.30: Frequency Adjustment of Piezoelectric Resonators

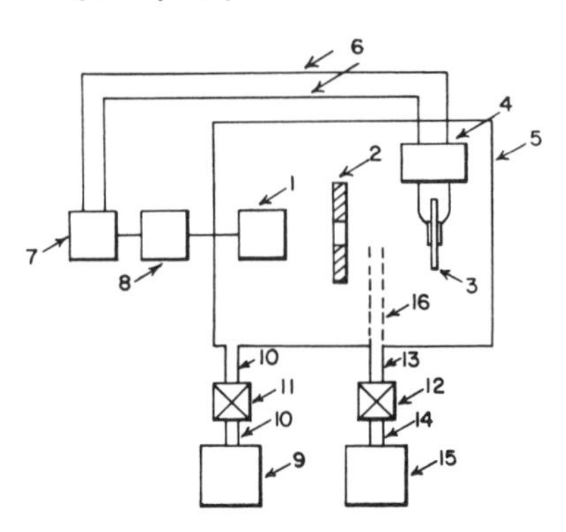

Source: U.S. Patent 4,112,134

A typical operating cycle according to this process comprises steps (a), (b) and (c) described above for the conventional approach and continues with (d) turn off valve **11**; (e) turn off the filament current when |FR - FT| reaches a given value, usually zero; (f) open vent valve **12** when |FR - FT| reaches another given value, usually zero; and (g) take out the finished resonator.

Upon venting the vacuum chamber in step (e), the deposition is stopped almost instantaneously. Venting with air is usually not desirable because it may oxidize the deposition material and the filament, both of which are still hot at the instant. For this reason, vessel **15** is preferably filled with an inert gas at pressure around or above atmospheric pressure. Even more immediate suppression of the deposition may be obtained by placing the venting orifice in the area between deposition source **1** and resonator **3**, as indicated in Figure 6.30.

MAGNETIC COATINGS

RECORDING MEDIA

Cobalt Oxide Protective Layer

Magnetic recording members such as are commonly utilized in conjunction with computers of various known types as memory discs or the like are conventionally constructed utilizing a disc shaped substrate of a metal such as aluminum or brass appearing somewhat like a common phonograph record. The surface of such a substrate is commonly coated with a magnetic recording alloy layer and such a layer in turn is normally covered with a protective covering of one sort or another.

In general effective magnetic recording alloys as are used in such magnetic recording layers will normally contain from about 50 to 97% by weight cobalt and from about 3 to 50% by weight nickel and, if other secondary ingredients are present in such alloys, from about 3 to 50% by weight of such secondary ingredients. A particularly suitable magnetic recording alloy contains about 70% by weight cobalt, about 5% by weight phosphorus and about 25% by weight nickel.

Such magnetic recording alloy layers can be from about 3 to 100 microinches thick, although generally they are from about 14 to 20 microinches thick. The precise composition of such a layer and the precise thickness of such a layer will depend upon an intended use of a magnetic recording member. As such members are used they are employed in connection with a transducer in such a manner that there is relative rotation between the transducer and the recording member.

In order to avoid damage to the recording alloy layer in a magnetic recording member a number of different expedients have been proposed and adopted to varying extents.

P.K. Patel, D.H. Johnston and J. Makaeff; U.S. Patent 4,124,736; November 7, 1978; assigned to Poly-Disc Systems, Inc. describe a magnetic recording member having a substrate, a magnetic recording alloy supported on the surface of the substrate, and a protective covering on the surface of the magnetic recording alloy which includes a barrier layer located on the surface of the alloy and an oxide layer located on the surface of the barrier layer, this oxide layer consisting essentially of a nonmagnetic oxide composition of one or more metals selected from the group consisting of metals within Groups IV-A, V-A, VI-A, VII-A, and VIII, and I-B of the Periodic Table. This barrier layer consisting of one or more nonmagnetic metals from within this group is inert relative to the oxide layer and the magnetic recording alloy under the conditions of use of the magnetic recording member and during the manufacture of the oxide layer.

In Figure 7.1 there is shown a magentic recording member **10** which utilizes a base or substrate **12**.

Figure 7.1: Surface Protected Magnetic Recording Member

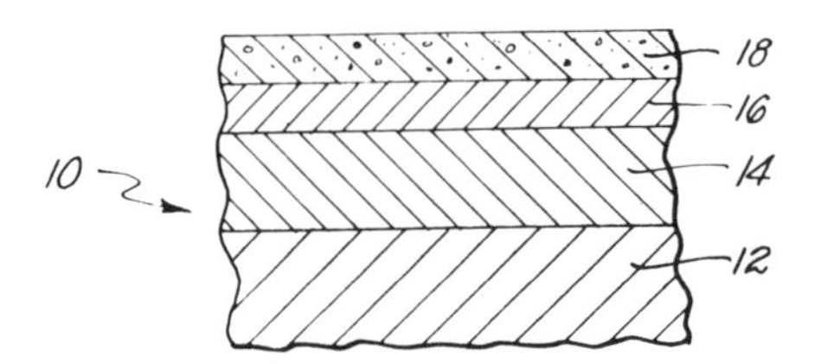

Source: U.S. Patent 4,124,736

Such a substrate **12** will normally be an aluminum disc or the like in accordance with conventional practice. This substrate **12** will, of course, be nonmagnetic in character. This substrate **12** supports a conventional magnetic recording alloy layer **14**. This layer **14** in turn supports a nonmagnetic barrier layer **16**, and this barrier layer **16**, in turn supports on oxide layer **18**.

It is presently considered that preferred results can be achieved if the layer **16** is from about 2 to 7 microinches thick. When the layer **16** is within this range of thickness, it is sufficiently thick so as to effectively serve as a physical barrier and there is very little danger of it containing surface discontinuities. Further, when the layer **16** is this thick it is sufficiently thin so that there is normally no danger of it affecting magnetic recording.

It is preferred to utilize the metal cobalt or an alloy such as a cobalt-phosphorus-nickel alloy, a cobalt-gold-phosphorus alloy or the like containing at least 85% by weight cobalt as the base metal to be used in forming a layer such as the layer **18**. This is because a cobalt layer can be readily oxidized to form an oxide mixture believed to contain predominantly Co_3O_4 which combines hardness satisfactory to provide resistance to abrasion in the complete member **10** in combination with frictional characteristics enabling the layer **18** to satisfactorily be used with transducers.

Similarly, certain metals within the groups as indicated may have more desirable frictional characteristics than cobalt oxides but will not possess the desired physical hardness. It is considered, however, that reasonably acceptable results can be achieved utilizing the metals tungsten, titanium, and tantalum. It is also considered that acceptable results can be achieved utilizing predominantly nickel alloys.

A layer of one or more metals as indicated to be utilized to form the layer **18** may be deposited in any desired conventional manner. The conversion of such a metal to an oxide so as to form the layer **18** may be conveniently carried out by heating a metal layer created in air, oxygen, or another oxygen-containing atmosphere. Such heating should be carried out at a sufficient temperature and for a sufficient time to convert all of the metal or metals present to the corresponding oxide or oxides so as to avoid the possibility of unoxidized metal remaining since any such unoxidized metal might interfere during the use of the member **10** in its intended manner.

As an example of this, cobalt or cobalt alloys as noted in the preceding can be satisfactorily oxidized in atmospheric oxygen or in pure oxygen at a temperature of from about 200°F to a temperature of about 800°F. At lower temperatures than within this range the oxidation takes place too slowly for practical purposes. At higher temperatures than within this range there is the danger of the various materials, other than in the metal or metals being used to create the layer **18** either melting or diffusing together and/or perhaps oxidizing.

The reaction involved here is essentially a normal time-temperature reaction. It is preferred to heat a cobalt alloy as indicated in an atmosphere as described at a temperature of from about 400°F for a period of about 3 hr to a temperature of about 530°F for a period of about 2 hr in order to obtain substantially complete conversion of the cobalt present to the oxide.

It is preferred to form the barrier layer **16** out of nickel-phosphorus alloy or copper because these metals are comparatively cheap and because as metals they are sufficiently hard enough to support or reinforce the layer **18** so as to render it relatively immune to surface abrasion. Such metals are also more resistant to oxidation than cobalt or cobalt alloys and, hence, can be utilized with the preferred cobalt and cobalt alloys.

Satisfactory results can also be achieved in forming the layer **16** out of metals such as gold, silver, platinum, chromium and the like. In general the higher a metal is located in the Periodic Table the more suitable it is for use in forming the layer **18** while the lower a metal is on the Periodic Table the more suitable it is in forming the layer **16**. This is due to a differential in the ease in forming oxides.

Addition of Barium Ion

R. Shirahata, T. Kitamoto and M. Suzuki; U.S. Patent 4,128,691; December 5, 1978; assigned to Fuji Photo Film Co., Ltd., Japan describe a process for the production of a magnetic recording medium by electroless plating, which comprises carrying out the plating in a plating bath containing a ferromagnetic metal ion, preferably, cobalt ion or cobalt ion plus nickel ion, barium ion and hypophosphite ion as a reducing agent and thus forming a magnetic metal thin film.

The electroless magnetic plating bath can contain ordinarily (a) cobalt, nickel, iron and/or other metal ions to form a ferromagnetic metal thin film, (b) a reducing agent such as hypophosphite, borohydride compound or hydrazine, (c) a complexing agent such as malonic acid, succinic acid, tartaric acid, citric acid or ammonium salts thereof, (d) a pH buffering agent such as formic acid, acetic acid, malonic acid, succinic acid or citric acid and (e) a pH regulator such as sodium hydroxide, ammonium hydroxide or sodium carbonate.

In a preferred embodiment of the process, the electroless plating bath consists of an aqueous solution containing 0.02 to 0.6 mol/ℓ of cobalt ion, 0 to 0.02 mol/ℓ of nickel ion, 0.0002 to 0.005 mol/ℓ of barium ion and 0.03 to 0.4 mol/ℓ of hypophosphite ion as a reducing agent.

The abovedescribed complexing agent, pH buffer and pH regulator can optionally be incorporated in this aqueous solution. The pH of the abovedescribed plating bath is preferably 6.5 to 9.0 in view of the use of hypophosphorous acid as a reducing agent and the temperature of the plating bath is preferably 60° to 95°C so as to advance the plating rapidly. The plating rate is generally 50 to 5,000 A/min, preferably 200 to 2,000 A/min.

The thickness of a magnetic thin film is determined so as to give a sufficient output as a magnetic recording medium and to effect satisfactorily a high density recording and, in general, 0.05 to 1.0 μ, preferably 0.1 to 0.4 μ. The substrate can be selected from those of nonconductive materials such as polyethylene terephthalate, polyimide, polyvinyl chloride, cellulose triacetate, polycarbonate, polyethylene naphthalate, glass and ceramics, and metallic materials such as aluminum and brass. As occasion demands, the substrate can be subjected to a pretreatment, for example, formation of metallic films by vapor deposition or by plating, or coating of adhesives. The magnetic thin film can be uniformly formed irrespective of the shapes of the substrate, for example, tapes, sheets, cards, disks and drums.

Example: A polyethylene terephthalate film of 22 μ in thickness was dipped in an aqueous solution of 5 mols/ℓ of sodium hydroxide, warmed at 80°C, for 5 min for the purpose of defatting, washed adequately with water and then subjected to a substrate surface activation treatment for 3 min using Catalyst 6F Solution and Accelerator 19 Solution (Shipley Co.) as a surface activation solution. Then the substrate was dipped in the following electroless plating solution and plated to give a plating film thickness of 0.12 μ.

	Mol per Liter
Cobalt chloride ($CoCl_2 \cdot 6H_2O$)	0.06
Sodium hypophosphite ($NaH_2PO_2 \cdot H_2O$)	0.10
Ammonium chloride (NH_4Cl)	0.20
Citric acid ($C_6H_8O_7 \cdot H_2O$)	0.09
Boric acid (H_3BO_3)	0.50
Barium chloride ($BaCl_2 \cdot 2H_2O$)	0-0.007

pH is 7.5, temperature is 80°C

The coercive forces and squareness ratios of magnetic plating films prepared with various concentrations of barium chloride and the plating rates were shown in the following table.

Barium Ion (mol/ℓ)	Coercive Force (Oc)	Squareness Ratio (Br/Bm)	Plating Rate (A/min)
0	535	0.71	600
0.001	790	0.73	500
0.002	825	0.75	480
0.003	850	0.79	450
0.004	870	0.74	440

(continued)

Barium Ion (mol/ℓ)	Coercive Force (Oc)	Squareness Ratio (Br/Bm)	Plating Rate (A/min)
0.005	885	0.70	400
0.006	890	0.64	300
0.007	880	0.65	220

As is evident from these results, the coercive force and squareness ratio are improved by the addition of barium ion. In particular, this effect is remarkable within a concentration range of 0.001 to 0.005 mol/ℓ and the plating speed lowers when the concentration is more than 0.006 mol/ℓ.

Cobalt-Doped Acicular Hyper-Magnetite Particles

T.M. Kanten; U.S. Patent 4,137,342; January 30, 1979; assigned to Minnesota Mining and Manufacturing Company describes magnetic recording media based on acicular iron oxide particles, each having a core of $(FeO)_xFe_2O_3$ where x is greater than one but not greater than 1.5 as determined by titration, and a surface layer comprising a cobalt compound. The cobalt in the surface layer provides 1 to 10% by weight of the particles.

The acicular hyper-magnetite core material can be produced by the single step of heating acicular alpha-FeOOH particles of submicron size in an inert atmosphere in the presence of a reducing agent such in hydrogen or a glyceride at about 300° to 600°C, preferably for about 15 min. Preferably a long-chain fatty acid glyceride or mixture of long-chain fatty acid glycerides is used at a temperature of at least 400°C.

The long chain may be straight or branched but should have at least 12 carbon atoms. Glycerides which are esters of acids having 16 to 18 carbon atoms are especially useful and may be employed in amount of 3 to 10% by weight of the alpha-FeOOH, preferably about 5%. When using an organic reducing agent, the resultant core material includes a carbon-containing residue which may provide an oily film when the particles are slurried in water. The carbon may comprise about 0.5 to 3% of the total weight of the particles.

While mixing a slurry of this core material, a soluble cobalt salt and a compound which will insolubilize the cobalt ions are added sequentially to deposit cobalt containing material onto the surfaces of the particles. The homogeneous mixture is filtered, and the cake is heated in an inert atmosphere at 80° to 200°C to fix the surface deposit. The heating is discontinued before substantial diffusing of the cobalt ions into the particle cores.

The heating temperature should not exceed 200°C. At temperatures below 80°C, the process is unduly slow and inefficient. By adjusting the amount of the cobalt salt in the slurry, the cobalt ions in the coating may be controlled within the desired 1 to 10% of the total weight of the particles.

Example: Raw materials are acicular alpha-FeOOH particles, average length of 0.5 micrometer and length to width ratio of 10:1 and triglyceride of C_{16-18} saturated carboxylic acids ("Neustrene" 059 of Humko Products Inc.). Forty pounds (18.2 kg) of the FeOOH particles and 2 lb (0.9 kg) of the triglyceride were charged to a 3 ft³ (0.085 m³) gas-fired rotary kiln. While maintaining a nitrogen purge of 3 cfm (0.085 m³/min), the contents were heated to 930°F (500°C) for 30 min and then cooled with water to room temperature. When cool, the contents were transferred, under a nitrogen atmosphere, to a steel container which was purged with argon. The container was sealed to maintain the inert atmosphere and placed in a glove box under nitrogen to prevent oxidation of the product. The Fe^{+2} content of the resulting acicular particles was 38%, as determined by titration.

Using a high-speed, high-shear homogenizing mixer, 20 lb (9.1 kg) of these acicular particles were slurried in 24 gal (91 liters) of deionized water of conductivity less than 20 micromhos. 6.0 lb (2.7 kg) of $CoSO_4 \cdot 7H_2O$ were dissolved in 10 gal (38 liters) deionized water and added to the slurry, and the mixing was continued for 30 min. 1,400 ml 29% aqueous NH_3 were then added. The pH rose to 9.1.

The slurry was then pumped into a plate-and-frame-type filter. The solids were washed by reslurrying in 50 gal (190 liters) deionized water and refiltering. The conductivity of the mother liquor after washing was less than 700 micromhos.

The solids were charged to a rotary kiln while purging with nitrogen at 3 cfm (0.085 m^3/min) and heating to 250°F (120°C). After 15 min at that temperature the kiln was cooled with water to room temperature while maintaining the nitrogen purge. The resultant particles had a carbon content of 1.8%, and each had a core consisting essentially of acicular $(FeO)_x Fe_2O_3$ where x was approximately 1.2, and a surface layer comprising a cobalt compound. The cobalt ions provided about 6% by weight of the particles. A slurry was made of:

8 g of the cobalt-containing particles;

50 g of the acetate salt of a polypropylene oxide quaternary amine of 2,200 molecular weight;

15 g of butyl myristate;

15 g of fine alumina particles;

416 g of a 30 wt % solution of a high molecular-weight polyester-polyurethane polymer synthesized from neopentyl glycol, poly-ϵ-caprolactonediol, and diphenylmethane diisocyanate dissolved in dimethylformamide;

836 g toluene and

50 g dimethylformamide.

The slurry was charged to a one-gallon-capacity (3.8 liter) sandmill and milled until smooth, which required 2 hr. This was coated onto a biaxially-oriented polyethylene terephthalate web one mil (25 micrometers) in thickness. The wet coating was passed through a magnetic field to orient the acicular particles in the longitudinal direction and heated to drive off the solvents. After polishing to a surface roughness of 4.0 microinches (0.1 micrometer) peak-to peak, the web was slit into tapes of standard tape widths. The thickness of its magnetizable layer was approximately 200 microinches (5 micrometers).

The magnetic properties of the tape, measured in the presence of a 3,000-oersted, 60-hertz field using an M versus H meter, were: ϕ_r is 0.400 line per ¼ inch (0.6 cm) width of tape, M_r/M_m is 0.81, H_c is 632 oersteds, B_r is 1,270 gauss.

After storing the tape for 100 days at room temperature, its H_c increased to 650 oersteds. Extrapolation indicated that the H_c should rise to 675 oersteds after 20 months. Such increase would not cause any problems on ordinary recording and reproducing apparatus.

Adjustment of pH

A.J. Kolk, Jr.; U.S. Patent 4,150,172; April 17, 1979; describes an electroless plating process for producing magnetic thin films having square hysteresis loop characteristics and a coercivity that can be selected to lie between 300 and 1,000 oersteds by adjusting the pH of the plating bath. The magnetic film of cobalt-phosphorus or cobalt-nickel-phosphorus can be applied directly over a layer of nonmagnetic nickel and the bath contains no volatile constituents that can rapidly alter the magnetic properties of magnetic films deposited during the same run.

As a result, the process may be used to produce multilayer structures of alternate magnetic and nonmagnetic layers and each magnetic layer therein will have substantially identical magnetic characteristics. The process is particularly suitable for the magnetic plating of computer memory discs which may be recorded at bit densities in excess of 5,000 bits per inch along each magnetic track.

Figure 7.2a is an elevation view illustrating the various layers plated on a substrate **10**.

Figure 7.2: Electroless Plated Magnetic Medium

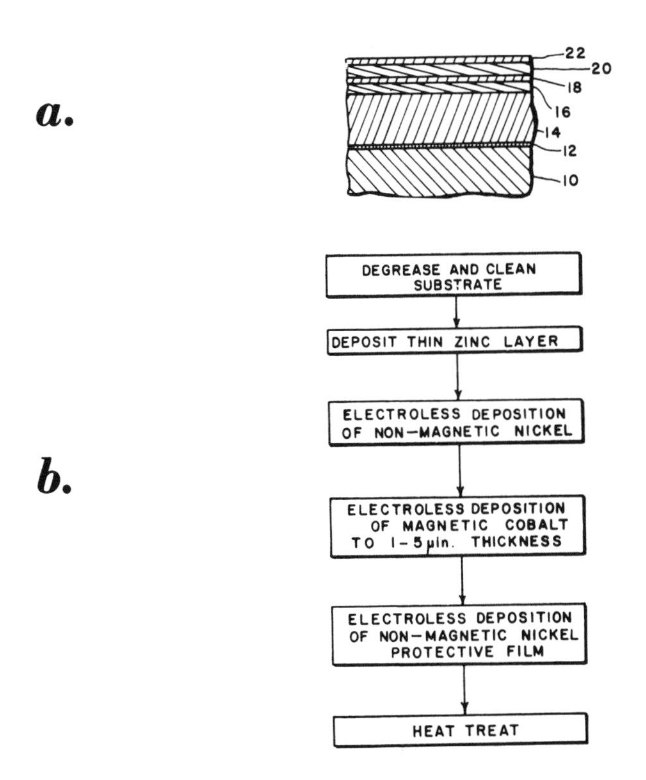

Source: U.S. Patent 4,150,172

In the description to follow, the substrate **10** will be assumed to be an aluminum alloy disc upon which will be plated a magnetic medium for use as a digital computer memory. It will be understood that the substrate **10** may, if desired, be of any other material, such as glass or plastic. Substrate **10** is suitable lapped, polished and then degreased and cleaned to receive a thin flashed layer of zinc **12** which, in turn, supports a layer **14** of nonmagnetic electroless nickel deposited to a thickness of 60 microinches or more.

A magnetic cobalt layer **16** is then deposited from an electroless bath to the surface of the nonmagnetic nickel layer **14**. The magnetic cobalt layer **16**, which is deposited to a thickness of between 1 and 5 microinches, supports a second layer **18** of nonmagnetic nickel which is deposited to a thickness of between 1 and 2 microinches as a protective covering for the cobalt layer **16**.

If it is desired to produce a thick film from several thin film layers, another equally thick layer of magnetic cobalt **20** along with a second protective nickel layer **22** may be applied to the nonmagnetic nickel layer **18**. Additional thin magnetic cobalt layers may be applied over thin nonmagnetic nickel layers to produce thick magnetic structures having the square hysteresis loop characteristics of the individual thin magnetic layers.

Figure 7.2b illustrates the various plating steps in the process of applying a magnetic coating to an aluminum disc to produce a computer memory disc capable of very high packing densities in the order of 5,000 bits per inch or better. As illustrated in Figure 7.2b the initial step in the process is to degrease the substrate with an organic solvent and then clean the disc with a nonetching aluminum cleaner followed by a spray rinse of distilled water.

After cleaning, the substrate is subjected to a zincating process to form a thin zinc layer on the substrate surface. There are many commercially available zincating materials and processes, such as the well-known process described in U.S. Patent 3,216,835.

The zinc-coated substrate is then plated with electroless nickel using processes such as described in U.S. Patent 2,876,116. Again, proprietary electroless nickel baths are commercially available for performing this step in the process. One variation from the standard practice in the performance of this step is that it is required that the pH be adjusted to a low level to maintain phosphorus content of the plate at a level above 10% so that the nickel is deposited in a nonmagnetic composition. This can be achieved if the pH is kept at a level of about 4.15 measured at the bath operating temperatures of 80°C. The nonmagnetic nickel layer is plated to a thickness of at least 60 microinches to both physically and chemically protect the aluminum substrate.

The nickel-plated substrate is followed by the formation of a magnetic cobalt layer from an electroless bath. The bath is prepared from appropriate sources of cobalt, hypophosphite, phosphate and citrate ions in approximately the following concentrations and ratios:

	Grams per Liter
Co^{+2}	2.5
$H_2PO_2^-$	3.5
PO_4^{-3}	5.5
$C_6H_5O_7^{-3}$	15.4

The above ions must be added to water of high purity level to achieve consistent magnetic properties in the deposited material. If desired, an ethanolamine or a mixture of ethanolamines in a quantity of approximately 0.2 g/ℓ to 2.0 g/ℓ may be added to the bath to increase the brightness of the deposit and to act as a wetting agent. Also, if desired, some of the cobalt ions may be replaced by nickel ions to modify the deposit by forming a cobalt-nickel-phosphorus alloy. The bath temperature is preferably kept in the range of 82° to 85°C and the pH measured at the operating temperature is generally kept between 8.3 and 8.55 depending upon the value of coercive force which is desired. In general, coercive force increases with increasing pH and decreasing thickness of the deposit.

The other ions which may be present in the bath are introduced as necessary in order to utilize the most economical and commercially available chemicals in preparing the bath. These ions could be any neutral species such, for example, as sodium, potassium, sulfate, acetate, or chloride. Phosphite ion is formed by the oxidation of hypophosphite ion during the reduction of the metal and by other side reactions, but it causes no problems until the concentration becomes high—well in excess of 10 g/ℓ. The phosphite ion may have favorable action of another buffer in the bath.

The correct pH to obtain the desired coercive force and the plating time should be determined by plating upon a dummy or sample substrate. The reason for this is that the plating rate and the coercive force are functions of the substrate geometry and the speed with which the substrate is rotated in the bath.

After the cobalt layer has been applied to the disc, the disc is given a protective coating of about 1 to 2 microinches of electroless nickel. The primary purpose of this coating is to protect the cobalt from oxidation. If the boundary layer lubricant, such as an oxide or chemical conversion coating, is desired to further protect the magnetic coating, the lubricant layer may be applied over the nickel.

The final step in the process is to bake the structure at a temperature of at least 135°C, but not higher than 280°C. The baking improves the adhesion of the deposited layers, increases the coercive forces and stabilizes the magnetic parameters.

Back Coating of Magnetic Layer

G. Akashi, M. Fujiyama and M. Utumi; U.S. Patent 4,135,031; January 16, 1979; assigned to Fuji Photo Film Co., Ltd., Japan describe a magnetic recording substance comprising a magnetic recording layer on one side of a nonmagnetic support and a back layer on the other side thereof, the back layer being provided by coating a powdered solid with a binder and having spike grains with a height of 0.8 to 5 μ and an interval of 200 μ or more on the surface thereof and a friction coefficient of 0.25 or more.

Useful examples of the materials for such spike grains are talc, titanium oxide, zinc oxide (ZnO), lithophone, complex of zinc sulfide and zinc sulfate, aluminum oxide, silicon carbide, kaolin, carbon, barium sulfate ($BaSO_4$), silicon dioxide (SiO_2), barium carbonate ($BaCO_3$), etc. Above all, talc, lithopone and complexes of zinc sulfide and zinc sulfate are preferable. These spike grains can be used individually or in combination. For the purpose of provision of spike grains to satisfy the abovedescribed conditions, there are several methods, for example, comprising (1) adding an inorganic powder having a grain size corresponding to a desired spike grain to a coating composition at the final time of dispersion, (2) adding an inorganic powder having a grain size of somewhat larger than that of a desired spike grain to a coating composition at the initial time of dispersion and dispersing the mixture while controlling the dispersion time to obtain a desired spike grain size, (3) causing pigment grains to be aggregated and (4) adding two kinds of inorganic powders differing in grain size to a coating composition simultaneously or separately while controlling the dispersion time to obtain a desired spike grain size. In the last case, one inorganic power has generally a grain size of 0.02 to 5 μ, preferably 0.1 to 3 μ and the other inorganic powder has generally a grain size of 2 to 40 μ, preferably 5 to 30 μ.

The magnetic coating compositions contain predominantly ferromagnetic powders, binders and coating solvents, optionally with dispersing agents, lubricants, abrasives and antistatic agents.

Formation of a magnetic recording layer is carried out by dissolving or dispersing the abovedescribed composition in an organic solvent and then coating the resulting composition onto a support.

Suitable materials which can be used for this support are various plastics, for example, polyesters such as polyethylene terephthalate, polyethylene-2,6-naphthalate and the like, polyolefins such as polypropylene and the like, cellulose derivatives such as cellulose triacetate, cellulose diacetate and the like, polycarbonates, etc., and nonmagnetic metals, for example, copper, aluminum, zinc, etc. Such a nonmagnetic support can have a thickness of about 3 to 100 μm, preferably 5 to 50 μm in the form of a film or sheet.

The magnetic layer coated onto a support is dried after, if necessary, the coating has been subjected to a treatment for orientating the magnetic powder in the layer. If desired, the magnetic layer can be subjected to a surface smoothing treatment or cut in a desired shape to thus form the magnetic recording substance of the process. In this case, the orientating magnetic field can be either an ac or dc magnetic field with field strength of about 500 to 2,000 gauss. The drying temperature can range from about 50° to 100°C and the drying time is about 3 to 10 min.

In the process, furthermore, the support is subjected to back coating on the reverse side to the magnetic layer.

Suitable solid powders which can be used are carbon black, graphite, kaolin, barium sulfate, silicon dioxide, titanium oxide, talc, zinc oxide, etc. and as a binder, there are used various thermoplastic resins and thermosetting resins. These components are mixed with an organic solvent to prepare a coating composition as in the case of the magnetic layer and then applied to the reverse side of a support to the magnetic layer to thus form a back layer.

According to the process, there are obtained the following advantages or merits:

Increase of the modulation noise can be minimized in spite of the presence of coarse grains on the reverse side of the magnetic layer.

The edges of a tape wound at a high tape speed are well-ordered, so the fluctuation of signal output due to tape deformation can be decreased.

There takes place no tape loosening even if the tape is suddenly stopped as the tape end is in a free state.

There takes place no step-out of the wound edges along tape near the core.

Increase of the drop out can markedly be suppressed.

The tolerance of the electric resistance of a back layer is increased; that is, the static charge build-up is lowered for a same electric resistance.

Surface Activation with Palladium Dibenzalacetone Complex

G. Wunsch, P. Deigner, R. Falk, K. Mahler, W. Loeser, F. Domas, P. Felleisen and W. Steck; U.S. Patent 4,128,672; December 5, 1978; assigned to BASF AG, Germany describe a process for the manufacture of rigid magnetic recording media in which a thin film of a cobalt-containing ferromagnetic metal is deposited by electroless deposition on an activated rigid base followed by the application of a less than 0.1 μ thick protective layer to the surface of the thin film of cobalt-containing ferromagnetic metal.

The rigid base includes a nonmagnetic metallic base disc at least one surface of which carries a layer (S) from 2 to 100 μm thick, which layer has been brought by machining to a peak-to-valley height of not more than 0.3 μ, is insoluble in, and virtually nonswellable by, organic solvents and consists of a dispersion of from 0.5 to 5 parts by weight of a finely divided nonmagnetic pigment of a Mohs hardness of at least 6, per part by weight of a cured binder which firmly bonds the pigments to the base disc.

The process further comprises: activating the surface of the base disc by treatment with a solution of a palladium(O) complex in an organic solvent, decomposing the complex, and thereafter depositing a thin film of ferromagnetic metal of about 0.05 to 1 μ thickness on the so-activated surface of the base disc by electroless deposition.

Suitable nonmagnetic pigments are, e.g., hard oxides, carbides or nitrides of metals, such as aluminum oxide (Al_2O_3), chromium(III) oxide (Cr_2O_3), quartz (SiO_2), cerium dioxide (CeO_2), α-iron(III) oxide (α-Fe_2O_3), and titanium dioxide (TiO_2). Acicular α-Fe_2O_3 which has been obtained from acicular γ-Fe_2O_3 by a heat treatment, i.e., by heating it above the Curie point, has proved very suitable for the production of the layers S. The particle size of the nonmagnetic pigments used in the layers S should in general not be substantially greater than the ultimate thickness of the layer S and is in general from about 0.1 to 15 μ and, in the preferred embodiment of the process, not more than from about 1.5 to 1 μ.

The pigments contained in the layers S are firmly embedded in a cured binder. Curable binders which can be used to produce the layers S include all binders which adhere sufficiently firmly to the metallic base discs used, e.g., the aluminum discs, possess adequate binding power for the nonmagnetic pigment particles and give, after baking a layer of the dispersion of the nonmagnetic pigments in the binder, a tough layer having a hard surface which can be ground and/or polished. The latter aspect is of great importance for obtaining a surface having a small peak-to-valley height.

Binders which, in the form of their solutions or solutions of their components, are particularly suitable for the production of the layers S are those based on polyurethanes, e.g., mixtures of polyisocyanates with relatively high molecular weight compounds having active hydrogen atoms which react with isocyanate groups, and mixtures of such polyurethane binders with other binders, e.g., curable aminoplast, melamine and phenolic resins; curable epoxy resins, polyaminoamides or other polymers which react with epoxy resins, or with low molecular weight curing agents; and mixtures of curable epoxy resins, such as polyglycidyl ethers with curable melamine-formaldehyde precondensates or appropriate phenol-formaldehyde precondensates which have been plasticized, with alcohols of 1 to 4 carbon atoms.

Mixtures of curable epoxy resins or polyepoxide compounds with curable precondensates of phenols or alkylphenols with formaldehyde, which have preferably been produced in an alkaline medium by reacting the phenolic compounds with a 1.5- to 3-fold molar amount of formaldehyde, and of which the hydroxyl groups have been etherified at least partially with aliphatic alcohols of 1 to 4 carbon atoms, are very suitable.

The weight ratio of finely divided pigment to binder in the dispersion for producing the layer S is in general from about 0.5:1 to 5:1 and preferably from about 1.2:1 to 3:1.

Solvents which can be used to produce the pigment/binder dispersion and apply it to the base disc are the volatile solvents and solvent mixtures conventionally used in the surface-coating industry, in which the binders or binder components used are soluble, e.g., aromatic hydrocarbons, such as benzene, toluene or xylenes; volatile alcohols and glycols, such as propanol, butanol or ethylene glycol or their ethers and/or esters, such as ethylene glycol monoethyl ether or ethylene glycol monoethyl ether monoacetate (also called ethyl glycol acetate); ketones, such as acetone, methyl ethyl ketone and cyclohexanone; amides, such as dimethylformamide; ethers, such as tetrahydrofuran and dioxane and their mixtures.

To produce the dispersion, the nonmagnetic pigment or a mixture of different nonmagnetic pigments is dispersed in the curable binder and a sufficient amount of solvent, using conventional equipment, for example, a ball mill or tube mill. The application of the dispersion to the metallic base disc may also be effected by conventional methods. A very suitable method has proved to be applying first a layer of the dispersion to the slowly rotating base discs (e.g., at a speed of from about 100 to 500 rpm), e.g., by spraying, to give a thickness of, e.g., from about 1 to 3 mm, and then to produce the desired thickness of the layer S, e.g., preferably from 2 to 15 μ, by rotating the disc at a higher speed (preferably at from about 1,000 to 3,000 rpm).

After the coating operation is over, the layers S are cured or baked. In this treatment, the coated base disc is advantageously heated at about 120° to 250°C for, e.g., from ¼ to 1 hr, the curing temperature and curing time depending on the binder system used.

The baking step is advantageously followed by machining, such as grinding and polishing, of the surface of the cured layers S to produce the desired slight surface roughness.

Prior to the activation, of the base disc provided with the layers S (hereinafter referred to as base T_S), it is advisable, to clean the base T_S, e.g, rinse it with the solvent which is used in the palladium(O) complex solution in the subsequent activation.

Preferred palladium(O) complexes are the complexes or systems of palladium(O) with dibenzalacetone

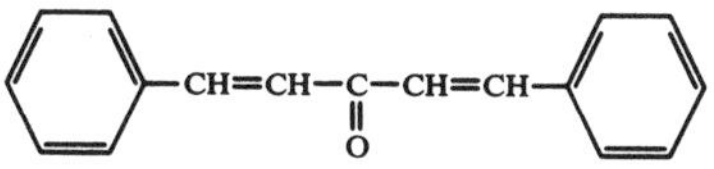

—CH=CH—C—CH=CH—
‖
O

which not only exhibit a good activating action but also particularly high stability in solution.

The stability of the solutions of the palladium(O) complexes may be increased by using benzene and, in particular, alkylated benzenes, e.g., toluene, as the solvent. The concentrations of the complexes in the solutions are from about 15 mg/ℓ to the saturation concentration at room temperature, and preferably from about 50 mg/ℓ to 2 g/ℓ, depending on the nature of the complex and of the solvent. Although halohydrocarbons and, e.g., acetonitrile, tetrahydrofuran or dimethylformamide are less suitable solvents because they may slightly decompose the metal complexes, this decomposing effect may be used to activate the substrate.

It was found possible, after dipping a base T_S into a solution of palladium dibenzalacetone complex in toluene at room temperature and then dipping the treated substrate into a chlorinated hydrocarbon, e.g., dichloroethylene, trichloroethylene or tetrachloroethylene, at room temperature, to electrolessly plate the base T_S, for example, in a cobalt salt bath.

Particularly suitable cobalt-containing ferromagnetic thin metal layers are composed of cobalt-phosphorus, cobalt-nickel-phosphorus and similar ferromagnetic alloys, the magnetic properties being greatly influenced by the Co/Ni ratio. The deposited magnetic films are from about 0.05 to 1 μ, and especially from 0.08 to 0.5 μ, thick and in general have a coercive force Hc of from about 23 to 75 kA/m. The coercive force of the magnetic layer can be increased by subsequent heating of the layer at from about 150° to 300°C under an inert gas.

The deposited magnetic films can be provided, by conventional methods, with a protective film which is in general less than 0.2 μ thick and in particular less than 0.05 μ thick, e.g., with a cobalt oxide film, a coating of rhodium or tungsten carbide, a silicone film, a wax film, a polymer film or a combination of such materials. Treatment of the surface of the thin layer of ferromagnetic cobalt-containing metal has proved best.

In this process, a solution of a film-forming organic synthetic resin in a volatile solvent is applied to the thin film of cobalt-containing ferromagnetic metal, the layer of solution in the dried, but not baked, state being not more than 0.3 μ thick and in particular not more than 0.05 μ thick, and the applied layer is dried and baked for from about 1 to 15 hr at from about 200° to 300°C in an atmosphere containing oxygen. Preferably, the applied layer is, after baking, no longer identifiable as a separate layer.

The process is more economical than prior art processes because, owing to the fact that the layers S compensate for defects in the metal base disc, and the base T_S provided with the layers S can be readily machined and is resistant to basic and acid metallizing baths, it reduces the number of base disc rejects and thus makes possible the production, in large numbers, of magnetic discs which are capable of withstanding repeated momentary contact or sustained direct contact with magnetic heads without exhibiting any signs of damage even after being in operation for several hours.

BUBBLE CIRCUITS

Nonuniform Spacing Layer

Magnetic bubble technology (MBT) involves the creation and manipulation of magnetic bubbles in specially prepared magnetic materials. The work bubble is intended to encompass any single-walled magnetic domain, defined as a domain having an outer boundary which closes on itself. The application of a static uniform magnetic bias field orthogonal to a sheet of magnetic material having suitable uniaxial anisotropy causes the normally random serpentine pattern of magnetic domains to shrink into isolated, short-cylindrical configurations or bubbles whose common polarity is opposite that of the bias field. The bubbles repell each other and can be moved or propagated by a magnetic field in the plane of the sheet.

R.M. Sandfort; U.S. Patent 4,104,422; August 1, 1978; assigned to Monsanto Company describes an efficient technique for fabricating bubble chips which permits a variation in the thickness of the spacing layer to accommodate different degrees of coupling between bubbles and the overlay circuit to facilitate logic operations without degrading the integrity of normal propagation or storage operations.

According to the process, a spacing layer between the bubble material and the circuit overlay is designed with a special topography providing raised plateaus or mesas at selected locations where it is desired to weaken the coupling between bubbles and a particular overlay element or series of elements. The nonuniform thickness of the spacing layer facilitates logic operations by making bubbles more prone to deflection at selected locations of logic interaction while preserving the optimum spacing for orderly propagation in other areas of the same bubble chip.

Figures 7.3a, 7.3b and 7.3c illustrate a specific technique for achieving a nonuniform spacing layer in a bubble chip. The technique will be described in terms of sequential steps or stages of fabrication. First the epitaxial layer **14**, usually garnet, is grown on the nonmagnetic substrate **12**. Next, a spacing layer **16** of nonmagnetic, electrically nonconductive material (e.g., silicon oxide) is deposited on the surface of the layer **14** to a thickness, greatly exaggerated in Figure 7.3a, which would provide optimum spacing for normal propagation, or shift register operation, with a given type of overlay circuit element. The spacing layer **16** may be of uniform thickness as in standard practice.

Departing from conventional procedures, a second spacing layer **48** of nonmagnetic, electrically conductive material is deposited over the original spacing layer **16**.

Figure 7.3: Bubble Chip with Stratified Spacing Layer

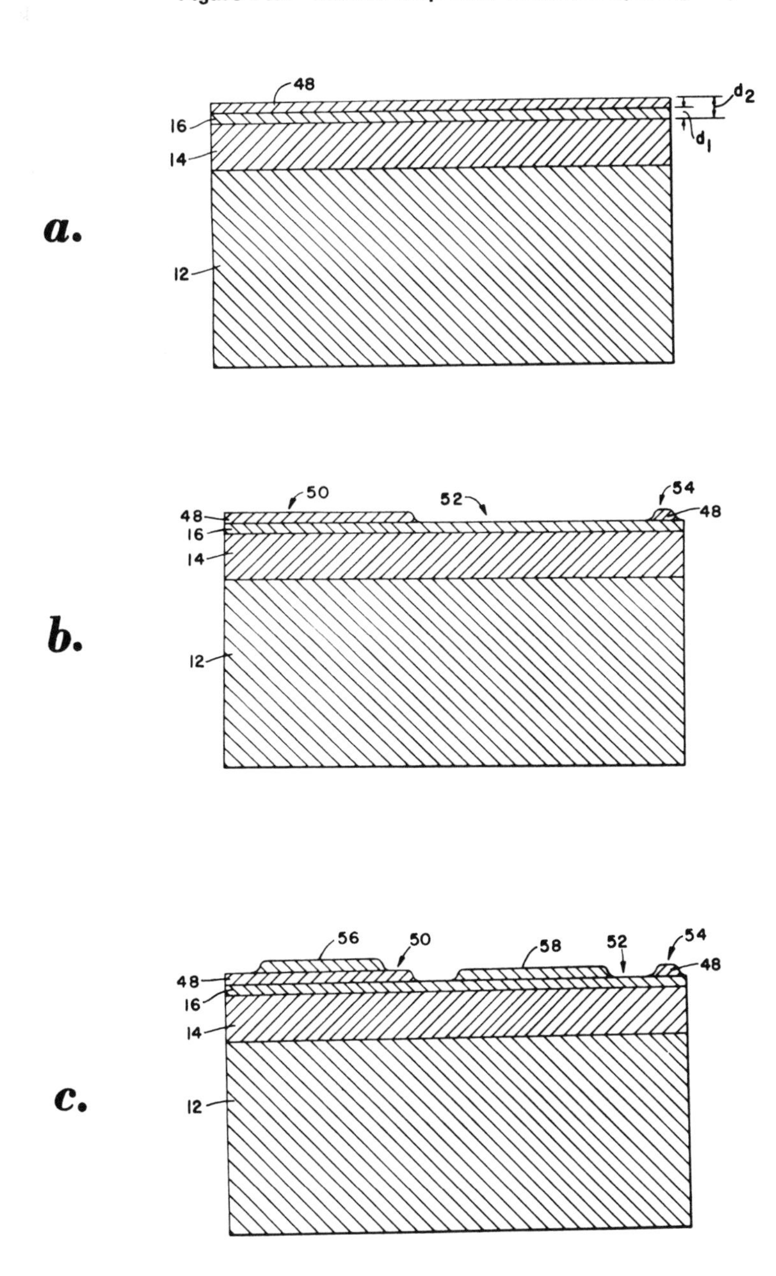

Source: U.S. Patent 4,104,422

Layers **16** and **48** together comprise a stratified spacing layer whose overall thickness d_2 represents a different optimum spacing for weakly coupling bubbles to an element from which bubbles are to be deflectible, while the thickness of the first spacing layer **16** would be suitable for normal shift register operation.

The next phase of this technique is to define a predetermined pattern on the surface of the second spacing layer **48** and to etch the layer **48** so as to remove whole portions thereof to leave a mesa topography. The word mesa refers, for example, to the raised plateau **50** remaining in Figure 7.3b after etching out the adjacent area **52**. The mesa topography may include a linear pattern of conductors or control leads, represented by ridge **54** in Figure 7.3b for use in transfer, generation annihilation and readout, for example.

The next step in the procedure is to deposit a thin layer of soft ferromagnetic material to form the circuit overlay directly on top of the composite spacing layer formed by the etched layer **48** and the planar layer **16**. A pattern of circuit elements is defined and etched in the ferromagnetic overlay by conventional techniques to leave circuit elements **56** and **58** on the exposed surface of the bubble chip as shown in Figure 7.3c. Circuit elements such as element **56** from which a bubble is to be deflectible, are prearranged to rest on top of the mesa **50**, for example, at the greater spacing d_2 between the overlay element and the bubble layer **14** in order to weaken the coupling. Circuit elements such as element **58** designed for normal propagation, or for carrying bubbles which will deflect bubbles from an element such as element **56**, are arranged to reside on the plains, directly on the conventional spacing layer **16**.

In practice it has been found that a suitable material for the electrically conductive nonmagnetic layer **48** is chrome-gold alloy. Other nonmagnetic metals like copper and aluminum may also be adapted for use in layer **48**.

Metal Masking Layer

D.C. Bullock and M.S. Shaikh; U.S. Patent 4,098,917; July 4, 1978; assigned to Texas Instruments Incorporated, describe a method of providing a substrate with a patterned metal layer disposed thereon, which method has particular application in magnetic bubble domain technology in providing a bubble propagation path of of magnetically soft material, such as permalloy on a magnetic bubble-supporting film.

In forming the pattern in the metal mask, a layer of a metallic masking material taken from the group consisting of vanadium, tantalum, titanium and a titanium-tungsten alloy is deposited onto the layer to be patterned. Thereafter, a layer of photosensitive material, i.e., photoresist, is deposited onto the layer of the metallic masking material, the layer of photosensitive material being selectively exposed to a source of light. After the photosensitive material is developed to define a pattern therein exposing selected regions of the layer of metallic masking material, the exposed regions of the layer of metallic masking material are subjected to a plasma etch to selectively remove these exposed regions down to the level of the layer to be patterned.

Thereafter, the remaining portion of the layer of photosensitive material is removed. In this way, the remaining portions of the layer of metallic masking material are exposed as a patterned metal mask protecting selected portions of the layer to be patterned.

In accordance with a specific embodiment of the process, the layer of metallic masking material is vanadium, which has the properties of being etched at a relatively fast rate in a plasma atmosphere while also being resistant to ion milling so as to be etched at a slow rate when subjected to this treatment. Thus, the method is employed to construct a magnetic bubble domain memory chip by forming a patterned bubble propagation path of magnetically soft material on a magnetic bubble-supporting film secured to a nonmagnetic substrate.

In this instance, the layer of material to be patterned which is deposited onto the substrate takes the form of a layer of magnetically soft material overlying a magnetic bubble-supporting film, with the layer of metallic masking material and the layer of photosensitive material being deposited thereon in sequence. After development of the layer of photosensitive material to define a pattern therein exposing selected regions of the layer of metallic masking material, these exposed regions of metallic masking material are subjected to a plasma etch for selective removal of same down to the level of the layer of magnetically soft material.

The remaining portion of the layer of photosensitive material is then stripped, exposing the remaining portions of the layer of metallic masking material as a patterned metal mask which protects selected portions of the layer of magnetically soft material disposed therebeneath. Ion milling is then employed to remove the exposed regions of the layer of magnetically soft material, thereby producing the bubble propagation path pattern in the remaining portions of the layer of magnetically soft material as protected by the overlying patterned layer of metallic masking material.

The patterned metal mask is then subjected to a plasma etch for stripping the metallic mask from the structure to expose the bubble propagation path pattern of magnetically soft material previously covered thereby.

In this technique, the layer of photoresist material employed may be relatively thin, since the actual patterned mask in forming the desired pattern in the layer to be patterned is provided by the metal mask. Use of a layer of photoresist material of relative thinness enables the method to provide extremely fine geometry patterns, wherein resolutions of 1 μ or smaller are achievable. The subsequent ion milling of the layer to be patterned enables the resolution achieved to approach 0.25 μ.

PHOTOACTIVE AND LUMINESCENT MATERIALS

PHOSPHOR LAYERS

Display Device Utilizing Superimposed Dual Masks

This process relates to display devices comprising a layer of electroluminescent material which is capable of conducting electric current and which is disposed between a pair of electrodes to which a voltage is to be applied to excite the electroluminescent material to luminescence.

P.J.F. Smith; U.S. Patent 4,119,745; October 10, 1978; assigned to Smiths Industries Limited and Phosphor Products Company Limited, England describes a method of manufacturing a display device of the kind specified, which comprises the steps of disposing, on an assembly having a substrate with one of the electrodes formed thereon, masking means with an aperture defining a region in which is to be provided electroluminescent material that is capable of conducting electric current, disposing the electroluminescent material in the aperture, forming the other of the electrodes on the exposed surface of the electroluminescent material when it has solidified, and removing the masking means. The masking means may be adhesively secured to the assembly, the bond between the masking means and the assembly being such as to permit ready removal.

Figure 8.1a is a plan view of a display device. Figure 8.1b is a sectional view on the line **II–II** of Figure 8.1a. Figures 8.1c and 8.1d are plan views of masks used. Figures 8.1e to 8.1g are a series of sectional side views used to explain the method. Referring to Figures 8.1a and 8.1b, the display device includes a rectangular substrate **10** of glass on which is formed by a conventional photoetching process a transparent, positive electrode **11**, and a conducting strip **12** at the left-hand edge of the substrate, to which are to be connected, respectively, the positive and negative terminals of a unidirectional voltage supply. The positive electrode **11** is rectangular and spaced from the conducting strip **12**, this space being indicated in Figure 8.1a by the dashed lines **13** and **14**. The two display regions are defined by respective rectangular sections **15** of the phosphor-binder mixture deposited on the positive electrode **11**, and the remainder of the upper surface (Figure 8.1b) of the positive electrode is coated with a layer **16** of electrically insulating lacquer by means of a conventional silk-screening or photoresist process. This layer **16** extends into and fills the space between the positive electrode **11** and the conducting strip **12**. Individual negative electrodes **17** of aluminum are provided on the upper surfaces of the phosphor display regions **15**. These electrodes **17** cover the whole of the upper surfaces of the display regions **15** and have conducting tracks **18** that connect the electrodes **17** to the conducting strip **12**. The upper surface of the display device is covered by a sheet **25** of glass that is mounted on two strips **26** of butyl rubber disposed adjacent the edges of the substrate **10**, the upper surface of the so-formed assembly then being covered by a layer **19** of a suitable encapsulation material.

Figure 8.1: Display Device Utilizing Superimposed Dual Masks

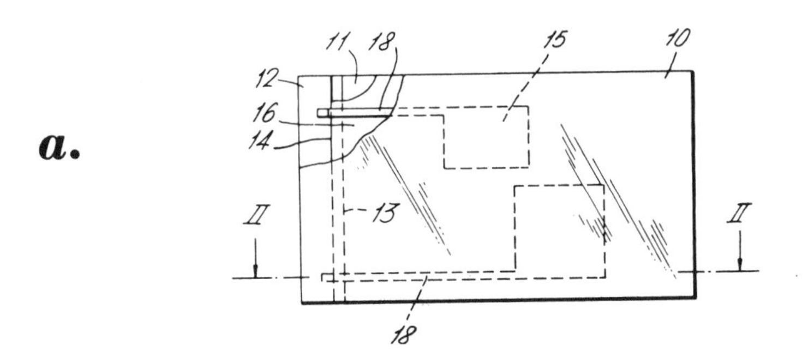

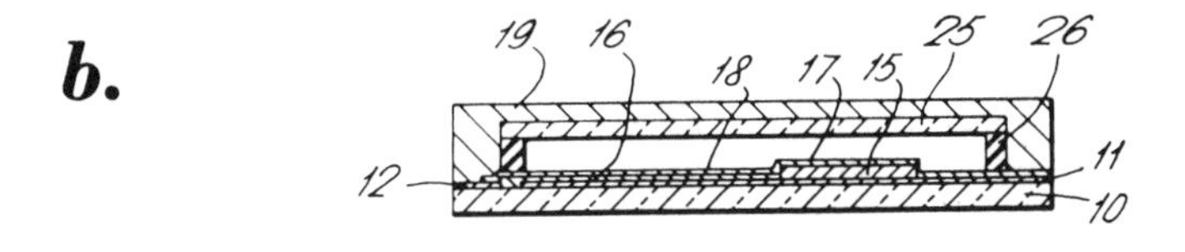

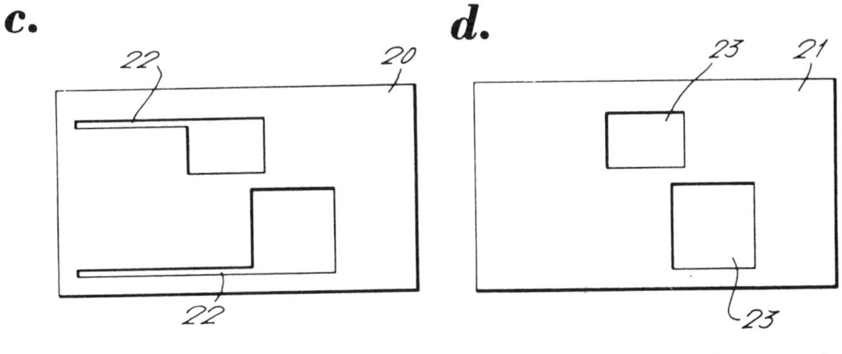

(continued)

Figure 8.1: (continued)

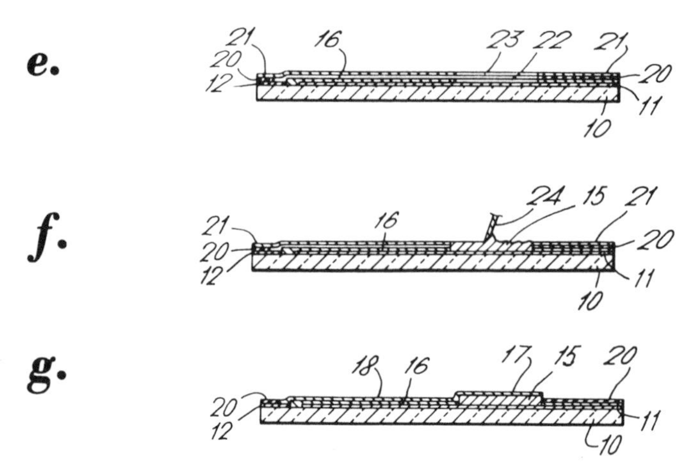

Source: U.S. Patent 4,119,745

When a undirectional voltage is applied between the positive electrode **11** and the conducting strip **12**, the phosphor display regions **15** are excited to luminescence, this light-emission being readily visible through the glass layer **10** and the transparent electrode **11**.

In accordance with this process, masking means are used in the formation of both the phosphor regions **15** and the negative electrodes **17**. This masking means comprises two masks **20** and **21** shown in Figures 8.1c and 8.1d, respectively, which are each of rectangular shape and of the same size as the glass substrate **10** of the display device. The mask **20** has apertures **22** therein shaped to correspond to the phosphor regions **15** and the conducting tracks **18** while the mask **21** has apertures **23** therein corresponding only to the phosphor regions **15**. These apertures **22** and **23** are formed by stamping.

The masks **20** and **21** may be of any suitable material such as plastics or paper and preferably have a thickness of the order of 50 to 100 μ. The masks **20** and **21** are each coated on their lower surface with adhesive (for example, a low-tack acrylic adhesive) and, if desired, the adhesive coatings may be covered by removable covering layers (not shown) having high-release surfaces to facilitate their removal. The covering layers are conveniently of paper coated with plastics.

Various materials have been found suitable for forming the masks **20** and **21**. For example, these masks may comprise transparent protection masks of Sellotape Type No. 1421. Alternatively, the masks **20** and **21** may comprise polyvinyl chloride tape such as Fablon.

The positive electrode **11**, the conducting strip **12** and the layer **16** are first formed on the glass panel **10** as previously described. The masks **20** and **21** are then adhesively secured to the layer **16** as shown in Figure 8.1e with the mask **20** disposed beneath the mask **21** and the apertures **22** and **23** aligned with one another and with the rectangular aperture formed in the layer **16**. The apertures **23** and the portions of the apertures **22** aligned therewith are then filled with the phosphor-binder mixture to form the display regions **15** (Figure 8.1f). The mixture is arranged to fill completely the aligned apertures, any excess phosphor mixture being removed by a scraping tool **24** so that the exposed surfaces of the regions of the phosphor mixture are flush with the upper surface of the mask **21**. This scraping tool **24** is wide enough to extend completely across the mask **21** so as to permit any excess phosphor mixture in the two regions to be removed in a single operation.

When the phosphor mixture has solidified by being cured or allowed to dry, the mask **21** is removed to expose the mask **20** (Figure 8.1g), and the negative electrodes **17** with their conducting tracks **18** of aluminum are formed by a conventional vacuum-deposition process on the exposed surfaces of the regions **15** and in the elongate slotted-portions of the apertures **22**. In an alternative arrangement, the negative electrodes **17** and the conducting tracks **18** are formed by a screen printing process.

Solid State Flat Panel Display Device

F.-C. Luo and T.P. Brody; U.S. Patent 4,135,959; January 23, 1979; assigned to Westinghouse Electric Corp. describe a large area flat panel solid-state display in which thin film transistor addressing and control circuitry are integrally connected to the display medium while being electrically isolated at noncontact portions to prevent extraneous excitation. An array of spaced apart interconnected rows and columns of thin film transistor control circuitry is disposed upon a substrate. The interconnecting of the circuitry is via switching signal, information signal, and power signal bus bars which define between their intersection unit display cells which are repeated over the entire panel. The method comprises:

(a) Depositing the interconnected array of thin film switching and control elements and signal buses, and the first electroluminescent electrode on the substrate by successive vacuum deposition of conductive metal, insulating metal oxide, and semiconductive material in patterns which define and interconnect the display elements;

(b) Applying a relatively thick laminated photopolymerizable insulator layer over the entire area of the panel over the deposited elements;

(c) Exposing to radiation the photopolymerizable insulator layer disposed over the thin film elements and the signal buses while not exposing the area over the first electroluminescent electrode to thereby polymerize the exposed portions;

(d) Removing the unexposed portions of the insulator layer to expose the first electroluminescent electrode;

(e) Depositing a layer of electroluminescent phosphor over the entire panel; and

(f) Depositing a light transmissive conductive top electrode onto the top surface of the electroluminescent phosphor layer.

Three-Layer Laminate

Phosphor screens for a cathode ray tube, for example, a tricolor mosaic screen for a color television picture tube, comprise phosphor dots or stripes arranged in a predetermined pattern on the surface of a glass face plate. The phosphor dots consist of dots of a blue-emitting phosphor, dots of a green-emitting phosphor and dots of a red-emitting phosphor. Formed on the surface of the phosphor screen is a so-called metal back, i.e., an electron-permeable metal layer like an aluminum layer. The metal layer acts as a light reflector and an electrical conductor. Since the surface of the phosphor screen is rough, the metal layer fails to have a specular smooth surface if formed directly on the surface of the phosphor screen, resulting in a decreased amount of luminescent light reflected by the metal layer toward the viewer. It follows that the light output of the cathode ray tube becomes insufficient, namely, the tube lacks in satisfactory brightness.

T. Ito and H. Tsukagoshi; U.S. Patent 4,122,213; October 24, 1978; assigned to Tokyo Shibaura Electric Company, Limited, Japan describe a process for forming a laminate of superposed layers respectively representing different organic substances on the surface of the phosphor screen for a cathode ray tube and, then, forming a metal layer on the laminate, followed by heating for volatilizing the laminate. The superposed layers constituting the laminate have different thermal decomposition points. The laminate may consist of an acrylic resin film or layer and a film of a water-soluble high molecular compound formed on the acrylic resin film. But, a more preferred laminate is of a three-layer type, which is prepared by forming an additional film of a silane compound beneath the acrylic resin film.

The laminate of this process is advantageous over the conventional acrylic resin film of one layer type in that the laminate permits forming a metal layer having a more smooth surface.

According to Figure 8.2, one preferred embodiment of this process is explained as follows. Reference numeral **11** denotes a portion of the glass face plate of a color television picture tube, numeral **12** representing dots of phosphor formed over the face plate **11**. A silane compound layer **13** is formed to cover the phosphor dots **12**. The layer **13** is formed by coating the phosphor dots **12** with an aqueous solution consisting essentially of a silane compound, followed by drying. It is possible to remove excess solution and level the thickness of the solution film by rotating the face plate **11** during the coating step. The silane compounds used include glycidoxypropyl trimethoxy silane (SH-6040), γ-glycidoxypropyl trimethoxy silane (A-187), γ-amino propyl triethoxy silane (A-1100) and γ-methacryloxypropyl trimethoxy silane (A-174). It is preferred that the aqueous solution contain about 0.1 to 10% by weight of the silane compound.

An acrylic resin layer **14** above the silane compound layer **13** is formed by coating the film **13** with an aqueous emulsion consisting essentially of an acrylic resin, followed by drying. The acrylic resin used is substantially the same as that used in U.S. Patent 3,582,390 and includes acrylate resin and acrylate resin copolymer. A preferred acrylic resin content of the emulsion ranges from about 10 to 30% by weight. As is the case with the formation of the silane compound layer, it is preferred to rotate the face plate **11** during the coating step of the acrylic resin emulsion.

Figure 8.2: Three-Layer Laminate

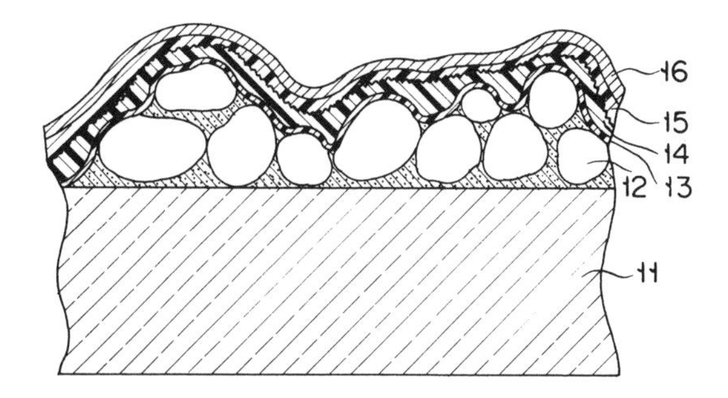

Source: U.S. Patent 4,122,213

Formed on the acrylic resin layer **14** is a layer **15** of a water-soluble high molecular compound. The layer **15** is formed by coating the acrylic resin layer **14** with an aqueous solution consisting essentially of a water-soluble polymer, followed by drying. The water-soluble polymers used include polyvinyl alcohol, polyvinyl pyrrolidone and polyacrylamide. The aqueous solution of the polymer may contain traces of colloidal silica. It is preferred that the aqueous solution contain about 0.05 to 20% by weight of the water-soluble polymer. It is also preferred to rotate the face plate **11** during the coating step of the water-soluble polymer. An aluminum film **16** is formed on the water-soluble polymer layer **15** by vacuum deposition.

The metallized laminate is then subjected to continuous heating to 430°C. Since the silane compound layer **13**, the acrylic resin layer **14** and the water-soluble polymer layer **15** are sequentially decomposed and volatilized with time in accordance with their thermal decomposition points, a large amount of gas is not evolved at a time. The gases evolved from the layer having a lower thermal decomposition point form passageways through which the gases evolved from the other films escape, thereby preventing blistering of the aluminum film.

The water-soluble polymer layer **15** serves to make smoother the surface of the acrylic resin layer **14**, resulting in the aluminum film **16** formed on the layer **15** being able to reflect the luminescent light more effectively than an aluminum film formed directly on the acrylic resin film.

The laminate made of a plurality of organic films produces an additional effect. Whether the acrylic resin layer is slightly thinner or thicker than desired, the blistering of the metal layer is not caused, leading to a prominently high productivity or yield of the product cathode ray tube.

Example: The following table shows the brightness of 18" color television picture tubes respectively provided with aluminized tricolor mosaic screens prepared according to the method of this process and a prior art method.

Color	This Process	Prior Art
Green	133.6 fL	119.3 fL
Blue	21.0 fL	18.5 fL
Red	32.0 fL	29.1 fL
White	213 μA	237 μA

In the case of this process, a silane compound film, an acrylic resin film and a water-soluble polymer film were successively formed over the phosphor dots supported by the glass face plate in the manner described. Specifically, the substances mentioned were coated while the glass face plate was rotated and then dried. For the formation of the silane compound film, an aqueous solution containing 5% by weight of A-187, γ-glycidoxypropyl trimethoxy silane, was used. An aqueous emulsion containing 14% by weight of acrylic resin was used for the formation of the acrylic resin film. On the other hand, 6 wt % aqueous solution of polyvinyl alcohol was used for the formation of the water-soluble polymer film. Finally, the organic laminate thus formed was aluminized by vacuum deposition and then baked so as to volatilize the films of the three organic substances.

In the case of the prior art, an aqueous solution containing 14% by weight of acrylic resin was coated over the phosphor dots, followed by drying so as to form an acrylic resin film. Then, an aluminum layer was formed on the acrylic resin film by vacuum deposition, followed by baking out the resin film.

The brightness in foot Lamberts (fL) for each of green, blue and red colors on the color television picture tube was determined under a fixed current and voltage. On the other hand, the brightness of the white screen was determined in terms of the total current (μA) flowing through the three electron guns required for keeping the brightness of the white screen, a specified tint of white, at 32 foot Lamberts.

The above table shows that the case of this process is brighter than the prior art case by about 10%.

Two-Layer Resin Substrate

K. Mitobe and Y. Uehara; U.S. Patent 4,123,563; October 31, 1978; assigned to Hitachi, Ltd., Japan describe a process for the production of color television picture tubes which comprises at least the step of forming a phosphor layer on the inner surface of a panel, the step of coating the inner surface of the panel including the phosphor layer with an aqueous emulsion of a water-insoluble film-forming resin to form a volatilizable substrate layer, the step of forming a metal film on the substrate layer and the step of volatilizing the organic substances, characterized by forming the substrate layer from at least two layers each of an emulsion, adding colloidal silica, an aqueous ammonium oxalate solution and aqueous hydrogen peroxide to the first layer emulsion, and adding polyvinyl alcohol-boric acid complex and a small amount of ammonium hydroxide to the emulsions of the layer contacting directly the metal film.

Example: A green phosphor slurry consisting of 30 parts by weight of a green phosphor, 2.5 parts by weight of polyvinyl alcohol, 0.2 part by weight of potassium bichromate, 0.05 part by weight of a surface active agent and 67.25 parts by weight of pure water is coated onto the inner surface of a faceplate by rotary coating method, heated by a heater and dried. A shadow mask is installed and the thus formed coating is exposed to light from an extra-high-pressure mercury lamp through the shadow mask for 60 seconds and developed by spraying with warm water to obtain a green phosphor picture element. Then, blue and red phosphor picture elements are formed in the same manner. The phosphor layer is coated with a filming emulsion consisting of 8.8 parts by weight of a 34% by weight copolymer of n-butyl methacrylate and methacrylic acid, 1.7 parts by weight of 30% by weight colloidal silica, 20.0 parts by weight of a 2.5% by weight aqueous ammonium oxalate solution, 1.4 parts by weight of 35% by weight aqueous hydrogen peroxide and 68.1 parts by weight of pure water, heated by a heater and dried to form a resin film as the first layer. Another filming emulsion consisting of 38.7 parts by weight of a 38% by weight acrylic resin Primal B74 (an aqueous emulsion of an acrylic resin), 30.0 parts by weight of a 2% by weight polyvinyl alcohol-boric acid complex, a small amount of ammonium hydroxide, 2.0 parts by weight of 100% by weight glycerol and 29.3 parts by weight of pure water is then coated and dried to form a resin film as the second layer. Aluminum is evaporated thereon to form a metal film and the organic substances are volatilized by bakeout.

The fluorescent screen thus produced is very good in that the blister or separation of the metal film does not occur on the surface of curvature of the panel, the panel skirt and the panel corner and the undecomposed resin which has turned brown does not remain. When a color television picture tube is assembled by the use of the fluorescent screen and a color picture image is produced, the brightness of the image is about 5% higher than that in prior art color television picture tubes.

In a similar process, *N. Watanabe; U.S. Patent 4,139,657; February 13, 1979; assigned to Hitachi, Ltd., Japan* describes a process for producing a color television picture tube which comprises at least the step of coating phosphor slurries onto the inner surface of a panel to form a phosphor layer, the step of forming two layers each consisting of an aqueous filming emulsion containing an acrylic resin, the step of forming a metal film on the second layer, and the step of baking out the organic substances, characterized by using the phosphor slurries containing an acrylic resin and using the first aqueous filming emulsion containing an acrylic resin having an elongation of 10% or more and using the second aqueous filming emulsion containing an acrylic resin having an elongation of less than 10%.

The value of elongation used herein means the breakdown elongation of the acrylic resin sheet (10 mm in width, 50 mm in length, and 1 mm in thickness) as made by pressing powder of the acrylic resin at 180°C and 100 kg/cm^2.

As shown in Figure 8.3, a phosphor slurry comprising a phosphor, polyvinyl alcohol, ammonium dichromate, an acrylic resin and a surface active agent is coated onto the inner surface of a faceplate **1**, and exposure to light and development are then repeated to form a phosphor layer **2** comprising green, blue and red phosphor picture elements. In this case, the amount of the acrylic resin contained in the phosphor slurry is 10 to 300 parts by weight, and preferably 20 to 100 parts by weight, per 100 parts by weight of the polyvinyl alcohol (solid basis).

A filming agent consisting mainly of an acrylic resin having an elongation of 10% or more, and preferably 40% or more, is then coated onto the phosphor layer **2**, heated and dried to form the resin film **31** of the first layer. Another filming agent comprising an acrylic resin having an elongation of less than 10%, boric acid ester of polyvinyl alcohol, etc. is coated, heated and dried to form the resin film **32** of the second layer. A metal film **4** is then formed thereon. Baking out is conducted to volatilize the organic substances. The fluorescent screen formed according to such a process is good in that the metal film **4** contains few pores and cracks and no blister is caused.

Figure 8.3: Two-Layer Resin Substrate

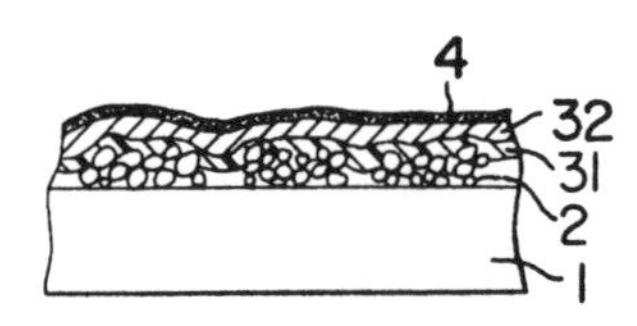

Source: U.S. Patent 4,139,657

Example: A green phosphor slurry consisting of 30 parts by weight of a green phosphor (ZnS:CuAl), 2.2 parts by weight of polyvinyl alcohol, 0.22 part by weight of ammonium dichromate, 1.1 parts by weight of an acrylic resin Primal C72, 0.05 part by weight of a surface active agent and 66.43 parts by weight of pure water is coated onto the inner surface of a faceplate by tilting and rotating the faceplate panel, heated by a heater and dried. A shadow mask is installed and the thus formed coating is exposed to light from an extra-high pressure mercury lamp at an appointed position through the shadow mask for 60 seconds and developed by spraying with warm water to obtain a green phosphor picture element.

Then, blue and red phosphor picture elements containing an acrylic resin are formed by the use of a blue phosphor (ZnS:Ag) and a red phosphor (Y_2O_2S:Eu) in the same manner. The phosphor layer is coated with a 6% aqueous emulsion of n-butyl methacrylate polymer (elongation 77%), heated by a heater and dried to form a resin film as the first layer. Another aqueous emulsion consisting of 15 parts by weight of an acrylic resin Primal B74 (elongation 1.3%), 0.5 part by weight of boric acid ester of polyvinyl alcohol, 0.5 part by weight of colloidal silica, 2 parts by weight of glycerol and 82 parts by weight of pure water is then coated and dried to form a resin film as the second layer. Aluminum is vapor-deposited thereon and the organic substances are volatilized by bakeout.

The fluorescent screen thus produced is very good in that the phosphor layer and the metal film contain few pores and cracks and no blister is caused. When a color television picture tube is assembled by the use of the fluorescent screen and a color picture image is produced, the brightness of the image is about 5% higher than the brightness of prior art color television picture tubes.

PHOTOGRAPHIC MATERIALS

X-Ray Recording Material

H. Dannert, H.-J. Hirsch, E. Klein, and K.-H. Panstruga; U.S. Patent 4,121,933; October 24, 1978; assigned to U.S. Philips Corporation describe an electrophotographic recording material, in particular for x-rays, in which at least a photoconductive layer of tetragonal lead monoxide and a binder is provided on a carrier.

The recording material is characterized in that the tetragonal lead monoxide has a grain size of from 1 to 50 μm and, in particular, from 5 to 20 μm. The re-

recording material is preferably manufactured by dispersing tetragonal lead monoxide with the aboveindicated grain size without mechanical force in the binder and sedimenting it on the layer carrier without mechanical force.

A particular advantageous embodiment of the method is characterized in that:

(A) Orthorhombic lead monoxide is manufactured from ultrapure lead acetate and ultrapure ammonia solution in very pure deionized water in a vessel of silicate-free material;

(B) The orthorhombic lead monoxide in a solution of 1.5 to 18% by weight of ammonia in water is kept at a temperature between 0° and 80°C from 2 to 2,000 hours, in which the reaction mixture may be kept in motion;

(C) The powdered tetragonal lead monoxide thus obtained is washed with an aqueous ammonia solution of 3½% by weight, dried at 100° to 120°C and, if desired, (a) is heated at a temperature between 110° and 350°C at normal pressure in air for 15 to 120 hours, and (b) if desired, after (a) is heated in an argon flow at 420° to 440°C, kept at this temperature for 1 to 5 hours and cooled to room temperature in a period of time of from 5 to 10 hours;

(D) The powdered tetragonal lead monoxide thus aftertreated in a solution of the binder of 2 to 25% by weight, preferably 3 to 10% by weight, which binder is selected from the group of lacquer synthetic resins and polyvinyl carbazole, in an organic solvent is provided on the layer carrier for sedimentation without force, the supernatant binder solution is decanted and the solvent is evaporated; and

(E) The recording material thus obtained is heated in air at 150° to 250°C for 1 to 2,000 hours, preferably 10 to 25 hours, and is then left to stand, if desired, for at least 20 to 30 hours under the influence of normal spatial illumination.

Example: 24 g of lead acetate (ultrapure, Merck) are dissolved in 50 ml of deionized water having a resistivity exceeding 10 Mohm cm. 240 ml of 10 N ammonia solution are added in 10 to 15 seconds while stirring rapidly. The initially white suspension is then stirred at the same speed for approximately 7 minutes. In this period, the conversion of the white basic lead acetate into the yellowish green orthorhombic lead oxide takes place. The precipitate is then washed decantingly 5 times, each time with 100 ml of 2 N ammonia solution. These treatments are carried out in polyethylene vessels.

The orthorhombic PbO powder thus obtained is converted into the tetragonal modification in the dark in 0.5 N ammonia solution at 25°C for a period of 5 days. X-ray examinations have proved that the conversion is complete. The obtained particle size was between 5 and 20 μm. The orthorhombic modification could no longer be established. The resulting red PbO powder was heated at 250°C in air at normal pressure for a period of 24 hours.

From a dispersion in 30 ml solution of 3% of a saturated polyester T203 (chemical composition: aromatic carboxylic acids and aliphatic diols; physical characteristic numbers: melting range 105° to 150°C, viscosity 0.6 to 0.7) in butanone, 2.2 g of PbO are sedimented on a surface of 27.5 cm^2 of aluminum as a layer carrier; the supernatant solvent is decanted. The layer is dried in air and then heated in the dark at 200°C for 24 hours. The sensitivity measurement was carried out repeatedly without exposing the plate to daylight. After this method, a sensitivity of 13 mr is obtained.

Encapsulated Dye Developer

W.J. McCune, Jr.; U.S. Patent 4,137,194; January 30, 1979; assigned to Polaroid Corporation describes photographic processes and products for obtaining color images by diffusion-transfer processes, wherein color-providing substances such as dye developers utilized to provide the colored images are encapsulated within minute capsules of alkali-permeable polymeric material, the capsules being coated with silver halide.

One particular applicability of the capsules is in certain products or assemblages useful in photographic transfer reversal processes capable of producing a color print. In processes for forming color images by transfer techniques, an imagewise distribution of one or more color-providing substances is formed in unexposed parts of a negative photosensitive element having one or more light-sensitive portions having silver halide therein and transferred to an image-receiving element. The imagewise distribution of each color-providing substance so transferred and deposited upon the image-receiving element arranged in superposed relation to the negative photosensitive element colors the image-receiving element a predetermined color to provide therein a monochromatic or multichromatic image comprising one or more positive images of negative latent color images formed by the exposure of the photosensitive element.

It has been found that color-providing substances such as mentioned above utilized to provide color images may be encapsulated or contained within a shell-like coating of a film-forming polymeric material. The nucleus of such capsules may comprise a color-providing substance in substantially solid form which is, for example, solubilized by the processing composition or the nucleus may comprise a suitable liquid solvent or medium in which a color-providing substance is dissolved or suspended.

In order to prevent loss of the nucleus liquid and/or to retard or prevent undesirable environmental materials from reacting with the nucleus materials, the outer surfaces of the capsule wall are coated with a thin continuous film such as, for example, of aluminum or silver. These capsules may be utilized in the photosensitive element, for example, in or behind the silver halide emulsion. Preferably, a coating or layer of the capsules containing a color-providing substance, e.g., a dye developer, is placed behind the silver halide emulsion, i.e., on the side of the emulsion adapted to be located most distant from the photographed subject when the emulsion is exposed and most distant from the image-receiving element when in superposed relationship therewith. Placing such capsule layer or stratum behind the emulsion has the advantage of providing increased contrast in the positive image.

When employing a layer or stratum of capsules which contain, for example, a dye developer, reducible dye or complete dye in the preferred manner, i.e., behind the silver halide emulsion, it has been found that the effective emulsion speed can be substantially increased by having the outer surfaces or portions of such capsules provided with a thin coating such as of a suitable metal adapted to reflect light of all visible wavelengths incident upon it. The increase in effective emulsion speed is apparently due to the fact that previously the colored dye would ordinarily absorb light whereas now the reflecting coating utilizes some of the otherwise wasted light to increase the absorption of light in the emulsion when it is exposed. Thus, the emulsion is effectively more sensitive since the exposure depends on the absorption of light.

In this process, there is thus employed capsules having the outer surfaces of the capsule wall coated with a thin, light-reflecting film or layer such as, for example, of silver or aluminum which not only increases the effective emulsion speed but also increases the impermeability of the capsule walls so as to prevent loss of encapsulated liquid material or to exclude environmental substances such as moisture, oxygen or the like from contact with the encapsulated materials. The thin continuous light-reflecting coating may be applied to the capsule walls by any one of several methods such as, for example, spraying, electroplating, vapor deposition and the like.

Bismuth Image-Forming Layer

H.H. Wacks, M. Izu and D.J. Sarrach; U.S. Patent 4,138,262; February 6, 1979; assigned to Energy Conversion Devices, Inc. describe imaging film which comprises a flexible, transparent, plastic substrate having deposited on a surface thereof a thin, continuous, high optical density layer of bismuth, or an alloy of bismuth. The layer of bismuth is highly opaque, and initially has a dull, grayish appearance. The outer surface of the bismuth layer is roughened and is uniquely adapted to receive a photoactive material which functions both as a photoresist and as a protective overlayer for the bismuth.

The bismuth may be deposited on the substrate by any of various techniques, including vapor deposition, vacuum deposition, or sputtering. In accordance with a preferred method, an initial layer of bismuth is deposited on the substrate by sputtering. A second or top layer of bismuth is then deposited on the first layer by evaporation. The sputtering and evaporation of the bismuth advantageously is carried out in-line, or successively, in a continuous operation in the same chamber.

The flexible plastic substrate employed in making the film desirably is transparent and advantageously is nonoutgassable, that is, it should not emit gas, usually air or water vapor, under vacuum conditions, or give off any volatile materials. In addition, the substrate should be free of voids and free of surface irregularities which can give rise to pinholes in the final structure. Over and above the foregoing considerations, the film should have good dimensional stability, and to this end, biaxially oriented films are preferred. Exemplary of plastic substances having utility for the purposes of this process are polyesters, particularly, polyethylene terephthalate, polystyrenes, polyethylenes, polypropylenes, and the like. Especially preferred is a polyester film available as Melinex O, 400 gauge.

This film is characterized by its exceptionally smooth surface, its clarity, its freedom from voids, and its excellent dimensional stability. The thickness of the substrate may range from 2 to 7 mils, with a thickness of 3 to 5 mils being preferred.

Example: A 1,200' roll of an outgassed polyester film (ICI Melinex O, 400 gauge) was placed in the chamber of a metallizing unit containing a radio frequency sputtering cathode comprising a 17.5" x 8" x 0.25" thick bismuth plate case directly on and bonded to a water-cooled backing plate, and an evaporation source comprising a single carbon resistance boat with dimensions of 15" x 1" x 0.25" containing 13 individual cavities measuring 1" x 0.75" x 0.5", each cavity having a 16 g preweighed bismuth pellet in it. A water-cooled drum was provided in the chamber. The chamber was maintained under a vacuum of 10^{-6} torr. High purity argon gas was introduced into the chamber to provide a background pressure of 5×10^{-3} torr.

The polyester film was fed at a speed of approximately 20 fpm past the bismuth sputtering cathode while maintaining the temperature of the film at 100°C. The forward power of the cathode was 400 W while its reflected power was 40 W. While maintaining the same rate of feed, the film is passed in a water-cooled drum through a bismuth vapor cloud emanating from the resistance boat which is heated to a temperature of 624°C. The bismuth coated film is rewound on a takeup roll in the chamber. Dry nitrogen gas is introduced into the chamber to 1 atm, after which the roll of bismuth coated film is removed. The bismuth layer on the film was about 2600 Å thick and had an OD of greater than 5.

A solution of a quinone diazide-type photoresist comprising a 50-50 mixture of Shipley's AZ 1350J and AZ 1375 positive photoresists, having a solids content of 35% and a viscosity of 80 cp, was applied to the bismuth coated film to a wet thickness of about 6 μ, at a web speed of 45 fpm, utilizing an offset gravure roller coater. The photoresist coated film was then passed, at the same web speed, through a flow out zone, an infra-red lamp drying zone and a convection drying zone, the latter two zones being maintained at a temperature to provide a film temperature of 100°C. The application of the photoresist layer and the drying of the layer were carried out under dust-free conditions.

The finished imaging film had a black, nonshiny finish on the photoresist side, and a shiny, highly reflective appearance on the substrate side. The crazing properties of the imaging film were tested by cutting a 35 mm strip, 12" long from the roll, and attaching a 500 g load to an end thereof. The thusly loaded strip was passed over a mandrel, of a given diameter, first in a downward direction and then upwardly. The unexposed film was then run through a standard etching solution two times, and examined with optical microscopy for material failure. If breaks or cracks developed in the photoresist materials, the etching solution would penetrate the film to the substrate. Any such breaks or cracks would be observed in transmitted light as a series of parallel lines perpendicular to the direction of travel of the film over the mandrel. No crazing was exhibited by the imaging film.

Silver Halide Vapor Deposition

Y.P. Malinovski, B.D. Mednikarov, and J.J. Assa; U.S. Patent 4,123,280; October 31, 1978; assigned to Zlafop pri BAN, Bulgaria describe a method for vacuum evaporation and deposition of pure silver halide layers on a continuously moving support to which has been applied a priming layer for enhancing silver halide substrate adhesion for a vacuum deposited silver halide layer, whereby uncontrollable sensitization of the silver halide layer by impurities in the vacuum chamber is avoided, so that the photographic material obtained has fully reproducible and controllable properties.

According to this process, there is provided a method for the production of a photographic material constituted by a silver halide coated substrate, which comprises placing a silver halide having a purity of not less than 99.999% in a crucible which is formed of a material which is inert when heated to an elevated temperature and which is disposed in or constitutes part of a chamber which is evacuated, heating the crucible by radiant heating means disposed outside the vacuum chamber and positioned and shaped to provide substantially uniform heating of the crucible and its contents and passing over the crucible a substrate primed with a substance which enhances the adhesion of solidified silver halide vapor to the substrate, the crucible being so shaped that a concentrated stream of silver halide vapor issues therefrom and impinges perpendicularly on the substrate which is positioned not more than 10 mm from wall means of the crucible between which passes the stream, and the heating of the crucible and shape of the crucible being such that a silver halide layer is deposited on the substrate at a rate of 200 to 2000 Å/sec.

Fully controllable and reproducible sensitization of the silver halide layer by, for example, gold-iridium treatment to obtain a negative imaging photographic material, or its fogging by the additional vacuum deposition of a monoatomic layer of silver or gold to obtain a positive imaging photographic material can then be effected. Other vapor deposition sensitization methods for producing both positive and negative imaging photographic materials are described in British Patent 1,154,741. Sensitization can also be effected by conventional methods used for sensitizing conventional emulsions.

Particularly suitable primer layers for application to substrates used in the process have been found to be the commercially available photoresist lacquers based on polyvinyl cinnamate (for example, KTFR) or on polyisoprene (for example, Copyrex RN 40). These lacquers enable particularly effective silver halide/substrate adhesion to occur, the adhesion being much better than when coating layers of gelatin or silicon monoxide are used and, at the same time, are suitable for use when sensitization is to be effected using aqueous solutions of sensitizers. These lacquers also cause an increase in the resolution of the final photographic material.

Especially when a direct positive photographic material with high sensitivity is to be obtained the priming layer is preferably constituted by a metal chromate layer up to 500 Å thick, deposited onto the heated substrate by vacuum deposition. Lead chromate, basic lead chromate, barium chromate, silver chromate, or bismuth chromate can be employed as metal chromate, lead chromate being preferred.

Lithographic Printing Plate

J.M. Kitteridge and R.J. Armstrong; U.S. Patent 4,126,468; November 21, 1978; assigned to Vickers Limited, England describe a radiation sensitive composition comprising: (1) a quaternary ammonium salt of the type which will accept at least one electron on exposure to radiation to form a substance capable of causing metal to be deposited onto the substance from an electroless plating solution in contact with the substance and comprising a salt of the metal and a reducing agent, and (2) a chemical sensitizer for improving the speed of photoreduction of the salt, the sensitizer having the general formula shown on the following page.

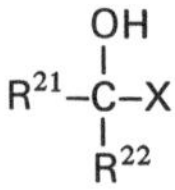

$$R^{21}-\overset{\displaystyle OH}{\underset{\displaystyle R^{22}}{C}}-X$$

In the above formula, R^{21} is a substituted or unsubstituted aryl group, an aryl alkenyl group, or an arylalkyl group; R^{22} is a substituted or unsubstituted aryl group, an alkyl group, an arylalkyl group, an arylalkenyl group, a carboxylic acid group, a carboxylic acid salt group, or a hydrogen atom; or R^{21} and R^{22} together represent a bivalent species; and X represents a carboxylic acid group or a carboxylic acid salt group.

The quaternary ammonium salt is preferably a diquaternary cyclic ammonium salt which, by the addition of one electron, forms Weitz radicals. Such salts are dicationic and contain nitrogen atoms in the molecule, at least two of the nitrogen atoms being quaternized and being constituents of ring structures which are linked and are at least partially aromatic, the link between the ring structures providing a chain of conjugated unsaturation between the nitrogen atoms. Particularly suitable diquaternary ammonium salts are the bipyridinium salts exemplified by:

(1) 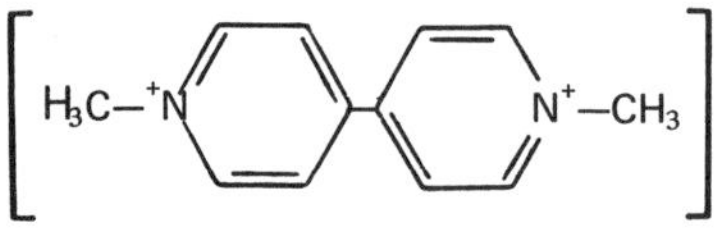 $[H_3C-{}^{+}N(C_5H_4)-(C_5H_4)N^{+}-CH_3]$ $2Cl^-$

(2) 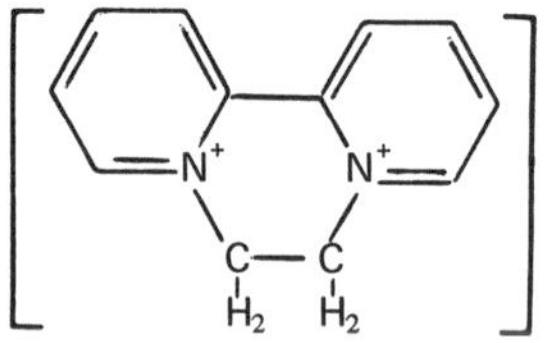 $2Br^-\cdot H_2O$

(3) 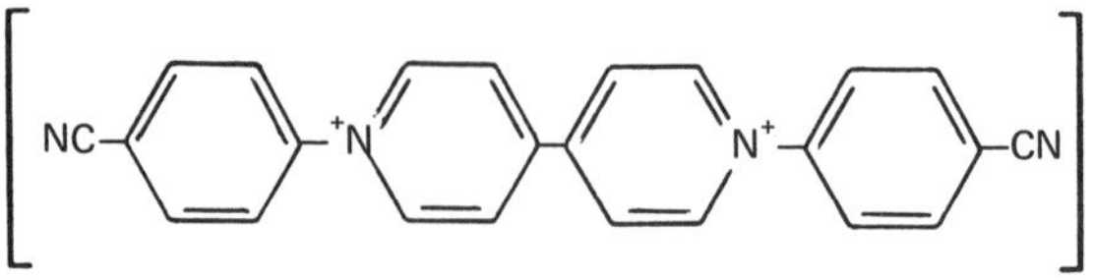 $[NC-C_6H_4-{}^{+}N(C_5H_4)-(C_5H_4)N^{+}-C_6H_4-CN]$ $2Cl^-$

Compounds (1) and (2) are Paraquat and Diquat, respectively. Compound (3) is N,N'-bis(p-cyanophenyl)-4,4'-bipyridinium dichloride and will hereinafter be referred to as CPP^{++}.

In one embodiment, in the formula for the chemical sensitizer, R^{21} is not a phenyl group when R^{22} is a phenyl group. Examples are mandelic acid, p-bromo mandelic acid, p-chloromandelic acid, p-fluoromandelic acid, m-methoxymandelic acid, o-methoxymandelic acid, p-methoxymandelic acid, m-nitromandelic acid, benzilic acid, 3-phenyllactic acid, atrolactic acid, 9-hydroxy-fluorene-9-carboxylic acid, 3,4-dimethoxy-2'-chlorobenzilic acid, di-p-methoxybenzilic acid, 1-naphthyl glycollic acid and styryl glycollic acid. It is particularly preferred to use water-soluble sensitizers such as sodium mandelate, ammonium mandelate, tetra-methyl ammonium p-fluoro mandelate, potassium atrolactate and hexamethylene tetramine mandelate. The chemical sensitizers appear to act as electron donors in the photoreduction action. The molar ratio of quaternary ammonium salt to chemical sensitizer may be, for example, from 10:1 to 1:50, respectively.

The inherent spectral sensitivity of the quaternary ammonium salts can be extended well into the visible region of the spectrum by the inclusion of spectral sensitizers and such sensitizers may be included. Suitable spectral sensitizers are riboflavin, 7-chloro-9-(N-methyl-2-diethylaminoethyl) isoalloxamine chloride (hereinafter referred to as compound RD) and other very similar compounds, and Arconol Yellow [3,6-dimethyl-2-(4-dimethylamino-phenyl) benzthiazolium chloride]. Other sensitizers which can be used include 3,3'-diethylthiacyanine iodide, proflavin, acridine orange, acriflavin, N-methylphenazinium methosulfate, 4-cyanoquinolinium methiodide and erythrosin.

The composition may additionally include a film-forming polymeric binder. The binder may be such that it is inert towards the substance formed on exposure of the quaternary salt or such that it has a stabilizing influence on that substance. Suitable water-soluble or swellable film forming polymers for use as the binder include poly(vinyl alcohol), poly(ammonium methacrylate), gelatin, alginates and maleic anhydride copolymers, e.g., a copolymer of maleic anhydride with styrene vinyl ether or ethylene. Soluble polysaccharides such as polysucrose may also be used as binder as may poly(N-vinylpyrrolidone) either alone or in admixture with poly(vinyl alcohol).

The radiation sensitive plate is prepared by applying to the substrate, preferably of grained and anodized aluminum, a coating solution comprising the radiation sensitive composition (e.g., including the quaternary ammonium salt and, optionally, chemical sensitizer, the binder material and any other desired additives). The amounts of quaternary ammonium salt and binder (if present) in the coating solution are not critical and are dictated solely by the desired sensitivity and by practical considerations. Adequate results may be obtained using a coating solution containing from 0 to 20 parts by weight of water-soluble polymeric binder, from 0.1 to 10 parts by weight of quaternary ammonium salt, and from 70 to 99.9 parts by weight of water. It is preferred to use a coating solution having a pH of from about 1 to 7. In general, the higher the pH, the more sensitive is the composition. The coating solution may be applied to the substrate in a variety of ways but dip or roller coating are the preferred methods.

In use of the radiation sensitive plate of the process in the photomechanical production of a lithographic printing plate, the radiation sensitive layer is image-wise exposed to actinic radiation, e.g., ultraviolet light. The quaternary ammonium salt accepts at least one electron in the areas struck by radiation to form an image which is a negative of the master.

The image-wise exposed plate is processed by contacting the same with an electroless plating solution, for example, a solution containing a silver salt and a reducing agent. Such solutions are well-known per se and are capable of depositing metal without the external application of an electrical field.

An adherent metallic image is thus deposited on the radiation struck areas of the layer. Once a trace of metal has been so deposited, it is capable of catalyzing further deposition of the same or a different metal from an appropriate solution. In this way, a metallic layer can be built up in the image areas.

Example: The following coating solution was prepared and roller-coated onto a sheet of aluminum which had been grained in nitric acid, desmutted in phosphoric acid and finally anodized in sulfuric acid.

	Parts by Weight
CPP^{++}*	1.0
Lemol 16-98	5.0
Riboflavin	0.5
Citric acid H_2O	1.3
Sodium hydroxide	0.5
Water	95.0

*N,N'-di(p-cyanophenyl)-4,4'-bipyridinium

Lemol 16-98 is a poly(vinyl alcohol) of the Borden Chemical Company.

The resultant radiation sensitive plate was exposed through a negative transparency in a printing down frame for 1.5 minutes to the radiation from four Philips 300 W MLU lamps suspended 600 mm above the frame. A green print-out image was obtained. The exposed plate was then immersed in the silver electroless plating solution for 3 minutes at room temperature. The green image was replaced by a black image of silver metal which could be polished to obtain lustrous silver which was in conductive contact with the aluminum substrate.

PHOTOCOPY

Arsenic-Selenium Alloy for Belt-Type Receptors

The formation and development of images on the imaging surfaces of photoconductive materials by electrostatic means is well-known (U.S. Patent 2,297,691). The best known of the commercial processes, more commonly known as xerography, forms a latent electrostatic image on the surface of an imaging layer by uniformly electrostatically charging the surface in the dark, and then exposing the charged surface to a light and shadow image. The light-struck areas of the imaging layer are thus made substantially more charge-conductive and the electrostatic charge is selectively dissipated in such areas. After light exposure, the latent electrostatic image remaining on the imaging surface (i.e., a positive electrostatic image) is made visible by contacting with finely divided colored or black electroscopic material, known in the art as toner. Toner is principally attracted to those areas on the image bearing surface which retain the original electrostatic charge and thereby form a visible positive image.

In structure, the conventional xerographic plate normally has a photoconductive insulating layer overlaying the conductive base or substrate and frequently an interface or charge blocking layer between the two.

The photoconductive layer may comprise a number of materials known in the art. For example, selenium-containing photoconductive material such as vitreous selenium, or selenium modified with varying amounts of arsenic are found very useful in commercial xerography. Generally speaking, the photoconductive layer should have a specific resistivity greater than about 10^{10} ohm-cm (preferably 10^{13} ohm-cm) in the absence of illumination. In addition, resistivity should drop at least several orders of magnitude in the presence of an activating energy source such as light. As practical matter, a photoconductor layer should support an electrical potential of at least about 100 V in the absence of light or other actinic radiation, and may usefully vary in thickness from about 10 to 200 μ.

A.J. Ciuffini; U.S. Patent 4,126,457; November 21, 1978; assigned to Xerox Corporation describes a process to obtain flexible, particularly belt-type photoreceptors having improved light sensitivity, stability and durability, utilizing arsenic-rich selenium alloys as photoconductors in successful working combination with flexible metal substrates having different coefficients of expansion than the photoconductor layer.

This is accomplished by effecting the evaporation and condensation of at least one arsenic-rich selenium photoconductor layer from a heated donor source in a profile of increasing arsenic concentration onto a prepared substrate or interface-substrate combination. Condensation is effected by utilizing as heated donor source during the evaporation and condensation, a plurality of receptacles containing selenium alloy of varying concentrations within the range of about 0 to 40% arsenic by weight. The receptacles can be individually or collectively heated by heating means to provide at least substantial partial evaporation from two or more receptacles simultaneously onto substrate or interface-substrate layers.

The evaporation-condensation step as above described is effected while simultaneously heating and maintaining the temperature of the substrate or interface-substrate layers during evaporation and condensation at a temperature no less than about the glass transition temperature of the selenium alloy of lowest arsenic concentration to be evaporated from the donor source and not less than about 85°C.

Where flexible metal substrates are likely to contain imperfections such as holes or pits capable of trapping liquids and gasses, it is preferred, but not mandatory, that the substrate or interface-substrate be pretreated. In this step, the substrate is preheated by conventional means such as glow discharge to at least 50°C before applying the photoconductive material. When used, the preheating step is usefully carried out for a period of about 15 to 45 minutes, under vacuum and/or in an inert gas such as argon and preferably at a temperature of about 50° to 150°C. This treatment helps to obtain the later even application of photoconductor material and avoids the formation of pits or weak areas on or between the selenium photoconductor and the substrate caused by the heating and escape of trapped gasses during condensation of the photoconductor. This degassing step is found particularly useful when the photoreceptor is an electrolytically-formed belt such as a nickel belt in combination with a polymeric interface layer.

While a variety of flexible metal or metal-covered substrates can be used, it is found that thin flexible belts of aluminum, steel, brass, nickel or the like are generally satisfactory.

Because of the brittleness of arsenic-rich photoconductor layers, it has been found useful, although not always necessary, to utilize belts of such materials which are first coated with a thin adhesion-promoting interface of about 0.5 to 3 μ. These can be applied to the substrate by usual means such as coating, draw coating, dip coating or flow coating. Suitable solvents such as cyclohexanone can later be conveniently removed by drying or washing. Such techniques and interfaces are described in Belgian Patent 784,453.

Polymeric resins acceptable for the above purpose include polybenzimidazoles, polycarbonates, polyesters, polyurethanes, etc., inclusive of blends. The interface layer can also optionally contain additives exemplified by small amounts of conductive or photoconductive pigments such as copper phthalocyanines, zinc oxides (electrography grade), cadmium sulfoselenide and metal-free phthalocyanines.

The heated donor source usefully consists of a plurality of open receptacles capable of holding selenium alloys of varying concentrations. Such receptacles can be of stainless steel and are usefully joined together as a multicompartmented boat for ease in introduction and removal from the vacuum chamber. It is useful if each individual receptacle is capable of holding a loading (i.e., amount of particular alloy) of at least about 60 g of alloy for each substrate to be coated in the vacuum chamber.

Alloy baths, suitable for use in this process, can include 0 to 40% arsenic plus 0 to 10,000 ppm of a halogen dopant such as chlorine or iodine and/or other additives as suggested, for instance, in U.S. Patents 3,312,548 and 2,822,300, and Belgian Patent 784,453. Such alloys are conveniently obtained in the manner disclosed in U.S. Patent 2,822,300, by heating desired amounts of selenium, arsenic, etc., in a sealed container to a temperature of about 825°C for at least about 30 minutes. In this general manner, it is possible to obtain any one of the desired selenium-arsenic alloys to carry out the process.

Barrier Layer of Aluminum Hydroxyoxide

G.L. Dorer; U.S. Patent 4,123,267; October 31, 1978; assigned to Minnesota Mining and Manufacturing Company describes a photoconductive element consisting essentially of: (a) an electrically conductive substrate, (b) a layer of aluminum hydroxyoxide crystallites on the substrate, and (c) a continuous photoconductive layer over the layer of crystallites wherein the photoconductive layer is selected from selenium, selenium compounds and alloys of selenium.

Figure 8.4 shows an electrically conductive substrate **10** to which a structured barrier layer **12** of aluminum hydroxyoxide crystallites (sometimes referred to hereinafter as boehmite) is bonded. Substrate **10** preferably has an electrical resistance several orders of magnitude less than the electrical resistivity of the photoconductive layer **12** after the layer has been illuminated. Generally, substrate **10** has a specific resistivity less than 10^{10} ohm-cm and usually less than 10^5 ohm-cm.

The structured layer **12** is formed by exposing the aluminum surface of substrate **10** to an oxidizing environment containing water so that crystallites **13** of hydrated aluminum oxide grow in situ thereon. Although this can be done by simply immersing substrate **10** in water for a period of time, it is preferable to expose it to a gaseous oxidizing environment that is essentially saturated with water vapor at about 20° to 150°C. The photoconductive materials, employed as layer **14**, are selected from selenium, selenium compounds and alloys of selenium.

Figure 8.4: Barrier Layer of Aluminum Hydroxyoxide

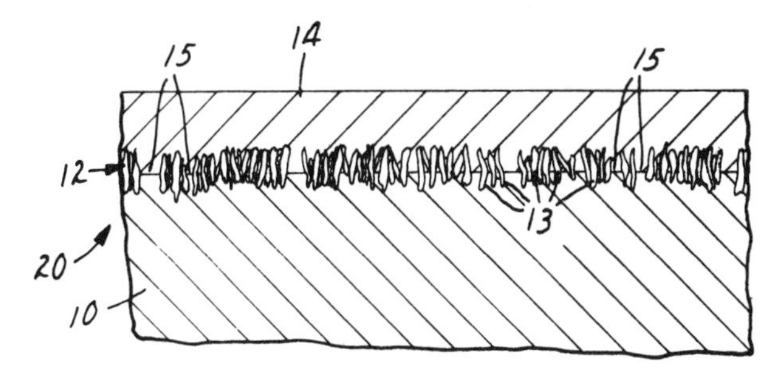

Source: U.S. Patent 4,123,267

When selenium is used, it may be in the amorphous or vitreous form. Representative useful compounds of selenium include arsenic selenide (As_2Se_3), cadmium selenide, tellurium selenide, etc. Representative useful selenium alloys include alloys of selenium with arsenic or tellurium in the vitreous form, arsenic tellurium doped selenium, etc. Preferably, the photoconductive material is selected from vitreous selenium, arsenic or tellurium alloys and arsenic selenide.

The photoconductive material is applied to the aluminum hydroxyoxide layer so as to provide a continuous surface thereon. The thickness of the photoconductive layer is not critical. However, it should be of sufficient thickness so as to provide a visible image when the photoconductive element is processed by electrophotographic techniques. Thus, the photoconductive layer is preferably in the range of about 40 to 60 μ thick.

The photoconductive layer may be applied to the aluminum hydroxyoxide layer by a variety of techniques. Typically, it is applied by evaporating it onto the aluminum hydroxyoxide layer by techniques known to the art.

Generation of Crystalline Trigonal Selenium

D.G. Marsh; U.S. Patent 4,115,115; September 19, 1978; assigned to Xerox Corporation describes a method for the preparation of an electrostatographic photoreceptor which comprises a layer of particulate trigonal selenium dispersed in a polymeric matrix, which comprises:

(a) Forming a solution containing:

(1) Dibenzoyl peroxide,

(2) An organo selenium compound which interacts with dibenzoyl peroxide to form zero valent selenium, and

(3) A matrix polymer which is substantially nonreactive with dibenzoyl peroxide and the organo selenium compound,

in a volatile solvent;

(b) Applying the solution to a substrate in the form of a thin film;

(c) Allowing the dibenzoyl peroxide and organo selenium compound to react thereby forming amorphous zero valent selenium particles dispersed in the matrix polymer; and

(d) Heating the matrix polymer/zero valent selenium combination to a temperature and for a time sufficient to convert the amorphous zero valent selenium to its crystalline, trigonal form.

Among those organic selenides, which will react with dibenzoyl peroxide to provide zero valent selenium, are the organo diselenides. Organo diselenides, useful in this process, are selected from those organo diselenides corresponding to the formula: R_1–Se–Se–R_2.

These compounds are capable of undergoing a decomposition reaction in response to activating radiation and yielding, as one of the products of such decomposition, elemental selenium. Typical of suitable compounds, corresponding to the above formula, which may be used, are those organo diselenides wherein R_1 and R_2 are independently selected from the group of benzyl, alkyl substituted benzyl, amino substituted benzyl, amido substituted benzyl, arylalkyl substituted benzyl, aryl substituted benzyl, alkoxy alkyl substituted benzyl, amino alkyl substituted benzyl, alkyl amino substituted benzyl, aryl amino substituted benzyl, alkyl carbonyl substituted benzyl, alkyl thio substituted benzyl, alkyl seleno substituted benzyl, carboxamido substituted benzyl, halogen substituted benzyl, carboxy substituted benzyl, cyano substituted benzyl and alkyl alkoxy, amino substituted alkyl, amido substituted alkyl, aryl alkyl, alkoxy alkyl, aryloxy alkyl, hydroxy substituted alkyl, carbonyl substituted alkyl, thio substituted alkyl, seleno substituted alkyl, carboxamido substituted alkyl, halogen substituted alkyl and nitro substituted alkyl; cyclo alkyl and substituted cyclo alkyl.

In any specific embodiment suitable for use as an electrostatographic photoreceptor structure, it is envisioned that a number of matrix polymers may be useful in this process. Poly(methylmethacrylate) (PMMA) is quite suitable. Preferred embodiments utilize polymers having glass transition temperatures, Tg, that are easily accessible by mild thermal treatments such as by heating to about 200°C. This is the case because exceeding the Tg will allow enhanced diffusion of selenium atoms or particles to nucleation sites for subsequent crystallization. Examples of matrix polymers include PMMA (Tg 72° to 100°C); poly(styrene) (Tg 100°C); poly(vinyl alcohol) (Tg 85°C); and poly(carbonate) (Tg 149°C). Polymers having Tgs below room temperature may also be used.

In fabricating photosensitive layers according to this process, dibenzoyl peroxide, the organo selenium compound and the matrix polymer are combined in a suitable solvent such as benzene, chloroform or tetrahydrofuran. The solution is spread onto a conductive substrate, e.g., aluminum sheet, by standard film coating techniques and the solvent evaporated optionally with the aid of gently heating.

Typical concentrations of the particular elements in the film before solvent removal based on weight percent per 100 ml of solvent are from 5 to 25% dibenzoyl peroxide, 5 to 15% organo selenium compound and from 5% matrix polymer up to its limit of solubility in the particular solvent being used.

After formation of the film, the interaction between the dibenzoyl peroxide and organo selenide will yield small particles of amorphous selenium. Heating the resulting film converts the amorphous selenium to trigonal selenium. The organic nucleation sites formed in situ during the process are responsible for the relative ease with which the amorphous selenium can be thermally converted to trigonal selenium. In general, heating temperatures of from 100° to 200°C for periods of 1 minute to 2 hours have been found sufficient to convert the amorphous selenium formed to its crystalline trigonal form. Too long a heating period at too high a temperature can be detrimental because the selenium will volatilize from the film. The optimum heating time and temperature for a particular composition can be readily determined by routine experimentation.

Example: A solution of 10% poly(methyl methacrylate), 5% benzyl diselenide and 10% dibenzoyl peroxide in methylene chloride is solvent cast onto a Mylar substrate using a doctor blade set at a 4-mil gap. The film is dried and assumes a pale yellow color typical of benzyl diselenide.

Upon drying the film in the dark overnight at room temperature, it is converted to a uniform brick-red color typical of amorphous selenium dispersed throughout the poly(methyl methacrylate) matrix.

Upon heating brick-red films containing amorphous selenium prepared as described above for 1 minute, 3 minutes or 10 minutes at 100°C, a mahogany to blackish colored film results. X-ray powder pattern analysis identifies the selenium in the film to be in the trigonal crystalline form.

Electroless Deposition from Selenious Acid

A.T. Ward, D.J. Teney, J.M. Ishler, and A. Damjanovic; U.S. Patent 4,098,655; July 4, 1978; assigned to Xerox Corporation describe a method for fabricating a photoreceptor comprising:

(a) Providing a substrate comprising a metal which is less positive than selenium in the electrochemical series, the substrate having a thin layer of an oxide of the metal on at least one surface thereof;

(b) Contacting the metal oxide layer with an aqueous selenious acid solution in the absence of applied electrical potential whereby the metal oxide is dissolved and a thin layer of selenium is formed on the surface of the substrate by electrochemical displacement; and

(c) Vacuum evaporating a photoconductive layer comprising selenium or its alloys over the thin selenium layer formed in step (b).

Example: The substrate of interest was an approximately 4" x 4" x 0.005" section of nickel sheet electroformed by the sulfur depolarized nickel anode process. This substrate was made one electrode of an electrochemical cell comprising a large open beaker containing 3 ℓ of 0.1 M aqueous sodium hydroxide solution. A similar piece of nickel sheet was made the counterelectrode. A dc pulse of 2.5 mA/cm^2 was applied first for 30 seconds with the substrate of interest as the cathode. Then the current direction was reversed so the substrate of interest was the anode of the cell and a pulse applied for 30 seconds. Reversal of the current direction was repeated subsequently at 30-second intervals until a total of 4 cathodic/anodic cycles and a final cathodic pulse were applied. The substrate of interest was then removed from the cell and rinsed thoroughly in a stream of deionized water.

The electrochemically oxidized substrate was then placed in an electroless deposition bath which was made up of an open beaker containing 1 ℓ of 0.1 M aqueous selenious acid maintained at a temperature of 95°C. After being immersed for about 2 minutes, the substrate was removed from the bath, rinsed thoroughly with a stream of deionized water and then allowed to dry in air for several hours.

The dried substrate was then mounted in a bell jar vacuum evaporator and a layer of a halogenated selenium-arsenic alloy was deposited thereon by conventional vapor deposition techniques to a thickness of about 60 μ. The substrate temperature was maintained at about 55°C during deposition.

The resulting photoreceptor was found to survive being flexed around a 2" diameter cylindrical mandrel indicating excellent adhesion between the vacuum deposited selenium alloy layer and the nickel substrate. The photoreceptor was then used to make reproductions of an original object using a Xerox Model D Processor. Excellent quality reproductions were obtained.

Electroconductive Transparency

J.C. Anderson; U.S. Patent 4,109,052; August 22, 1978; assigned to E.I. Du Pont de Nemours and Company describes a laminate film, having a light transparence greater than 40% and a surface resistivity of at most 10^7 ohms per square, the laminate comprising a substrate layer of a substantially optically transparent organic, polymeric material, a coupling layer of crosslinked acrylic, polymeric material bonded to one surface of the substrate layer, and a metal layer bonded to the coupling polymer layer. A top coating of a substantially optically transparent polymer, often having a photoconductive material dispersed therein, normally is applied over the metal layer.

Suitable substrate materials include, for example, polyesters, such as polycarbonates, poly(ethylene terephthalate), polyethylene 2,6-naphthalene dicarboxylate, and cyclohexanedimethanol terephthalate; polystyrene, cellulose acetate, cellulose butyrate, poly(vinyl chloride), poly(vinylidene chloride), poly(methyl methacrylate), polyethylene, poly(vinyl acetate), polycaprolactam, and poly(hexamethyleneadipamide). Poly(ethylene terephthalate) is the preferred substrate material because of its good mechanical strength, flexibility, and general handling character which makes it suitable for photographic use.

The substrate film usually will have a thickness of about 75 to 175 μ. Various additives may be incorporated into the substrate material, for example, dyes, slip additives, surface treatment agents, and the like, so long as those additives do not decrease optical transparency or surface resistivity or otherwise interfere with the intended use of the film.

The coupling polymer layer, which is applied to the substrate polymer film, is an acrylic copolymer having functional groups which allow crosslinking, such as hydroxyl, carboxyl, amine, amide or oxirane. Typically, these copolymers will be formed from acrylic esters, acids, and amides taken in suitable proportions so that the final crosslinked product will possess the necessary mechanical and optical properties. Both acrylic and methacrylic acids and their derivatives can be used as comonomers. Suitable esters include methyl, ethyl, propyl, isopropyl, butyl, hexyl, heptyl, octyl, 2-ethylhexyl, glycidyl, and hydroxyethyl acrylate and methacrylate. The crosslinking agent is normally added to the coating bath or coating composition. Suitable crosslinking agents can be readily chosen by one skilled in the art from a variety of commercially available polymer crosslinking agents capable of reacting with the functional groups of the acrylic polymer. Most frequently, these will be condensation products of formaldehyde with melamine, urea, diazines, and their derivatives. The amount of the crosslinking agent must be sufficient to produce, on reaction with the functional groups of the acrylic polymer, a coating having good mechanical and optical properties. The substrate polymer film often is oriented by stretching, either monoaxially or biaxially. When oriented film is used, the coating can be applied either prior to or after stretching, as well as between two stretching operations.

Typical acrylic copolymers will be an ethyl acrylate/methyl methacrylate/methacrylamide or a methyl methacrylate/ethyl acrylate/methacrylic acid terpolymer. These and other similar acrylic copolymers are available commercially from several sources. A commonplace crosslinking agent is a melamine/formaldehyde condensate.

The amount of coupling polymer applied to the substrate polymer is normally about 0.01 to 0.25 g/m^2. A dry polymer coating is about 0.01 to 0.25 μ thick.

The metal layer is applied to the coupling polymer coating by any convenient technique, such as, for example, by vacuum deposition, sputtering, or chemical deposition. The particular technique which should be used in each case depends on the metal being applied. Suitable metals include, among others, aluminum, nickel, palladium, chromium, gold, or a nickel-chrome alloy. It is necessary to deposit a sufficient amount of metal to obtain a surface conductivity value of less than 10^7 ohms per square. However, the amount of the metal should be sufficiently small to maintain the optical transparency of the metallized sheet at at least 40%. The preferred metal is palladium because it can be deposited by continuous vaporization in very thin layers, which are at least 75% transparent, while providing sufficient conductivity. Typically, the conducting metal layer is about 10 to 150 Å thick.

The process provides excellent adhesion, uniformity, and abrasion resistance. It is not necessary to preclean the substrate film, for example, by electron discharge, flame treatment, glow discharge, or argon plasma backsputtering.

When a photoconductive-material-containing top coating is applied over the metal layer, the laminate is suitable for use as an electrophotographic transparency. The top coating also provides mechanical protection to the metal layer. Many photoconductive materials are known to the art. Certain organic, photoconductive polymers have been previously reported. They include, for example, polyvinylcarbazole, polyvinylpyrene, polyvinylacridine, polyvinyldibromocarbazole, and polypyrenyl methylvinyl ether. However, a photoconductive material is not always present in the top layer. Transparencies, which can be used for other electrical, not photoelectric, imaging processes, will have a polymeric top coating containing no photosensitive additives. The top coating may also contain various stabilizers, slip agents, plasticizers, and other additives, so long as they do not interfere with the intended application of the finished article. This layer will usually be about

5 to 25 μ thick and is applied by any convenient method, including doctor knife, spray, and roll. While any polymeric material base containing, if desired, a suitable photopolymer and having sufficient adhesion, optical clarity, and mechanical properties can be used, the preferred top coating material is a vinyl acetate/vinyl chloride copolymer, which is available commercially as VAGH. This polymer can be applied from solution in organic solvents. Residual solvent is heat-evaporated.

Dielectric Resin Film

H. Burwasser and J.R. Wyhof; U.S. Patent 4,112,172; September 5, 1978; assigned to GAF Corporation describe an improved dielectric imaging member comprising a transparent dielectric substrate having a thickness of between about 75 to 175 μm, a conductive layer on the substrate and a dielectric material having a thickness of less than about 15 μm on the conductive layer. An imaging process is also provided comprising generating an electrostatic latent image on the dielectric imaging member and developing the image by contact with a developer material.

Referring to Figure 8.5, wherein like numerals indicate like elements, the process is illustrated. A typical 75 to 175 μm transparent dielectric substrate **10** has a thin transparent electrically conductive layer **12** applied to its surface. Conductive layer **12** has a resistivity value of less than 10^{10} Ω/sq, and preferably less than 10^8 Ω/sq. The conductive layer **12** can be any conductive material which is typically applied to paper such as quarternary ammonium salts, sulfonated polystyrenes, polyacrylic acid salts and the like. In addition, provided a reasonable amount of transparency is maintained, metallized films can also be employed.

A coating of a dielectric resin **14**, i.e., a resin that has electrical insulating properties, is applied over the conductive coating **12**. The dielectric resin coating thickness is less than about 15 μm, preferably 4 to 5 μm, and should adhere to the conductive substrate.

Dielectric resins suitable for use in either or both the dielectric substrate **10** or dielectric coating **14** include polyvinyl acetates, acrylics, styrenated acrylics, polyesters, polyvinyl butyral, polycarbonates and other high dielectric resins.

In effect, by virtue of introducing the conductive layer **12**, the capacitance of the substrate **10** is changed relative to a photoconductor **16** having a metallic substrate **18** and bearing an electrostatic latent image **20** formed thereon by conventional techniques. For the observer, however, the substrate **10** is virtually unchanged in thickness and transparency.

Because of the capacitative change provided by conductive layer **12** and dielectric coating **14**, it has been found that the image **20** on the surface of photoconductor **16** can be transferred to the dielectric coating **14** on substrate **10** by bringing the photoconductor surface **16** into intimate contact with the surface of dielectric coating **14**. The transferred electrostatic latent image **20'** on dielectric coating **14** will have sufficiently high charge density to permit development by immersion of the dielectric imaging member in a bath of liquid developer **22** or by other conventional development techniques.

Figure 8.5: Dielectric Resin Film

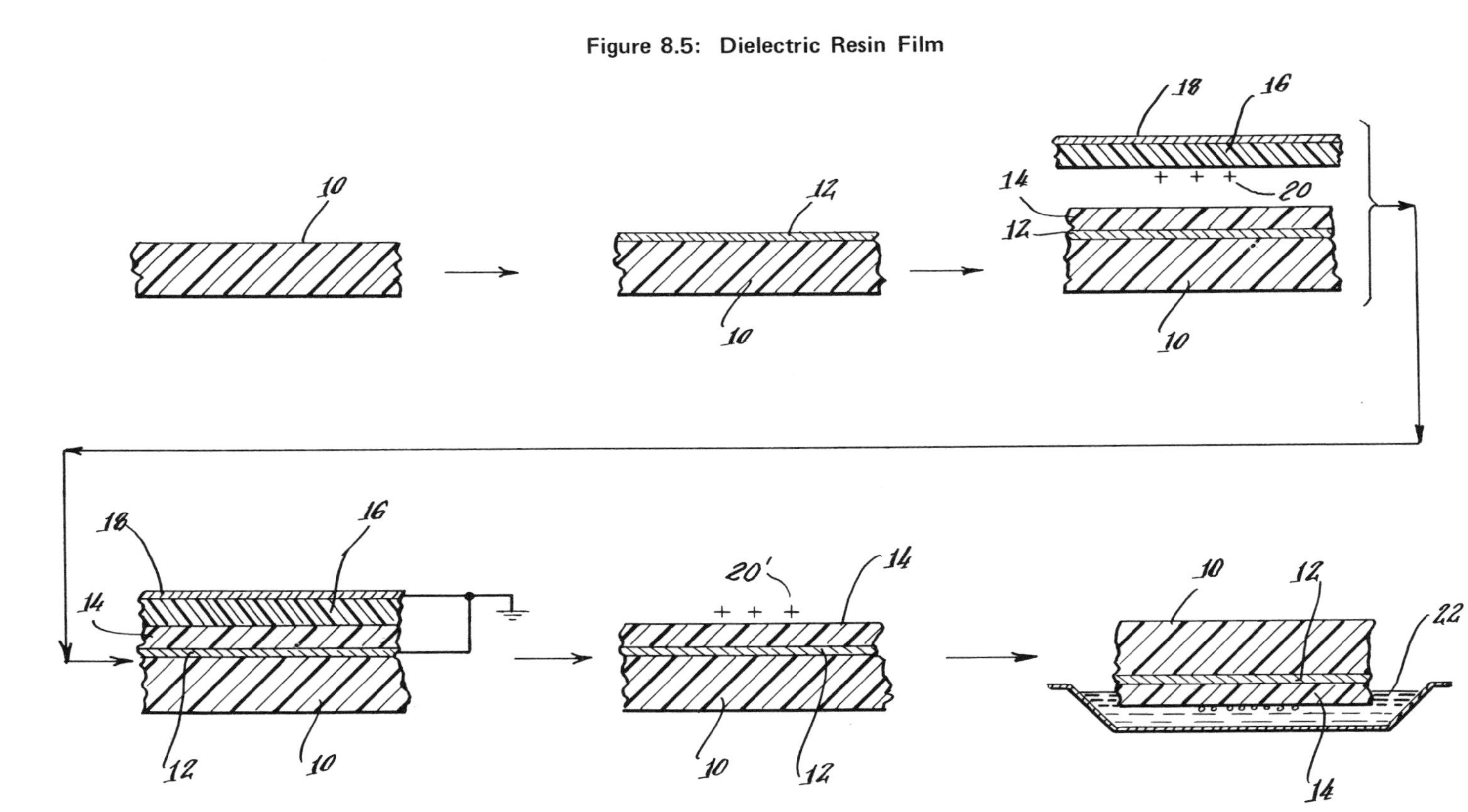

Source: U.S. Patent 4,112,172

Example: Dielectric imaging members were prepared by coating a polyester substrate (Melinex 505 preprimed available from ICI) having a thickness of 100 μm with a conductive layer of sulfonated polystyrene obtained from National Starch. The surface conductivity was 10^7 ohm-cm. The resulting conductive substrate was overcoated with a dielectric coating of styrenated acrylic from De Soto, Inc. in thicknesses ranging from 1.4 to 14.2 μm. Thickness measurements were made using a recording spectrophotometer and by mechanical means. An electrostatic latent image was transferred to the resulting dielectric imaging member by intimate contact with an electrostatic latent image-bearing photoconductive plate which itself has been charged by a -5,500 V corona. The 24 μm thick photoconductive plate had an initial surface voltage of 600 V. The photoconductor composition was a charge transfer complex of polyvinyl carbazole and trinitrofluorenone, of the type disclosed in U.S. Patent 3,484,237. The electrostatic latent image formed on the dielectric imaging member was developed by immersion in a liquid developer composition as disclosed in U.S. Patent 3,542,682.

Method of Positioning Recording Member

F.C. Gross; U.S. Patent 4,120,720; October 17, 1978; assigned to Scott Paper Company describes an improved technique for making effective electrical contact with the intermediate electrically-conductive layer of electrostatographic recording elements, including both electrophotographic recording members and electrographic recording members, combined with a means for accurately positioning the recording member during imaging thereof in an apparatus designed for this purpose.

This process accomplishes the above objectives by coating the inner surface of at least one of the holes located in the nonimage areas of the recording member which are used to accurately position the recording member during imaging with a conductive lacquer coating. The result is a means which simultaneously allows good electrical contact to be made with the intermediate electrically-conductive layer of the recording element and provides for accurate positioning of the recording member during imaging in an apparatus designed for this purpose.

This process, therefore, enables the recording element to be positioned accurately for imaging repeatedly within close tolerances and also provides a means for effectively grounding the intermediate conductive layer of the recording member during imaging and developing.

Figure 8.6a shows a conventional electrostatographic recording member, known as a fiche. The bulk of the body thereof comprises a plurality of imaging areas which are separate and distinct from each other and which are each adapted to receive information thereon. This type of fiche is best known as a microfilm card, each imaging area thereon being capable of receiving an image with the entire recording member being capable of receiving a plurality of such images depending upon the number of imaging areas contained therein.

An area is provided along the top of such recording elements normally designed to receive a title or some other kind of identification of the subject matter contained in the imaging areas of the recording member. A handling tab is provided in the center of the top of the recording element to permit one to handle the recording element without fear of damaging the information contained in the imaged areas. Such recording elements are also normally provided with end tabs or ears as is shown in Figure 8.6a.

Figure 8.6: Method of Positioning Recording Member

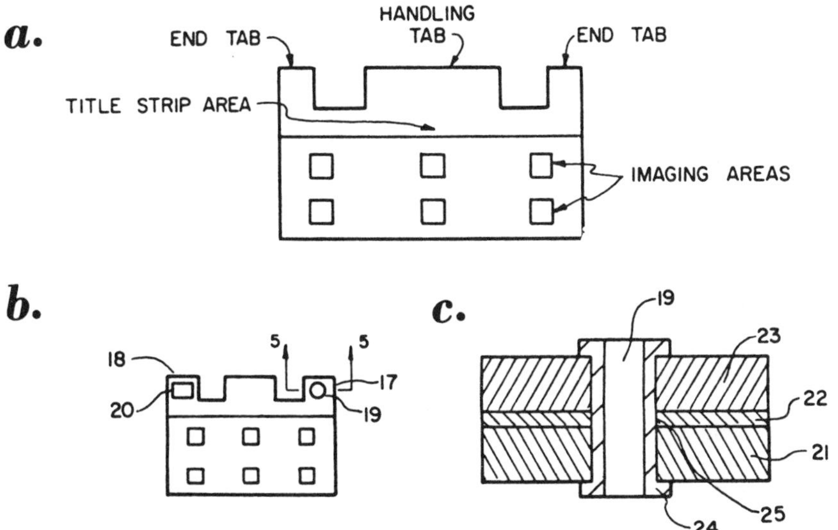

Source: U.S. Patent 4,120,720

The preferred embodiment of the process is shown in Figures 8.6b and 8.6c. Referring to the former figure, two holes **19** and **20** are provided in the respective end tabs **17** and **18**, these holes extending through the entire thickness of the recording member. The inner surface of at least one, preferably both, of the holes in the end tabs (which, of course, are in nonimage areas of the recording member) is coated with a conductive lacquer composition to thereby establish good electrical contact with the intermediate conductive layer of the recording element which is exposed on the inner surface of the holes. Normally, it is preferable to provide two holes in the recording element, although more than two holes may be provided if desired, in which case all of the holes may have their inner surfaces coated with the conductive lacquer.

Figure 8.6b also indicated the preferred embodiment of one hole (**19**) being circular and the other being elongated (**20**), in order to accurately position the recording element in the apparatus designed to selectively image the respective imaging areas contained therein. Use of the elongated hole **20** enables larger tolerance in placing the holes.

Figure 8.6c shows, in cross section, the recording element shown in Figure 8.6b, Figure 8.6c being taken along the line **5–5** in Figure 8.6b. Referring to Figure 8.6c, a support **21** is shown carrying thereon an intermediate conductive layer **22** onto which is superposed a photoconductive layer **23** (it will be apparent that the following description will also hold for electrographic recording members wherein photoconductive layer **23** will be replaced by a dielectric layer of high volume resistivity). From Figure 8.6c, it is seen that hole **19** has been covered on its inner surface with a conductive lacquer **24** establishing electrical connection with the intermediate conductive layer **22** at interface **25** between lacquer

24 and the conductive layer **22**. It will be appreciated that the arrangement shown in Figure 8.6c enables any buildup of potential in conductive layer **22** to be dissipated through the conductive lacquer **24** to ground (such as by contact with the apparatus used in the imaging thereof) thereby assuring a high quality image.

Any suitable coating technique may be employed to coat the conductive lacquer onto the inner surfaces of the holes provided in the recording element. In addition, the composition of the conducting lacquer employed is not particularly critical.

The electrically-conductive lacquer composition can generally be defined as a dispersion of an electrically-conductive pigment in a binder therefor. The electrically-conductive material or pigment dispersed in the binder can be any finely divided particulate material having good electrical conducting properties. Suitable conducting materials include carbon blacks, graphite, acetylene black, metal particles such as nickel or silver, semiconductive materials, etc. The particular electrically-conductive pigment or material used is not critical. It is present in the composition in an amount of at least the minimum required to render the composition electrically conductive.

In addition, the particle size of the electrically-conductive pigment is not critical and can vary as desired depending upon the particular material used and the end use required. Generally, the particle size of these materials will vary from a range of from 5 to 200 mμ. Those skilled in the art can determine the optimum particle size with a minimum degree of experimentation using methods well-known to those skilled in the art.

REFLECTIVE COATINGS

Light-Scattering Surface for Liquid Crystal Displays

R. Doriguzzi, M. Egloff, M. Kaufmann, J. Nehring, and T.J. Scheffer; U.S. Patent 4,106,859; August 15, 1978; assigned to BBC Brown Boveri & Company Limited, Switzerland describe a light-scattering reflector for a liquid crystal display which is simple and economical to manufacture and which exhibits an optimal light diffusion characteristic which may be matched to a particular display element.

Referring to Figure 8.7, there are represented the different steps of the method for the manufacture of the reflector, which is designated by **1** in Figure 8.7f. For this purpose, a major surface **2** of a substrate **3** (Figure 8.7a), which in this case may consist of a small glass plate of dimensions 20.0 x 9.4 x 0.4 mm is subjected to a grinding operation until the glass plate has the desired thickness and its surface **2'** has a uniformly roughened structure (Figure 8.7b) caused by the grinding medium (a corresponding structure may be obtained by sand blasting the glass surface). Silicon-carbide powder of grain sizes 400, 800, 1,000 or 1,200 (such as is supplied by the Struers firm in Denmark) has proved particularly suitable.

The grinding process itself, especially in the mass production of reflectors, is preferably effected with the help of a lapping machine such as is employed in semiconductor technology for the lapping of silicon wafers. Less than 5 minutes are required in order to obtain a substantially uniformly roughened surface.

Figure 8.7: Light-Scattering Surface for Liquid Crystal Displays

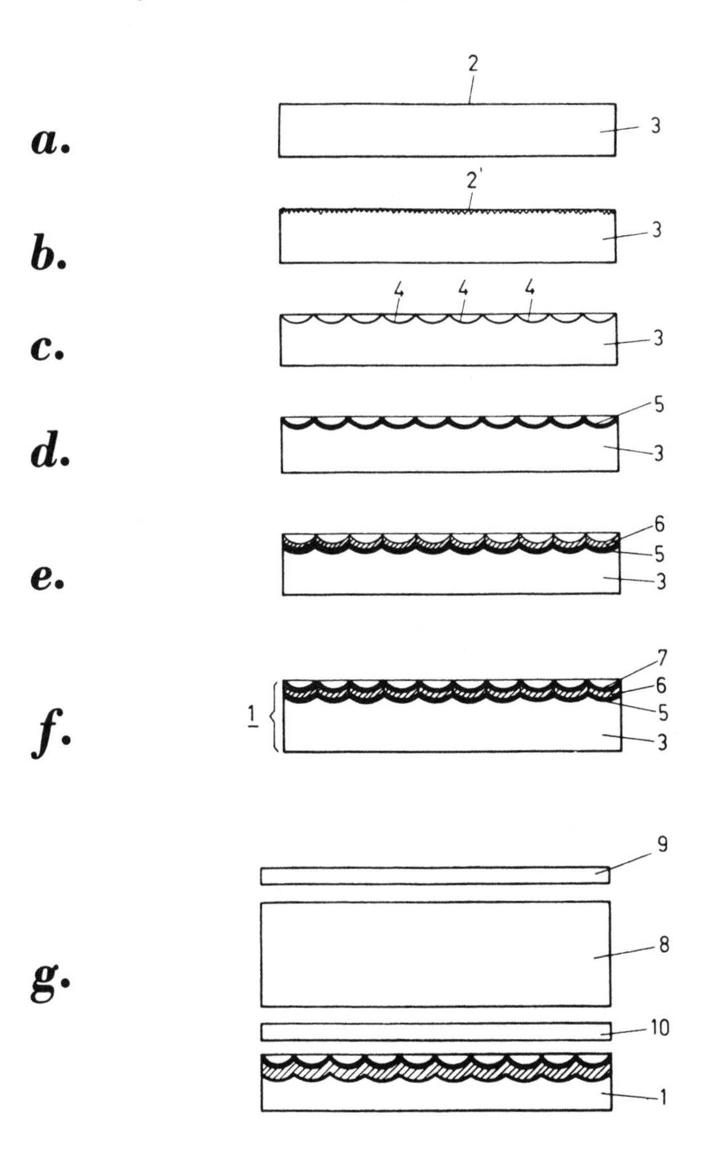

(a)-(f) Schematic illustrations of the different steps in the manufacture of the reflectors of the process

(g) Illustration of a liquid crystal display device with a reflector of the type shown in Figure 8.7f

Source: U.S. Patent 4,106,859

Following the grinding or sand-blasting operation, the small glass plate is cleaned several times in distilled water, for example, in an ultrasonic bath, and finally immersed for 3 to 10 minutes in an etching bath, that preferably consists of a 1:1 mixture of H_2O and HF, and moved to and fro in the acid. Finally, the thus-treated glass plate is again rinsed with distilled water and then dried.

In Figure 8.7c, there is schematically represented a section of such a glass plate. The depressions **4** are here shown in an idealized form.

The highly reflective metal layer is applied upon the surface which has been etched and cleaned with water. In order to achieve a better attachment of this metal layer to the glass plate, it has proved to be advantageous first to evaporate upon the glass surface a chromium layer **5** (Figure 8.7d) some 10 to 50 Å thick and then to deposit upon this chromium layer **5** the reflective metal layer **6**, which may, for example, consist of silver (Figure 8.7e). For protection of the silver layer **6** against ambient influences, there is finally applied upon this layer a layer **7** of silicon dioxide about 250 Å thick (Figure 8.7f).

Figure 8.7g shows schematically a liquid-crystal display device, in which the new reflector **1** is employed. The actual liquid-crystal cell **8** (which may in practice be a twisted cell) is situated between two polarizers **9, 10**. The reflector **1**, which is situated behind the polarizer **10** on the side remote from the liquid crystal cell **8**, is either stuck directly to the polarizer **10** or is positioned behind the polarizer. Any epoxy cement of high quality (e.g., Lens Bond M 62) may be employed for adhesion.

For liquid-crystal displays, which are to be employed both in daytime and also in the night (wrist watches), the use of semitransparent, diffuse metal reflectors has proved especially advantageous. In nighttime operation, a light source that is situated at the rear of the reflector supplies the necessary illumination.

Multifaceted Optical Scanners

Multifaceted scanners usually comprising multifaceted rotating mirrors are employed in well-known techniques for effecting optical scanning between a light source and a photocell. Typically, a light illuminates a silvered mirror, for example, at an angle of 45° to direct light toward a facet that is reflected from the facet toward the object being scanned. Normally, the object reflects this light back along the same path upon a photocell. The duration of the scan corresponds to the time for a facet to pass the light beam along the object being scanned. It is usually preferred that the object path scanned be independent of which facet is then in the light beam path.

T. Fisli; U.S. Patent 4,101,365; July 18, 1978; assigned to Xerox Corporation describes multifaceted scanner systems which may be precisely machined inexpensively and with greater facility than known scanner systems, enabling these multifaceted scanners to be considered for employment in a vast number of applications other than military or development laboratories where the exorbitant costs of available scanner systems can only be justified.

This process is accomplished, generally speaking, by providing acrylic high speed multifaceted scanners by injection molding. Injection molded acrylic has been used for the production of low cost, medium quality lenses. However, it has not been practical to employ acrylics in reflecting optics due to its low adhesion to thin film coatings such as aluminum. Since the advent of the application of magnesium fluoride as an overcoating to acrylic substrates, it has been made possible to properly adhere surfaces which possess the proper reflecting optics or reflectivity to the acrylic substrates.

Two methods, for example, that may be employed in providing injection molded high speed multifaceted scanners include providing an aluminum hub which is placed into a die cavity having the proper facet geometry. The aluminum hub is sized so as to provide a suitably dimensioned injection ring having suitable mechanical properties and optical properties for high speed scanning applications. Acrylic is injection molded into the gap between the aluminum hub and the die cavity. In order that the resulting member represent a stress-free ring at room temperature, the aluminum discs have to be preheated to approximately 500°F, for example, or roughly 100°F above molding temperature to accommodate the slightly high acrylic shrink rate. The effect of the preheat will be minimal on the physical properties of the aluminum alloy employed of which aluminum 7075-T651 is preferred. A variation of this first method, which is less complicated and expensive, eliminates the aluminum resulting in a solid acrylic injection molded scanner.

In another embodiment, the acrylic ring is molded separately and then cemented onto an aluminum hub fabricated of the aluminum alloy mentioned above. In both cases, the acrylic is coated with magnesium fluoride and then a mirror-like finish of aluminum is applied with a protective coating of, usually, silicon monoxide to complete the process if desired.

Typical applications of this system include deflection of a light beam, such as laser, in such a manner that it produces a flying spot. When this bright spot is moved across an object-document having high and low density areas by rotation of the scanner, a light detector (placed in the vicinity) provides an electronic signal which is low or nonexistent when the spot is in a dark area, and high when the spot is in a light area of the document. This type of scanning system is used in facsimile devices and in optical character readers. Since this system can be used (in conjunction with other hard and software) to decode alpha numerics, it is also known as a reader.

Another system, which also uses multifaceted scanners, is the so-called write system. The overall arrangement in general is the same except that in the stationary path of the beam (before the scanner) a light switch, known as a modulator, is used to write the image on a xerographic photoreceptor. The signal going into the modulator can come either from the light detector of the read station, or from a character generator which is the case with computer printers.

Example: An aluminum 7075-T651 preformed disc having the following dimensions: 2" o.d., ¾" i.d., ⅜" thick is held in place in the mold while the molded acrylic is injected into the die cavity having the proper facet geometry. After the injection die is cooled, the part is removed. The acrylic ring is firmly held on the aluminum disc by the force developed from the greater shrinkage of the acrylic deposition. A 100 Å layer of magnesium fluoride is then uniformly applied over the acrylic by vacuum evaporating and then 100 Å of aluminum is deposited over the magnesium fluoride adhesive coating by vapor deposition. 200 Å of SiO is then vacuum deposited over the mirror-like aluminum facets to protect them. Several discs are assembled on a rod and placed in a vacuum chamber, along with MgF_2, aluminum and SiO. The chamber is pumped down to the required vacuum and the MgF_2, aluminum and SiO are evaporated in that order by heating them above their melting temperatures. The parts are rotated during evaporation.

Disc-Shaped Information Carrier

H. Soeding; U.S. Patent 4,126,727; November 21, 1978; assigned to Polygram GmbH, Germany describes an information carrier for high-density storage, in particular for the storage of video signals, in which the information is provided on both sides of the carrier and stored in the form of a beam-reflecting surface structure which is covered, in each case, by a layer of translucent material.

An information carrier of the type mentioned above is known in the art, for example, reference may be made to the British Patent 1,446,009 and the Australian Patent 5,974,573. This prior information carrier comprises two separate discs of transparent material which are connected together by an intermediate layer and having their respective sides provided with a beam-reflecting surface structure each of which faces the other. With this structure, the information is read by an optical scanning beam which penetrates from the outside through the transparent carrier layer to the beam reflecting surface, where the beam is reflected and modulated according to the profile of the beam-reflecting surface. Here, the thickness of the transparent carrier layer is not particularly critical because the scanning beam is focused as precisely as possible in the information plane.

However, considerable demands are made on the homogeneity of the transparent carrier material and, in this respect, certain difficulties arise in the production of such information carriers because material residues, called tags, stick to the pressing die and cause corresponding fault zones in the information carriers which are made from these dies. Other material discrepancies are caused by internal material stresses, bubbling or other harmful phenomena which arise during the production process.

It is the primary object of the process to provide an information carrier for signals that can be read optically and in which the aforementioned discrepancies or fault zones do not have an adverse effect on the reading process.

The information carrier essentially comprises a disc-shaped base element constructed of an opaque material. On each side of the base element, there is a surface structure corresponding to the information stored, the surface structure being a profile which is produced by stamping, pressing, injection molding or other suitable techniques. A beam-reflecting metallic coating of roughly 300 to 1000 Å thickness is provided over each surface profile of the base element by vapor deposition or chemical processes. It will be appreciated that the metallic coating reproduces and maintains the surface profile and, therefore, the stored information.

Finally, a transparent lacquer coating of about 200 to 400 μ thickness is applied over each of the metallic coatings in order to protect the information against scratches, dust and the like. The lacquer coatings form the entry and exit faces for the optical scanning beam. In comparison with the carrier molded under the influence of pressure and heat as known in the prior art, these lacquer coatings offer considerable advantages with regard to their optical properties because the chance of potential fault zones, caused by internal material stresses, bubbling or other harmful phenomena, are substantially reduced. Advantageously, the thickness of the lacquer coatings is selected such that dust particles, scratches, etc., on the surface of the lacquer are so far outside the depth of focus of the optical reproduction system that such surface faults are no longer a nuisance. The total thickness of the information carrier is about 1 to 1.5 mm.

Multilayer Dielectric Laser Mirror

Multilayer dielectric mirrors have long been used in lasers and for other applications because of their high reflectivity. Commonly, such mirrors have included a substrate of fused silica or glass. A mirror coating comprising a plurality of layers of selected dielectric materials is deposited on a polished surface of the substrate. Mirror substrates of fused silica or glass have most commonly been polished by optical (or mechanical) polishing techniques.

Inhomogenities in the dielectric layers of the mirror coating on such prior art laser mirrors have caused difficulties in some applications. For example, a cause of phase locking in ring laser gyros is believed to be back scattering from inhomogeneities in the laser mirrors used in these gyros.

R.D. Henry; U.S. Patent 4,101,707; July 18, 1978; assigned to Rockwell International Corporation describes a multilayer dielectric laser mirror wherein homogeneity of individual dielectric layers of a mirror coating is greatly improved over that of the prior art. The improvement results from the use of a mirror substrate of garnet. The surface of the garnet substrate on which the mirror coating is deposited is preferably final polished by chemical-mechanical polishing techniques. A surface of a garnet substrate subjected to chemical-mechanical polishing is relatively free of inhomogeneous strain. It has been discovered that the equivalent mean surface roughness of the mirror coatings deposited on polished garnet surfaces free of inhomogeneous strain is significantly reduced from that of mirror coatings deposited on prior art polished fused silica or glass substrates.

Figure 8.8 shows a portion of a multilayer dielectric mirror **10** having a substrate **12** fabricated of a suitable garnet. The substrate **12** typically comprises a monocrystalline nonmagnetic garnet. As used here, the term nonmagnetic garnet refers to garnet materials containing no more than an insignificant amount of iron. However, magnetic garnets may also be used. The nonmagnetic garnets are considered to be metal oxides designated by the general formula: $J_3Q_5O_{12}$, where J is at least one element selected from the lanthanide series of Periodic Table, lanthanum, yttrium, magnesium, calcium, strontium, barium, lead, cadmium, lithium, sodium, and potassium. The Q constituent is at least one element selected from gallium, indium, scandium, titanium, vanadium, chromium, manganese, rhodium, zirconium, hafnium, molybdenum, niobium, tantalum, tungsten and aluminum.

The substrate **12** has a substantially rectangular cross section. One surface of the substrate **12** is a polished surface **14** upon which is deposited the mirror coating **16**. The mirror coating **16** comprises a plurality of individual layers **20a-20d** of selected nonmagnetic dielectric materials cumulatively deposited in a stacked relationship, one upon the other. The individual layers **20a-20d** are deposited coatings of such dielectric materials as titanium dioxide, magnesium fluoride, silicon dioxide, or other nonmagnetic dielectric materials suitable for this purpose.

Figure 8.8: Multilayer Dielectric Laser Mirror

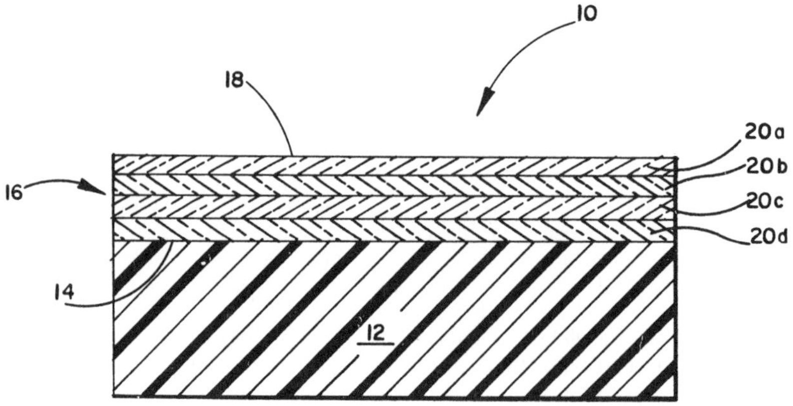

Source: U.S. Patent 4,101,707

In the preferred embodiment, the garnet surface **14** is final polished by standard chemical-mechanical polishing techniques known to be useful for preparing silicon and garnet substrates for epitaxial deposition thereon.

When a dielectric mirror coating **16** is subsequently deposited on the polished garnet substrate surface **14**, the surface roughness due to physical thickness inhomogeneity of the mirror coating surface **18** is reduced to significantly less than 100 Å and no appreciable optical thickness inhomogeneity resulting from index of refraction variation occurs.

Retroreflectorization of Fabrics

W.K. Bingham and T.R. Bailey; U.S. Patent 4,103,060; July 25, 1978; assigned to Minnesota Mining and Manufacturing Company describe a retroreflective treatment for fabrics that is so inconspicuous in daylight and of so little effect on hand, feel, and breathability, that garments made from the fabric will be widely worn by pedestrians; and that yet is so brightly retroreflective that the pedestrians will be readily visible at night for several hundred feet and more under illumination from oncoming motorists.

Briefly, the retroreflective treatment for fabrics that are to be worn comprises discrete retroreflective areas applied in a spaced, sparse manner over the surface of a base fabric. These retroreflective areas include a thin layer of binder material adhered to the base fabric and transparent microspheres supported or held in the binder material. At least about one-third of the microspheres have reflective means between them and the fabric whereby the microspheres are made retroreflective, and the surface that faces away from the fabric is optically exposed for receiving and returning the light rays. On the average, there are less than about 2,000 microspheres, and preferably less than about 500 microspheres, in any square centimeter of the surface of the fabric, and the smallest surface dimension (that is, a dimension along the surface of the fabric) of the continuous portions of the coating is no greater than about 0.5 centimeter.

A retroreflective treatment as described may be provided in different ways, but one method is at present very much preferred. This method uses a unique retroreflectorizing material, namely, a free-flowing mass of minute retroreflectorization particles. These minute retroreflectorization particles each comprise one or more transparent microspheres arranged in a closely packed monolayer; a solid binder layer in which the microspheres are supported and which may at least in part be softened to adhere the particles to a substrate; and specular reflective means underlying the microspheres and supported by the binder material in optical connection with the microspheres to make the microspheres retroreflective. The surface of the microspheres opposite from the reflective means is optically exposed to receive and return light rays.

These retroreflectorization particles are generally applied by cascading, metering, or otherwise depositing them onto a base fabric under conditions that soften the binder layer. At least a portion of the cascaded particles become adhered to the base fabric with the optically exposed surface of the microspheres facing away from the fabric.

Example: The steps in this example are discussed with reference to the illustrative drawings in Figures 8.9a through 8.9d. Visibly transparent glass microspheres **10**, which averaged 60 μm in diameter and had a refractive index of 1.92, were coated in a monolayer onto a composite web **11** which comprised a Kraft paper **12** and a polyethylene layer **13**. The web was heated so that the microspheres sank into the polyethylene layer **13** to approximately 30% of their diameters. The exposed microspheres were then vapor-coated with a layer **14** of aluminum in a thickness of approximately 250 Å to produce a web **15** as shown in Figure 8.9a.

Figure 8.9: Retroreflectorization of Fabrics

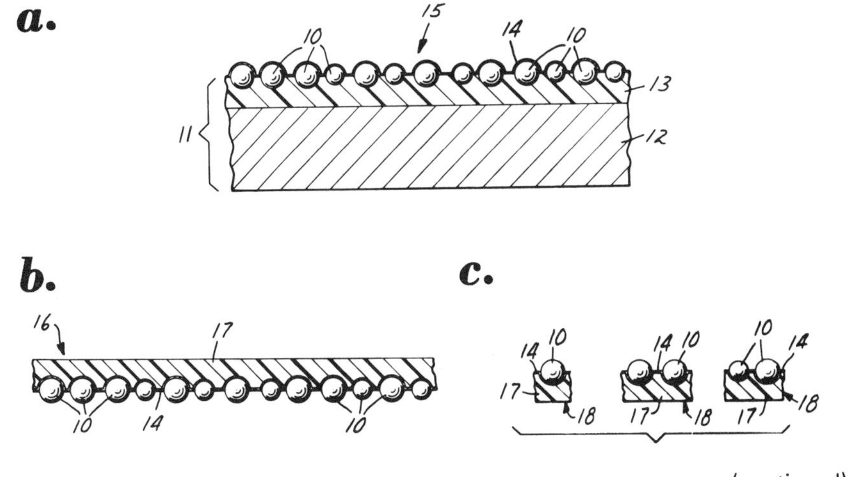

(continued)

Figure 8.9: (continued)

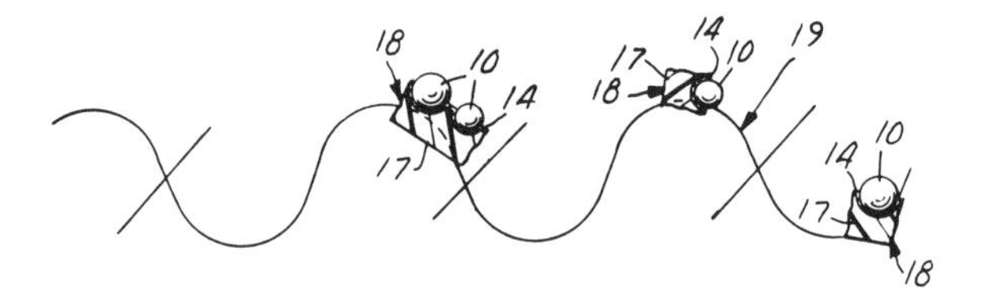

(a)(b) Enlarged sectional views of illustrative sheet materials prepared in the course of manufacturing retroflectorization particles
(c) Enlarged sectional view of illustrative retroreflectorization particles
(d) Enlarged schematic sectional view of a retroreflectorized fabric

Source: U.S. Patent 4,103,060

A linear saturated hot-melt polyester adhesive material (Bostik 7979, a typical useful polyester, is the reaction product of terephthalic acid, isophthalic acid, ethylene glycol and neopentyl glycol) was dissolved in a solvent that was a mixture of equal parts of toluene and methyl ethyl ketone to give a 60% solids solution.

This solution was knife-coated over the aluminum layer **14** of the microspheres **10** of the web **15** at a wet thickness of 0.004" (100 μm) and then dried 10 minutes at 150°F (66°C) and 20 minutes at 200°F (93°C). The polyethylene-coated paper **1** was then stripped away leaving sheet material **16** as shown in Figure 8.9b, which comprised a monolayer of glass microspheres **10** each approximately hemispherically reflectorized by the aluminum **14** and partially embedded in a binder layer **17** of the polyester. The sheet material **16** was then placed in a Waring Blender and chopped to a fine particle size with the aid of a small amount of dry ice. The resulting retroreflectorization particles **18**, as illustrated in Figure 8.9c were retained on a 200-mesh (U.S. Standard) screen but passed an 80-mesh screen, meaning that their size was between 74 and 180 μm. In such particles, the number of microspheres in the particles ranges from 1 to about 11.

Particles **18**, prepared in the above manner, were uniformly dispersed over the surface of a dark blue denim fabric made from a blend of cotton and polyester fibers by pulling the fabric through a curtain of the particles as they were dropped from a vibrating inclined plane. The coated cloth was then placed in an oven heated to 350°F (177°C) for 2 minutes, whereupon the particles were bonded to the yarns and filaments of the cloth. A fabric **19**, as illustrated in Figure 8.9d, resulted. To further improve nestling of the particles into crevices of the fabric, the fabric can be briefly ironed with a conventional laundry iron heated to about 300°F (150°C).

This fabric was tailored into jackets and pants and viewed under automobile headlamps. A person wearing the jacket and pants was quite visible at 300' with low beams and over 500' with high beams. The retroreflective efficiency varied over the surface of the fabric between 1.75 and 2.75 candelas per square meter of fabric per lux of incident light. The concentration of microspheres was counted and measured as approximately 550 microspheres per square centimeter over the treated surface of the fabric. The microspheres were randomly oriented, meaning that their optically exposed surfaces faced in a variety of directions. At least about one-third of them were oriented so as to retroreflect light that is perpendicular to the surface of the fabric.

The garments prepared were laundered 50 times and found to show a retention of about 50% of their original retroreflective brightness. There were no other apparent changes to the garments after laundering in comparison with a control fabric that had not been treated. The garments had essentially the same hand, feel, and breathability as control garments that had not been treated, and the glass microspheres could be discovered only by a careful scrutiny of the garments. The overall appearance of the treated garments under ordinary daylight viewing was almost identical to that of the control garments.

MISCELLANEOUS PROCESSES

Photomask Preparation and Repair

Photomasks for the electronics industry must be perfect and have high resolution. For TV picture tube shadow masks, which are large in scale, about 2' x 3' (about 61 x 91 cm) and which have hundreds of thousands of metal dots deposited on the glass surface, the photomask must have no missing or malformed dots, since these result in corresponding imperfections on the cathode ray viewing screen which are visible and annoying to the viewer.

E.A. James and P. Kuznetzoff; U.S. Patent 4,107,351; August 15, 1978; assigned to RCA Corporation have found that a pattern of metal can be applied to a substrate in a wholly additive process by applying a sensitizer and activator for subsequent electroless metal deposition onto the substrate, coating the surface with a photoresist, exposing and developing the photoresist to form the desired pattern of openings in the resist, then redeveloping the resist with a developer containing the activator and finally electrolessly depositing the metal in the openings and removing the remaining photoresist.

The process can be employed for many applications where a patterned metal layer is to be applied to an insulating substrate, but will be illustrated in detail with reference to the repair or preparation of a metal-on-glass photomask.

A glass photomask to be repaired is first cleaned thoroughly and dried. If the missing or imperfect dots to be repaired are on a localized area of the photomask, those areas alone can be repaired by covering the remainder of the mask or by surrounding the defective area with a dam of wax or other insoluble material. Defective or malformed dots can be removed, if desired, prior to repair.

The cleaned surface is then sensitized by treating with a solution of a tin salt, such as tin chloride. A suitable sensitizing bath contains 70 g/ℓ of stannous chloride and 35 g/ℓ of hydrochloric acid. The sensitized surface is rinsed with deionized water. Then, the sensitized surface is activated by treating with a solution of noble metal salt such as gold chloride, potassium gold chloride or palladium chloride. Suitably, a solution containing 0.5 g/ℓ of palladium chloride and 0.5 ml/ℓ of hydrochloric acid is employed. The activated surface is then rinsed, first with deionized water, then with acetone, and dried.

A layer of positive photoresist is then applied. The Shipley Company's AZ 1350 or AZ 111 photoresists are suitable. The photoresist can be dried and hardened by heating at about 70° to 75°C, if desired. The plate is then exposed to ultraviolet light through a mask of the desired pattern, followed by conventional developing, using, for example, Shipley AZ developer.

The photomask is subjected to a second development step, using Shipley AZ developer or a 0.15 to 0.2 N solution of sodium hydroxide, to which has been added the activator employed in the previous activation step. A concentration of about 0.5 g/ℓ of palladium chloride and 0.5 g/ℓ of concentrated hydrochloric acid is suitable. A 15-second spray or immersion is generally adequate.

The substrate is then immersed into an electroless plating solution for 15 to 20 minutes. Metal will be plated directly over the defective dots and missing dots. The photoresist is then removed in conventional manner, as by immersing or spraying with acetone or other suitable solvent. The repaired photomask is inspected, and if satisfactorily repaired, the metal dots can be hardened, as by baking at about 375°C for 1 hour.

Photoconductor with Interlaying Electrodes

Light conducting structures, which have a pair of light conductors on a substrate with an electrode disposed therebetween are known. Such structures are used as electrically controllable directional couplers in optical communication technology and act as on/off or crossover switches. In addition, such structures are used as electrical-optical modulators.

A common feature of these structures is that the two light conductors possess a zone in which they are very closely adjacent to one another. Electrodes are arranged between the light conductors and beside each of the light conductors. In this zone, a typical value for the spacing between a pair of light conductors is 3 μm. This means that the electrodes must be precisely aligned in their position and that permissible tolerances in the location of the electrodes must be less than 1 μm.

F. Auracher and G. Bell; U.S. Patent 4,136,212; January 23, 1979; assigned to Siemens AG, Germany describe a method for forming a light conducting structure having interlying electrodes which are extremely accurately dimensioned within narrow tolerances.

This process is directed to a method for forming a light conductor structure having a pair of light conductors embedded in one surface of a substrate of an electrooptical material having its c-axis extending parallel to the surface and at right angles to the light conductors and having electrodes extending between and along the light conductors, the structure being particularly adapted for use as an electrically controllable coupler. The method comprises providing a substrate of the electrooptical material having the one abovedescribed surface; applying a layer of polycrystalline silicon on the one surface of the substrate; removing a portion of the silicon layer from the one surface by etching to provide silicon-free portions or zones on the one surface having a configuration of the light conductors which are to be formed with the remaining portions of the silicon layer forming a mask; applying a layer of diffusion material on the silicon layer and the silicon-free portions of the one surface of the substrate, diffusing the diffusion material into the surface of the substrate at the silicon-free portions to form light conductors having an index of refraction greater than the index of refraction of the substrate by heating to an elevated temperature; cooling the substrate; applying a layer of negative-acting photo or light sensitive lacquer to the light conductor and the layer of diffusion material; exposing the layer of negative-acting photo or light sensitive lacquer by projecting light through the substrate with the remaining portion of the layer of silicon acting as a mask; developing the layer of the photolacquer so that the unexposed portions are removed and the remaining lacquer layer covers the light conductors; removing the remaining portions of the silicon layer with diffusion material thereon from the one surface; applying a metal layer on the one surface to form the electrodes; and then removing the photo-lacquer with the metal disposed thereon from the light conductor.

Preferably, prior to applying the metal layer to form the electrodes, a thin dielectric layer, whose index of refraction is lower than that of the substrate, is applied to the substrate and is then followed by the application of the metal layer forming the electrodes. Preferably, prior to the step of etching the polycrystalline silicon, the layer of silicon is subjected to ion bombarding so that beveled etching edges are formed during the subsequent etching process. The substrate is preferably either a lithium tantalate crystal or a lithium niobate crystal.

DC-Excitable Fluorescent Film

Conventional methods of producing thin fluorescent films for electroluminescence generally employ the embedding process or other heat-treating processes, wherein activators, such as, for example, Al, Ag, Au, Cu, Mn and Pb, and coactivators, such as, for example, NaCl, $ZnCl_2$, NH_4Cl, NH_4I, Al, Cu and Ga, are added to base materials, such as, for example, ZnS, ZnSe and CdS, so as to form evaporated films. However, such conventional methods have several disadvantages in that they cannot produce fluorescent films for electroluminescence which emit light by the action of direct current, unless coactivators are in fact added thereto as mentioned above.

K. Morimoto, Y. Utamura, and T. Takagi; U.S. Patent 4,098,919; July 4, 1978; assigned to Futaba Denshi Kogyo KK, Japan describe a method of producing an excellent fluorescent film which can emit light as a result of the application of dc voltage thereto, without using coactivators. This is achieved through the provision of a method of producing a fluorescent film for electroluminescence which comprises the steps of heating raw materials, composed of at least a base material and an activator or components thereof, within an enclosed crucible so as to generate a mixed vapor, injecting the mixed vapor into a vacuum zone through means of at least one nozzle provided upon the crucible, radiating an electron beam so as to ionize at least a portion of the injected mixed vapor, accelerating the ionized mixed vapor within an electric field, and projecting the accelerated ionized mixed vapor onto a base plate so as to form a vapor-deposited fluorescent film thereon.

Referring to Figure 8.10, the primary portion of an apparatus for practicing the process is illustrated, and the same is to be operated under high vacuum conditions, such as, for example, 10^{-2} torr or less, and preferably, 10^{-4} or less. An enclosed crucible **1** is provided with one or more nozzles **2**, for projecting vapor therefrom and into the vacuum zone, and crucible **1** is constructed such that the contents thereof are heated by means of electric heating, electronic impact, high frequency induction heating or the like. The crucible **1** contains raw material **3** composed of a base material, such as, for example, ZnS and an activator, such as, for example, Mn at a ratio of 0.01 to 10% by weight, and preferably at a ratio of 0.03 to 0.1% by weight.

Figure 8.10: DC-Excitable Fluorescent Film

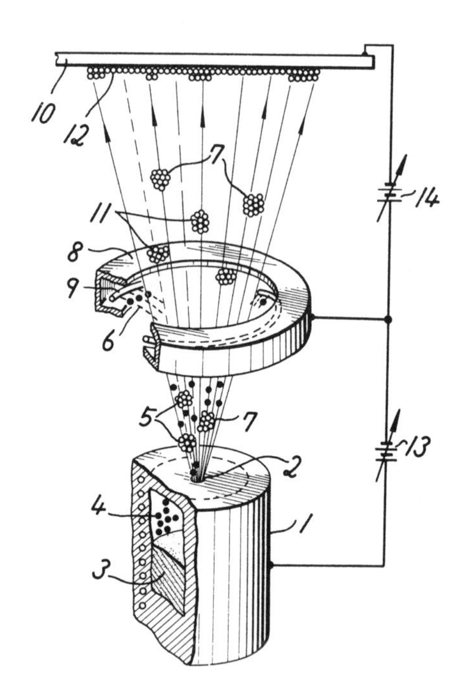

Source: U.S. Patent 4,098,919

Within this process, a compound, such as, for example, ZnS, is used as the base material, however, the raw material **3** may be composed of simple substances, such as, for example, the elements of the compounds. Accordingly, the raw material **3** may be composed of a mixture of simple substances of three kinds of elements, such as, for example, Zn, S and Mn. In the cases where Zn, S and Mn are used, the mixing ratios thereof are 1 to 99% by weight, 1 to 99% by weight, and 0.01 to 50% by weight, respectively, and preferably, mixing ratios of 20 to 80% by weight, 20 to 80% by weight, and 0.02 to 10% by weight, respectively.

The abovementioned crucible **1** is heated to a temperature of approximately 1000°C or more, and preferably at a temperature of 1200°C or more, and thus, a mixed vapor **4** of the raw material **3**, which vapor **4** has a temperature corresponding to the aforenoted heating temperature and a high vapor pressure, prevails within the space of the crucible **1** disposed above material **3**. The mixed vapor **4** of the raw material **3** is then injected into a high vacuum zone through at least one nozzle **2** provided within the upper surface of crucible **1**.

The mixed vapor **4** is, during this injection phase of the process, adiabatically expanded so as to attain a supercooled state, and within such state, the thermal energy of the mixed vapor particles is extremely decreased and, therefore, attractive forces between the particles work effectively to form massive atom groups or clusters **5**.

An electron beam **6** is projected onto the clusters **5** so as to ionize at least one of the atoms thereof and thereby form ionized clusters **7**. Within this state, the sectional area for collision of electrons is proportional to the mass of the clusters **5**. Therefore, the clusters **5** composed of hundreds of atoms have large sectional areas for collision of electrons, and accordingly the process is carried out very effectively and efficiently.

One of the various possible forms of the electron source mechanism for obtaining the electron beam **6** for ionizing the clusters **5** is shown within Figure 8.10 as an example, and the same is seen to comprise a generally hollow doughnut type acceleration electrode assembly **8** disposed above the injection nozzle **2** of the crucible **1** so as to accelerate the ionized clusters **7**. The acceleration electrode assembly **8** contains therein a filament coil **9** for electron emission, and the same is adapted to focus the electron beam **6** onto the injection nozzle **2** and the mixed vapor **4** of the raw material prevailing within the neighborhood of the injection nozzle **2**, which mixed vapor has been projected from the injection nozzle **2** and contains the clusters **5**.

If the acceleration electrode assembly **8** is maintained, relative to the crucible **1**, at a high negative voltage of -100 V or more, and preferably at a voltage of -500 V to 10 kV and, in addition, if the filament coil **9** is heated, then the electron beam **6** projected from the filament coil **9** is accelerated towards and focused within the neighborhood of the injection nozzle **2** so as to effectively collide with the clusters **5** and thereby ionize the same.

The ionized clusters **7** are then accelerated in the direction opposite that of the electron beam **6** by means of the action of the same electric field of negative voltage as that given by the acceleration electrode assembly **8** for accelerating the electron beam **6**, and consequently, clusters **7** move at high speed towards a base plate **10**, such as, for example, a glass and plastic plate after passing through the center of the hollow doughnut type acceleration assembly **8**.

Neutral clusters **11** which have not been subjected to the electron beam **6** also move towards the base plate **10** as a result of the kinetic energy thereof which they have when they are projected from the nozzle **2**. The ratio between the numbers of the neutral clusters **11** and the ionized clusters **7** can be adjusted by optionally selecting the strength of the electron beam **6**, the acceleration energy thereof, and the like. The strength of the electron beam **6** may be selected to be 10 mA or more, and preferably to be 10 to 300 mA. The acceleration energy of the electron beam **6**, namely, the negative voltage imparted to the hollow doughnut type acceleration assembly **8**, is preferably between -100 to -300 V from the standpoint of ionization efficiency; however, it is preferably -500 V or less from the standpoint of the desirable acceleration of the electrons as well as that of the ions.

The ionized clusters **7** and the neutral clusters **11** moving towards the base plate **10** both collide with the base plate **10** so as to vapor-deposit thereon and thereby form a vapor-deposited film **12**. At this time, a greater portion of the large kinetic energy of the ionized clusters **7** is converted into thermal energy so as to heat the base plate **10**, however, the base plate **10** is not heated uniformly, but on the contrary, is heated locally within those portions, and the adjacent areas thereof, upon which the vapor-deposition takes place. Thus, the vapor-deposited film **12** rapidly becomes a single crystal.

The bonding strength between the particles of the clusters depends upon van der Waals forces and, therefore, is weak. For this reason, when the clusters collide with the base plate **10**, the same are disintegrated into individual atoms which move onto the base plate **10** and condense at stable points thereof so as to form the vapor-deposited film **12** having good crystallization properties. In addition, when the ionized clusters **7** collide with the base plate **10**, the surface of the base plate **10** is subjected to spattering and, therefore, is purified owing to the kinetic energy of the clusters **7**. Thus, the vapor-deposited film **12** exhibits a high adhesion force. The abovementioned processes occurring simultaneously with the collision between the base plate **10** and the neutral clusters **11** thus create the vapor-deposited film **12** upon the base plate **10** which film has a high adhesion force and good crystallization properties.

The clusters **7** are also accelerated by means of a negative voltage, provided by means of a battery type power source **13**, which is applied to the acceleration electrode **8** and, if necessary, also by means of an additional negative voltage, provided by means of another battery type power source **14**, which is applied to the base plate **10**, and, consequently, the same are provided with a large kinetic energy with which they collide with the base plate **10**. Therefore, an effect analogous to ion injection can be expected, namely, the same process as that carried out when Mn ions are injected into ZnS material is expected simultaneously with the formation of the vapor-deposited film **12**. Accordingly, ions of an activator, such as, for example, Mn, are injected deeply into the base material, such as, for example, ZnS. They effectively become substituent atoms, which act as light-emissive nuclei, by annealing after the formation of the vapor-deposited film, and thus, light emission becomes possible by means of a dc excitation without utilizing any coactivators.

MISCELLANEOUS PROCESSES

THERMAL COATINGS

Porous Metallic Interface

Heat exchanger devices of tubular, planar and other configurations with a porous metallic surface layer have been proposed in the art. Such a layer is known to provide a highly efficient thermal interface by virtue of the extended effective surface areas for transferring the heat compared with smooth and conventional finned thermal interfaces. Furthermore, with the porous interface in contact with a boiling liquid, the individual pores or cavities when properly sized and distributed serve to provide highly effective sites for bubble nucleation and thus promote the nucleate bubbling (bubble forming and growing) process. As a result, effective heat transfer coefficients 10 times greater than conventional fin type members or even more can be obtained.

K. Inoue; U.S. Patent 4,120,994; October 17, 1978; assigned to Inoue-Japax Research Incorporated, Japan describes a method of preparing a heat transfer member having a porous metallic heat transfer interface which basically comprises the steps of disposing at least one surface of a substrate in contact with a solution containing a salt of a thermally conductive metal and depositing the metal from the solution upon the substrate chemically or electochemically so as to form a dendritic metallic layer thereon constituting the porous heat transfer interface.

It is desirable that the surface of the substrate be first mechanically roughened or flawed so that a multitude of minute protrusions are formed thereon in closely spaced relationship and in a uniform distribution to provide sites on which deposition should take place preferentially. It has been found that this preparatory treatment is highly effective to form dendrites uniformly over the entire surface and thus to enhance the evenness of the porous structure throughout the desired surface. To this end, flaws should advantageously be produced by scratching throughout the surface with a depth of cut around or greater than 0.1 mm and a spacing between adjacent cuts of 0.1 to 0.3 mm.

For the sake of insuring the effective formation of the desired dendritic structure and the uniformity thereof, it has also been found to be desirable that the solution in contact with the substrate during the depositing operation be held at an elevated temperature ranging between 60° and 110°C, preferably higher than 90°C near its boiling point.

When the internal surface of a tube is treated by this method, it has been found that the porous layer thereby provided throughout the inside wall serves as a highly efficient heat exchanger interface former at its heat input and output regions and in addition as an excellent wicking or capillary layer across its intermediate region.

Base Metal-Braze Metal Coating

R.C. Borchert; U.S. Patent 4,101,691; July 18, 1978; assigned to Union Carbide Corporation describes a method for manufacturing an enhanced heat transfer device consisting of a metal substrate and randomly distributed metal bodies bonded to the substrate.

In this method base metal powder with particles of major dimensions less than 0.1" is provided and mixed with first liquid binder in proportion such that the weight ratio of base metal powder to first liquid binder is between 20:1 and 30:1, so as to form an adherent mass. Braze metal powder having a melting point lower than the base metal powder is also provided having particles of major dimensions such that the major dimension ratio of braze metal powder to base metal powder is between 1:60 and 1:3.

The braze metal powder and the aforedescribed adherent mass are mixed in weight proportion such that the braze metal powder is between 10 and 30% of the braze metal powder plus the base metal powder, so as to form braze metal-coated mass. A second liquid binder is applied on the metal substrate and the aforedescribed braze metal-coated mass is then applied on the second liquid binder coated metal substrate. The so-formed braze metal-coated mass metal substrate is heated sufficiently to remove the first and second binders, melt the braze metal and metal bond the base metal to the metal substrate thereby forming the metal bodies.

The base metal powder constitutes the bulk structure of the final metal bodies and the powder is preliminarily sized to provide the desired major or greatest dimension of the individual particles. The major dimension of the base metal powder particles should not exceed 0.10" and preferably should be within the range of 0.006 to 0.060". Particle dimensions for these purposes correspond generally to the opening size of U.S. Standard series screen through which the particles will pass.

The adherent mass comprises base metal powder particles individually covered with a thin layer of first liquid binder. The first liquid binder may be a single component material, but preferably is a plural component material. A required component is a low volatility organic compound capable of wetting the base metal particles and capable of being removed from the base metal particles by vaporization and/or chemical decomposition without leaving an undesirable residue on the metal surface.

A preferred low volatility organic compound is an isobutylene polymer having a molecular weight of at least about 90,000. A higher volatility compound which is a solvent for the low volatile organic compound may be included as another component of the first liquid binder. A preferred higher volatility compound is kerosene.

The braze metal-coated mass is produced by admixture of the braze metal powder with the adherent mass thereby obtaining a relatively dry flowable powderlike material. The braze metal powder is a metal or alloy material having a melting point lower than the base metal powder and capable of forming a metallic bond therewith and with the substrate. Its particle size is substantially smaller than that of the base metal powder, such particle size being defined as the major dimension of the particles. Preferably, the major dimension ratio of braze metal powder to base metal powder is between 1:30 and 1:3.

"Partially rigidizing" is a procedure for stiffening or setting the first liquid binder after formation of the braze metal-coated mass. The braze metal particles are thereby firmly attached to the base metal particles. A preferred method of partially rigidizing is by heating the braze metal-coated mass at temperatures of 150° to 200°F in order to vaporize higher volatility component from the first liquid binder.

Figure 9.1 is a block diagram showing the several steps of the manufacturing method of this process.

Suitable materials for the base metal powder include copper, copper-nickel alloy (e.g., 70 to 90% copper), iron, stainless steel (e.g., Type 304 or 316) and aluminum.

Examples of suitable braze metal powders are copper alloy containing 8% phosphorus, nickel alloy containing 11% phosphorus, nickel alloy containing 13% chromium and 10% phosphorus, and aluminum alloy containing 12% silicon. It is to be understood that the selection of the braze metal powder must be compatible with the base metal powder and with the substrate so as to produce a structurally sound metallic bond therewith.

Figure 9.1: Base Metal-Braze Metal Coating

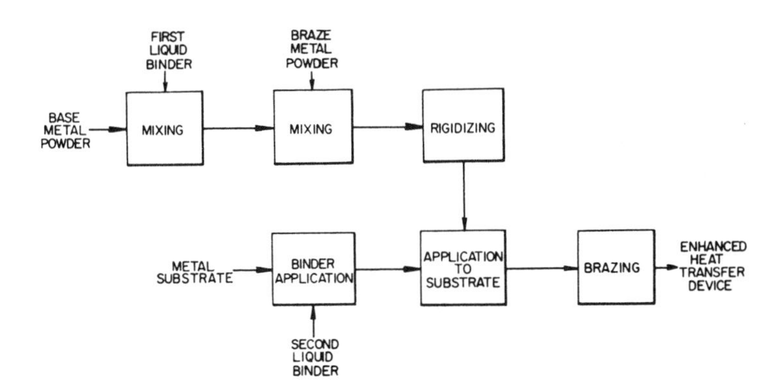

Source: U.S. Patent 4,101,691.

Many combinations of base metal powder, braze metal powder and metal substrate may be employed usefully and effectively. The table lists several such combinations by way of illustration. One combination of the table which is particularly useful is a copper bearing base metal powder, a phosphorus-nickel braze metal powder and a copper bearing metal substrate.

Base Metal Powder	+	Braze Metal Powder	+	Metal Substrate
Cu or Cu + 5-30% Ni		Cu + 8% P or Ni + 11% P		Cu or Cu + 1-2% Fe or Cu + 5-30% Ni
Steel or Cu		Ni + 11% P or Ni + 10% P + 13% Cr		carbon steel or low alloy steel (3.5-9% Ni)
Steel or SS 304 or SS 316		Ni + 11% P or Ni + 10% P + 13% Cr		SS 304 or SS 316
Al		Al + 12% Si		Al

Example: A single layer of randomly distributed metal bodies was bonded to the inner wall of a tubular substrate. This single layer surface was prepared by first screening copper powder to obtain a graded cut, i.e., through 50 and retained on 60 U.S. Standard mesh screen from a commercially available copper powder, Amax "O" (American Metals Climax, Inc.). A portion of this graded cut weighing 2,070 g was placed into a large evaporating dish and subsequently slurried with 285 g of a 6% by wt isobutylene polymer (Vistanex LM-MS), 6% kerosene and 88% benzene solution. After thorough mixing, sufficient of the benzene was evaporated on a hot plate to result in an adherent mass of copper particles and first liquid binder.

Phos-copper brazing alloy of 92% copper, 8% phosphorus by weight weighing 517 g which had been screened from phos-copper powder 1501 (New Jersey Zinc Company) to remove all particles larger than 325 mesh (U.S. Standard series sieves) was added to the adherent mass of copper base particles to formulate a ratio of 4 pbw to 1 part phos-copper. Thus, the major dimension ratio of braze metal powder to base metal powder was about 1:7. This is based on a base metal powder dimension of 0.0117" corresponding to No. 50 U.S. Standard mesh screen opening and a braze metal powder dimension of 0.0017" corresponding to No. 325 U.S. Standard screen opening. After thorough blending, the resultant dry mix of phos-copper coated copper base metal particles was allowed to stand at ambient temperature overnight. So treated, the particles of phos-copper brazing alloy were evenly disposed on and secured by the polyisobutylene coating to the surface of the copper particles. The powder was dry to the touch and free-flowing.

A CDA-192 copper alloy tube with a 0.679" i.d. and a 0.735" o.d. was coated with a second liquid binder composed of 30% polyisobutylene in kerosene by filling the tube with the binder followed by draining it from the tube to leave a thin adherent internal film of the binder on the internal tube wall. Next, the phos-copper coated copper base metal powder particles were poured through the tube, coating the internal tube surface substrate with a uniformly spaced single layer of braze metal-coated mass.

The external surface of the tube also was coated with a multiple layer of stacked copper particles integrally bonded together to form interconnected pores of capillary size in a manner described in U.S. Patent 3,384,154 to R.R. Milton (porous boiling layer). The tube was then furnaced at 1600°F for 15 minutes in an atmosphere of disassociated ammonia, cooled and then tested for heat transfer characteristics as an enhanced heat transfer device.

The sensible heat transfer enhancement of the described test device was determined by boiling Refrigerant 12 (dichlorodifluoromethane) at 48 psia on the exterior tube surface and by flowing water at higher temperature at 9 ft/sec through the internal surface covered with the single layer of bonded randomly distributed bodies. The boiling side heat transfer coefficient was already known, having been determined on a separate but similar boiling surface under the same conditions. The data of the instant test was reduced by extracting the known boiling side heat transfer resistance and the wall resistance in order to determine by difference the water side sensible heat transfer coefficient. The coefficient was found to be 2.55 times higher than obtained on a smooth surface metal substrate.

Heat Radiation Element

T. Kuze, T. Matsuki, K. Nagaoka and N. Iwai; U.S. Patent 4,119,761; assigned to Tokyo Shibaura Electric Co., Ltd., Japan describe a heat radiation element which comprises a substrate and a heat radiation layer formed on the substrate. Preferred materials for the substrate include, for example, metals such as iron (including steel), nickel, chromium, copper, aluminum and silver and various alloys such as iron-chromium, nickel-chromium and iron-nickel-chromium alloys. On the other hand, the heat radiation layer is constituted by oxides of a chromium-containing alloy formed in direct contact with the substrate. The oxides constituting the heat radiation layer contain at least 35% by weight, and preferably from 60 to 99% by weight, of chromium based on the total weight of the metals contained in the oxides.

Among the various chromium-containing alloys providing precursor substances of the oxides constituting the heat radiation layer, preferred alloys are iron-chromium alloy, nickel-chromium alloy and iron-nickel-chromium alloy.

The surface roughness of the heat radiation layer should preferably range in general from 0.05 to 30 μ, especially from 0.3 to 5 μ, as measured by the Japanese Industrial Standard (JIS) B 0601 in order to facilitate the heat dissipation of the heat radiation element. Further, the density ratio of the heat radiation layer to the theoretical density thereof i.e., the ratio of the actual density of the heat radiation layer to the theoretical density thereof, preferably ranges in general from 0.6 to $<$1.0, especially from 0.7 to 0.8.

The element can be prepared by various methods. One convenient method is to oxidize the substrate constituted by chromium-containing alloy. In order to meet the requirement that the oxides constituting the heat radiation layer should contain at least 35% by weight of chromium, the chromium content of the iron-chromium alloy should be at least 2% by weight, preferably 10% by weight or more. The chromium content of the nickel-chromium alloy should also be at least 2% by weight, preferably 5% by weight or more. On the other hand, the required chromium content of the iron-nickel-chromium alloy is 3% by weight or more, preferably 10% by weight or more.

An excessive chromium content in a chromium-containing alloy which constitutes the substrate should also be avoided in terms of the machining property of the substrate. In the case of the iron-chromium alloy, the chromium content should not exceed 35% by weight. Otherwise, σ-phase is deposited, rendering the alloy brittle. In the case of the nickel-chromium alloy, the chromium content should preferably be less than 60% by weight. When it comes to the iron-nickel-chromium alloy, the chromium content should preferably be less than 40% by weight, and in addition, the nickel content should preferably fall within the range of from 3 to 85% by weight.

The substrate oxidation is effected in general by heating the substrate in the air at 400° to 1300°C for several seconds to several minutes, though the heating conditions will vary depending on the chromium content of the substrate alloy. Alternatively, the oxidation may be effected by heating the substrate at 800° to 1350°C for one minute to several hours in wet hydrogen, i.e., a mixture of hydrogen gas and water vapor, having a dew point ranging from −10° to 40°C.

The oxidation treatment described above permits forming a heat radiation layer consisting of black metal oxides very tightly attached to the substrate. Generally, the heat radiation layer thus formed is 4000 Å to 10,000 Å thick.

A heat radiation element may also be produced by coating a substrate with chromium-containing alloy, followed by oxidizing the alloy. In this case, the coating may be effected by vapor deposition sputtering, plating, cladding or spraying. As is the case with the alloy substrate described previously, preferred alloys to be coated on the substrate are iron-chromium alloy, nickel-chromium alloy and iron-nickel-chromium alloy. The upper limit of the chromium content of the coating layer alloy need not be considered as far as the coating layer constitutes an alloy in the ordinary sense. Specifically, the chromium content should be at least 2% by weight, preferably 10% by weight or more if the coating layer is made of iron-chromium alloy. If the coating layer is formed of nickel-chromium alloy, the chromium content should be at least 2% by weight and preferably 5% by weight or more. When it comes to iron-nickel-chromium alloy, the chromium content should be 3% by weight or more, preferably 10% by weight or more.

The chromium-containing alloy coated on the substrate is oxidized in the air or in wet hydrogen under the conditions as described previously.

The table below summarizes preferred conditions for oxidizing chromium-containing alloy.

		In Wet Hydrogen		In Air	
	Dew	Heating			
Cr Content of Alloy	Point (°C)	Temperature (°C)	Time (min)	Temperature (°C)	Time (min)
Up to 12% by wt	20-40	1100-1350	>1	–	–
12 to 17% by wt	10-40	1000-1350	>1	600-1100	>0.5
>17% by wt	0-40	800-1350	>1	700-1300	>0.5

The total emissivity σ of the heat radiation element is increased up to at least about 0.90 if at least one of vanadium, titanium, zirconium and niobium acting as an emissivity-improving agent is present in the heat radiation layer.

In order to allow the heat radiation layer to contain the emissivity-improving agent mentioned above it suffices to add the emissivity-improving agent to a chromium-containing alloy such as iron-chromium alloy, nickel-chromium alloy or iron-nickel-chromium alloy, followed by the heat treatment under the conditions described previously. The additive content of the alloy should be at least 0.03% by weight, preferably 0.07% by weight or more. No detrimental effect is produced if an excessive amount of the emissivity-improving agent has been added. But, an appreciable improvement in emissivity is not recognized when the additive content exceeds 5% by weight.

The heat radiation element according to this process can be extensively used as members requiring a good heat radiation including, for example, an anode of an electron tube, a heating wire, a boiler shell, etc. It should also be noted that a material capable of a good heat radiation is also good in heat absorption capability. In this sense, the heat radiation element of the process can effectively be used as a heat absorption member of, for example, a solar heat absorption apparatus.

ANTIPLATING AGENTS

Removable Film

Recently, in the field of the steel sheet to be used for automobile, domestic electrical equipment, building material and the like, it has been eagerly demanded to produce a so-called one-side plated steel sheet by plating only one side of a steel sheet to give the sheet a sufficiently high corrosion resistance and at the same time to improve the weldability of the sheet.

M. Goto and H. Komura; U.S. Patent 4,125,647; November 14, 1978; assigned to Kawasaki Steel Corporation, Japan describe a method of producing one-side plated steel sheets by the use of an antiplating agent.

It was found that it is effective to use an antiplating agent which forms a crystalline or amorphous film consisting of four components of SiO_2, B_2O_3, MgO and M_2O, wherein M represents an alkali metal.

In the antiplating agent, as the SiO_2 component, alkali silicate ($M_2O \cdot nSiO_2 \cdot mH_2O$) can be used; as the B_2O_3 component, boric acid (H_3BO_3), boric anhydride (B_2O_3) or sodium borate ($Na_2B_4O_7$) can be used; as the MgO component, magnesium oxide (MgO) or magnesium hydroxide [$Mg(OH)_2$] can be used; and as the M_2O component, sodium hydroxide (NaOH), lithium hydroxide (LiOH) or potassium hydroxide (KOH) can be used. These compounds are added to the proper amount of water and used.

Of course, the above described M_2O component can be partly or wholly replaced by the M_2O component of alkali silicate or Na_2O component of sodium borate. Among alkali silicates, water glass ($Na_2O \cdot nSiO \cdot mH_2O$, $2 \leqslant n \leqslant 4$) is inexpensive. As the alkali, NaOH is inexpensive.

It has been found that the above described four components are essential and that, when any one of them is not present, satisfactory one-side plated steel sheets cannot be obtained.

An aqueous slurry having the above described composition is coated on one side of a sufficiently degreased steel sheet, and dried at low temperature, preferably at a temperature of not higher than about 200°C. The coating can be effected by any one of roll-coating, spray-coating, brush-coating or other optional methods. The proper amount of the aqueous slurry to be coated is 20 to 50 g/m^2 in dry weight. When the coating amount is smaller than 20 g/m^2, the coating slurry film cannot completely cover the steel sheet surface. while, when the coating amount is larger than 150 g/m^2, cracks are apt to occur in the coating slurry film during the drying of the film. Therefore, the resulting film cannot prevent completely the coated side of the steel sheet from being plated.

The water contained in the coated slurry film is evaporated by a low-temperature drying. This low-temperature drying is a necessary step in order that the atmosphere in the following annealing step is kept to a reducing atmosphere and that the breakage and exfoliation of the coating film due to rapid heating up to high temperature are prevented.

After the dried uniform film is formed on one side of the steel sheet, the steel sheet is annealed at a temperature (usually at about 700°C) of not lower than the recrystallization temperature of the steel sheet, cooled to a temperature which is near the plating bath temperature (in the zinc plating, about 460°C), and dipped in the plating bath in the same manner as that of commonly known continuous hot-dip plating process.

The steel sheet brought up from the plating bath is plated at its one side only, and another side of the sheet is covered with the film of the antiplating agent without being plated.

After the plating, the film must be removed. However it has been found that the film can be easily removed by quenching the plated steel sheet, after the sheet is brought up from the plating bath, from a temperature of not lower than about 100°C to room temperature.

The quenching may be carried out before the plating metal is solidified in order to regulate simultaneously the spangle size or after the plating metal is solidified. Alternatively, after a plated steel sheet may be once cooled gradually, the sheet may be again heated at any convenient time and then quenched.

The quenching is easily and effectively carried out by immersing the plated steel sheet in water. It has been found as the result of experiments that, when the plated steel sheet is quenched in water, the film can be completely peeled off from the steel sheet surface without changing the original cold rolled surface thereof.

Reactive Type Maskant

R.C. Elam; U.S. Patent 4,128,522; December 5, 1978; assigned to Gulf and Western Industries, Inc. describes a controllable, reactive type maskant for use principally in masking nickel, cobalt and iron based substrates during surface coating with aluminum based coatings.

The maskant is comprised principally of the following ingredients: (1) a reducible material such as, for example, titanium dioxide; (2) an inhibitor such as nickel; (3) an inert material such as aluminum oxide; and (4) a vehicle such as a mixture of methyl ethyl ketone and an acrylic resin.

A reactive type maskant functions to prevent the deposition of the coating on the substrate by reacting with the coating being applied.

Examples of reducible materials include zirconium oxide, ferric oxide, titanium sesquioxide, nickelous oxide and other oxides which have a higher free energy of formation than aluminum oxide and are accordingly less stable than aluminum oxide. A preferred reducible material is titanium dioxide. When titanium dioxide is used, it is reduced by the aluminum-freeing elemental titanium which is coated with aluminum to form a titanium aluminide. This prevents deposition of the pack coating on the masked portion of the substrate since the underlying substrate is prevented from reacting with the aluminum containing vapor generated during pack coating.

A particularly preferred amount of reducible material is between about 8 and 12% by weight of the solids portion of the composition.

The maskant should also contain an inhibitor which serves to control the rate of reaction between the reducible material and the pack coating. Preferred inhibitors are, for example, refractory metal including, but not limited to, tungsten, molybdenum and tantalum, as well as cobalt, chromium and iron. A particularly preferred inhibitor material is nickel.

A most preferred amount of inhibitor is within the range of from about 8 to 12% by weight of the solids portion since an amount within this range results in a fully controllable maskant.

The addition of an inert oxide material to the composition is also preferred since such a material aids, by dilution, in controlling the reaction between the reducible material and the pack coating. Inert oxide materials selected should have a free energy of formation no higher than the free energy of formation of aluminum oxide and they may have a lower free energy of formation than aluminum oxide. Preferred inert materials include, but are not limited to, the oxides of most rare earth metals such as, for example, scandium, yttrium and lanthanum, and the oxides of refractory elements such as, for example, hafnium, thorium and tantalum. A particularly preferred inert oxide material is aluminum oxide.

A preferred amount of inert material is within the range of from about 60 to 90% by weight, and a most preferred amount is between about 75 and 85% by weight of the solids portion. The vehicle in which the solids portion is suspended may constitute any suitable binder and preferably includes a mixture of a resinous material and a solvent.

The maskant may be applied to the portion of the substrate to be masked by conventional methods including, for example, by dipping the substrate into the maskant, by an applicator such as a brush or spatula or by spraying. The maskant should be permitted to dry prior to coating.

After the coating is applied to the unmasked portions of the substrate, the maskant may be removed by fracturing it with an appropriate tool such as a hammer or plastic mallet and then peeling the fractured portion of the maskant away from the substrate.

The maskant can also be removed using a scraping device, wire brush, or by a light grit blasting.

Example: In order to more fully illustrate the preparation of a maskant according to the process, a maskant composition comprising the following ingredients was admixed at ambient temperatures.

Ingredients	Percent by Weight	Percent by Volume
Solids		
Nickel	10	–
Titanium dioxide	10	–
Aluminum oxide	80	–
Vehicle		
Methyl ethyl ketone	–	50
Acryloid resin	–	50

The ingredients comprising the solids portion of the maskant were blended together in a mixing device and added to the vehicle in a volumetric ratio of two parts solids to one part vehicle thus producing a viscous blend which was spread out on a nickel alloy substrate and allowed to dry. An aluminide pack coating was applied and, after coating, the maskant was removed and the substrate tested both metallographically and with a heat tint test. The tests indicated complete masking of the masked area and virtually no migration of the uncoated area beyond the previously masked area indicating that the reactive type maskant was totally effective and completely controllable.

Oxidized Surface

K. Asakawa and M. Yoshida; U.S. Patent 4,107,357; August 15, 1978; assigned to Nippon Steel Corp., Japan describe a method for effecting one side molten metal plating in which a material metal is subjected to activation treatment on its surface to be plated by heating in a reducing atmosphere and then dipped into a molten metal bath for continuous plating, which comprises blowing an oxygen-containing gas to only one side of the material after the activation treatment and maintaining a nonoxidizing atmosphere at the outlet side of the bath whereby preventing adherence of the plating metal to the surface of the material on which the film has been formed.

In Figure 9.2, a strip (S) is uncoiled in turn from an uncoiler **1** and passed into a preliminary treatment unit **2**. This unit may be an oxidizing furnace, a nonoxidizing furnace, or a degreasing pickling tub, etc. according to the particular plating system used, where the strip is cleaned on its surface. Then it is carried to a reducing furnace **3** which has been filled with a reducing atmosphere, where it is reduced and activated on its surface and adjusted to a temperature suitable for plating. The strip is thereafter blown with an oxidizing gas on one side thereof for oxidation just before it is dipped into a metal plating bath **5**, for example, at a snout part **4**. It is thus dipped into the plating bath without contacting the outside atmosphere and finally carried upward from the bath **5**. In this case, the outlet side of the bath is filled with a reducing atmosphere by means of a seal box **6**, whereby any plating metal adhering to the oxide surface of the strip is removed. The strip is then cooled in the air and coiled by a coiler **7**.

Figure 9.2: Continuous Molten Metal Plating Line

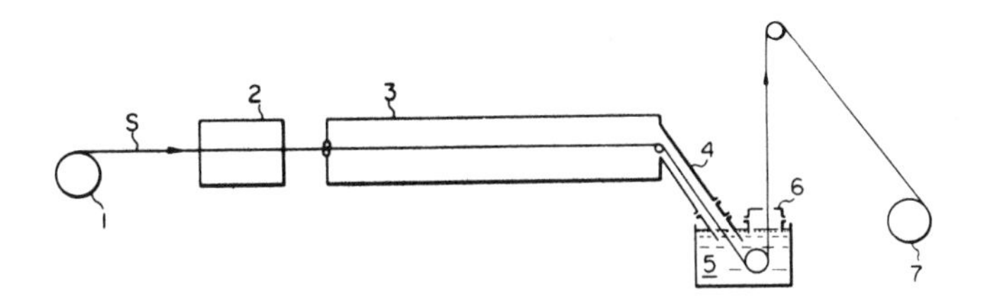

Source: U.S. Patent 4,107,357

In other words, this process may be practiced in the conventional continuous plating line by oxidizing one side of a strip at the terminal end of the furnace **3** filled with a reducing atmosphere, for example, at the snout part **4**; dipping the same into the plating bath **5** while the opposite or reverse side thereof is kept activated so that a formation of an alloy layer is prevented; and providing a seal box **6** at the outlet side of the bath whereby the repelling of the plating metal adhering to the oxide surface is accelerated and the plating on only the nonoxidized surface is carried out. In this way, a one side plated steel sheet of lower cost and higher quality than the conventional one can be obtained.

OTHER PROCESSES

Homogeneous Precipitation Deposition on Carrier Materials

J.W. Geus; U.S. Patent 4,113,658; September 12, 1978; assigned to Stamicarbon NV, Netherlands describes a process for the preparation of a particulate, at least inchoately catalytically active crystalline composition, affixed to the surfaces of a solid particulate supporting material which comprises the combination of steps of:

(A) Forming a body of an aqueous solution-suspension by admixing:

(1) A solution of at least one metal compound wherein said metal is Bi(III), Co(II), Cr(III), Fe(II), Mo(VI), Ni(II), Sn(IV) or Pt(IV), which compound is:

(a) Substantially soluble in said aqueous medium within at least a range of pH value below a predetermined pH value of less than about 7; and

(b) At least one metal is capable of forming an insoluble compound at a pH above said range; and wherein said solution of said compound contains 5 to 15% by weight of the metal ion of said metal compound; and

(2) A finely divided stable particulate solid, nucleating surface providing supporting material which is SiO_2,

Al_2O_3, or TiO_2 and which has a sedimentation rate sufficiently low to permit homogeneous distribution thereof in said solution under the agitated conditions during the process and to form a suspension under said agitated conditions in said solution; the particles thereof having a surface which, under the conditions of the process is a nucleating surface with respect to the insoluble compounds; and

(B) Initially controlling and adjusting the pH of said aqueous medium to a value within said pH range within which said compound is substantially soluble in said aqueous medium;

(C) Sufficiently vigorously agitating the resulting aqueous medium so that said suspension particles are homogeneously distributed throughout said body, along with said dissolved at least one metal compound; while also

(D) Increasing the total hydroxyl ion concentration within said agitated suspension solution wherein the total hydroxyl ion concentration is

(a) At a level sufficiently high to initiate nucleation and deposition precipitation of said insoluble compound on and over the said surface of the said particulate supporting material itself, but

(b) Below that concentration at which the total hydroxide ion concentration and the total concentration of ions of said metal exceed the solubility product of that insoluble compound, of step (A) (1) (b), in said suspension-solution and thus at a level insufficient to induce any substantial free nucleation of said insoluble compound within the solution phase per se and with rapid dilution and distribution of any local momentary incremental increase of hydroxyl ion wherein the step of increasing the hydroxyl ion concentration is effected by introducing a hydrolizable agent into said solution, in an amount of 1 to 10 times the theoretical quantity required to convert said metal compound to said insoluble compound which hydrolizable agent is urea, acetamide or hydrolyzable derivatives thereof and will, under the conditions of the process, hydrolyze to form hydroxyl ions, and hydrolyzing said agent to form said hydroxyl ions during the process under controlled and gradual conditions; wherein the control of said hydrolysis is essentially effected by controlling the temperature of said solution at a level whereat the required and desired rate of hydrolysis, under the conditions of the process, take place, wherein said temperature can be up to 100°C, and

(E) Continuing increasing total hydroxyl ion concentration until the desired amount of insoluble metal compound has been transferred from said solution to said supporting material, wherein the rate of said further increases in said total hydroxyl ion concentration being maintained substantially at the level of the metal migration rate to the surface of the supporting material, thereafter separating these solid supporting materials, having this insoluble compound substantially homogeneously deposited thereon, from the remaining aqueous solution of the aqueous medium.

Example 1: *Illustrating the In Situ Generation of Hydroxyl Ion* – Preparation of a Catalyst Mass Containing Nickel as the Catalytically Active Material: 102 g of $Ni(NO_3)_2 \cdot 6\ H_2O$, 49 g of $NaNO_2$, 43 g of urea, and 20 g of "Aerosil" (a silica preparation obtained by flame hydrolysis of silicon tetrachloride) were added to about 4 liters of water.

The suspension was heated for 44 hours at 88°C with vigorous agitation, and passed through a glass filter. The mass was washed with distilled water and subsequently dried for 84 hours at 120°C. Upon analysis the catalyst mass proved to contain 27.2% of Ni in the form of NiO, 0.14% of NH_3, 0.26% of NO_3, and 0.19% of Na. After reduction in a flow of hydrogen, this catalyst can be converted into a nickel metal catalyst containing 60 pbw of Ni to 40 pbw of SiO_2. The nickel surface of this catalyst, measured by hydrogen chemisorption, amounted to 184 m^2 of Ni/g, from which it follows that the average particle size was 36 Å.

Example 2: *The Preparation of Zinc Catalysts* – 31.6 g of $ZnCl_2$, 0.8 cm^3 of concentrated HCl, 20 g of Aerosil, and 800 cm^3 of distilled water were added together. With continuous intensive stirring, a 1 N ammonia solution was injected below the liquid surface at room temperature, at the rate of 218 cm^3 per hour. The injection continued for 2 hrs, 10 min, until the pH of the suspension had reached a value of 7.8. At this value, the experiment was stopped. After the carrier material, now loaded, had settled, it was filtered off on a normal paper filter, washed with 800 cm^3 of distilled water, and dried at 120°C for 19 hrs.

Analysis of the resulting material gave a Zn to SiO_2 ratio of 1:1. The mean rise of the pH value during the precipitation amounted to 0.03 units per min. Precipitation is complete only in the range of pH value under 8. The method as applied here is particularly suitable for abruptly terminating the rise of the pH when the value of 8 has been reached.

X-ray diffraction testing yielded a pattern for $Zn_5(OH)_6Cl_2$. The broadening of the x-ray diffractions indicated that the mean particle size was 450 Å units. However, electron microscopy showed that only a few big zinc-containing crystallites were present, but that the majority of the zinc oxide was dispersed in a finely divided form over the carrier material. The size of these small zinc oxide particles was about 30 Å units.

Example 3: *The Homogeneous Reduction of the Valency of the Metal to Be Deposition Precipitated* – Preparation of Copper Catalysts: A solution of 38.0 g of $Cu(NO_3)_2 \cdot 3\ H_2O$ in 500 ml of water and a solution of 44.5 g of potassium sodium tartrate, 90 g of sodium bicarbonate and 22.5 g of sodium carbonate in 500 ml of water were put together. The resulting deep blue solution was made up to 1.5 liters, whereupon 10 g of silica Aerosil 220 V (surface area 200 m^2/g) were suspended in it. Finally, 31.0 g of glucose were added to this suspension.

The temperature was raised to 74°C in about 6 hours, while the suspension was thoroughly agitated. The suspension was kept at this temperature for 20 hours, during which period the pH increased from 8.0 to 8.3. Finally the suspension, which in the meantime had turned yellow, was cooled to room temperature with stirring. It was found that the suspension could very easily be filtered. The filter residue had an ochre color and contained a small quantity of an orange material which had deposited on the wall of the glass reaction vessel during the precipitation process. The filtrate had a pale green color, which resulted from the yellow color of partially decomposed glucose and a very small quantity of blue nonreduced cupric tartrate.

After drying at 120°C the x-ray diffraction pattern indicated the presence of cuprous oxide; the broadening of the reflections indicated the presence of particles of 110 Å. An electron microscopic investigation of the dried precipitate showed that it contained very little cuprous oxide particles of larger dimensions (approximately 100 Å). These strongly agglomerated particles, which caused the broadening of the x-ray reflections, were those that had deposited on the glass wall. The remainder of the preparation consisted of the carrier, homogeneously coated with extremely fine cuprous oxide particles measuring approximately 20 Å.

Example 4: *The In Situ Generation of the Supporting Material* – Preparation of a Cobalt Nickel Alloy Dispersed on Silica: 152 g of $Co(NO_3)_2 \cdot 6\ H_2O$ and 82 g of $Ni(NO_3)_2 \cdot 6\ H_2O$ were dissolved in 2 liters of distilled water. 45 ml of 1 N HCl were added to this solution. The temperature of the solution was raised to 90°C, whereupon 25 ml of a potassium silicate solution, containing 2.5 g of SiO_2, was injected into the agitated solution in 20 minutes. At the end of the injection the pH of the suspension thus obtained was 2.85. Next, a solution of 69.3 g of NH_4HCO_3 and 60 ml of a 25% ammonia solution in 500 ml of water were injected into the thoroughly agitated suspension at the same temperature, in 50 minutes. The injection was stopped when the pH had reached a value of 7.25. The violet precipitate could be readily filtered off; the filtrate (a total of 3 liters, including the wash water) contained 51 mg Co/ℓ and 409 mg Ni/ℓ.

After drying for 16 hours at 120°C the material was pressed to pellets with a diameter of 2.8 mm and a height of 2 mm. These pellets were reduced in a stream of hydrogen at 400°C for 68 hours. Next, a column of approximately 20 cm length was formed from the pellets in an inert atmosphere and the hysteresis loop of the column was determined. The coercive force was 370 oersteds.

Example 5: *Deposition Precipitation by the Decomplexation of Metal Ions* – Preparation of an Iron(III) Oxide-on-Silica Catalyst System: 48 g of $FeCl_3 \cdot 6\ H_2O$ and 66 g of Komplexon III (the disodium salt of ethylenediaminotetraacetic acid) were dissolved in 1.5 ℓ of water. Thereafter, 18 g of NaOH and 11.4 ml of glacial acetic acid were added by way of buffer, and 10 g of silica Aerosil 200 (specific surface area 204 m^2/g) was suspended in the solution. The suspension was yellowish brown; pH value, 4.6. Next, the suspension temperature was raised to 70°C after which a 30% hydrogen peroxide solution was injected by means of a plunger pump under the liquid level into the vigorously agitated suspension, at 200 ml/hr. After ∿300 ml of hydrogen peroxide had been injected, the pump was stopped; the pH value rose to 5.6 and the suspension turned brown.

Thereafter the loaded carrier material was isolated by filtration and washed. The filtrate has a light yellow color. The filter residue was dried for 18 hours at 120°C. The x-ray diffraction pattern of the dried material showed three very broad unidentifiable bands, in addition to the lines characteristic of sodium chloride. The broadening of the reflections indicated that the particle size was about 10 Å units. This was confirmed upon examination of the material in the electron microscope. These extremely small particles were found to be very homogeneously distributed over the carrier material.

Example 6: *Injection of a Solution Containing the Metal Ions to Be Precipitated* – Preparation of a Permanently Magnetizable Mass Containing Iron(III) and Cobalt: 25 g of Aerosil (specific surface 375 m^2/g) were suspended in 2 ℓ of water. The temperature was raised to 94°C and the pH value adjusted to 5.0 by means of ammonium bicarbonate. Under vigorous agitation 500 ml of a somewhat acidulated solution containing 96.5 g $FeCl_3 \cdot 6\ H_2O$ was injected under the liquid level during a period of 60 minutes (0.003 g ion per minute per liter suspension). By simultaneous injection of a solution of ammonia and ammonium carbonate the pH value of the suspension was kept between 5.0 and 5.8. After the injection of the ferric chloride, the suspension was cooled to room temperature, and the pH value was then brought to a value of 4 by means of nitric acid. At this pH value Fe(III) is insoluble but Co is still soluble.

Now, 24.3 g of $Co(NO_3)_2 \cdot 6\ H_2O$ and 100 g of urea were added to the suspension. During vigorous agitation, the temperature of the solution was raised and kept at 100°C for 17 hours. According to the principles of the process, a cobalt compound was precipitated onto the iron loaded carrier material.

The loaded carrier material could rapidly be separated from the liquid by filtration; the filtrate was clear and colorless. The precipitate was washed and dried at 120°C for 24 hours.

The composition of the mass obtained was 28.9% by weight Fe, 7.4% by weight Co and 31.0% by weight SiO_2. Good mixing of the metal ions as oxide particles was provided by calcining the mass at 900°C for 42 hours, and a cobalt ferrite was formed as could be concluded from a magnetic moment of 40.5 $g/cm^3/g$ of the material obtained.

Example 7: *Homogeneous Oxidation to a Higher Valence State* – Preparation of a Permanently Magnetizable Powder Containing Cobalt Oxide and Iron Oxide: 2.5 g of the high porosity silica known as Aerosil 380 (specific surface area 380 m^2/g) was suspended in a solution of 36.8 g of $Co(NO_3)_2 \cdot 5\ H_2O$ (containing 7.5 g of cobalt), 80 g of urea and 60 g of NH_4NO_3 in 3 liters of water. The suspension was boiled to remove the dissolved oxygen.

After the suspension had been cooled to room temperature in a nitrogen atmosphere, the pH was adjusted to 2 with nitric acid; next 540 ml of a ferrous chloride solution, containing 40 g of iron, was added. The suspension was then heated to 100°C while being thoroughly stirred. The pH value of the suspension was recorded automatically; this value increased quickly to 5.2 and subsequently dropped to 4.8, which latter value was reached 132 minutes after the start of heating. After heating for 16 hours, the pH had increased again to 5.8 and a deep black precipitate had formed on the carrier material.

To precipitate the last residues of cobalt from the suspension, the pH of the suspension was increased to 7.3 by injecting a solution of NH_4HCO_3 and NH_3, whereupon the precipitate was filtered off. The filtration proceeded extremely fast and the filter residue could be readily washed out.

After drying at 120°C the material showed an x-ray diffraction pattern analogous to that of magnetite (Fe_3O_4). The broadening of the lines indicated a particle size of 325 Å units.

The powder dried at 120°C was mixed with a solution of 20% by wt of polyurethane in dimethyl formamide and then spread on a polyester film. The amount of powder in the coating was 60% by wt. On the film thus obtained, the hysteresis loop was measured. The maximum field strength was 2000 oersteds, and the coercive force was approximately 800 oersteds.

Maintenance of Nitrogen Atmosphere in Preparation Furnace

F. Byrd and J.A. Fisher; U.S. Patent 4,148,946; April 10, 1979; assigned to Armco Steel Corporation describe a method for pretreating a ferrous base metal strip in a metallic coating line, and more particularly a method and means for maintaining a nonoxidizing atmosphere at positive pressure within the strip preparation furnace of the metallic coating line during line stops.

The process is applicable to that type of strip preparation furnace of a coating line which incorporates a direct fired furnace, a controlled atmosphere heating furnace, one or more cooling chambers and a snout, the free end of which extends below the surface of the molten coating metal bath. All of these elements are joined together in sealed relationship. The direct fired furnace is connected by conduit means to an exhaust fan which draws the products of combustion from the direct fired furnace to the atmosphere or to pollution control means, the nature of which are well known in the art.

The conduit connecting the direct fired furnace and its exhaust fan will normally be provided with an air dilution opening so that the products of combustion or exhaust gases withdrawn from the direct fired furnace by the exhaust fan during line operation will be diluted with respect to their heat content.

A refractory lined door means is provided in association with the duct connecting the direct fired furnace with its exhaust fan. The door is shiftable between a normal open or retracted position wherein the products of combustion are free to be drawn through the conduit by the exhaust fan and an extended or closed position during line stops wherein it seals the direct fired furnace from the exhaust fan and the air dilution openings in the duct. The door means is refractory lined since the gases within the direct fired furnace are normally at a temperature of at least 2300°F (1260°C). The door means is preferably located in association with the air dilution openings in the conduit so that when the door means is in its extended or closed position, the exhaust fan can continue to draw air above the door means to cool the door means.

At the time of a line stop, the process shown in Figure 9.3 uses excess flow of a safe, non-oxidizing atmosphere into the strip preparation furnace. Nitrogen, for example, may be used for this purpose. The excess nitrogen serves as a safety factor. In addition, it prevents the formation of oxide and scale on the surfaces of that portion of the ferrous base metal strip 9 located within the strip preparation furnace, as well as detrimental effects on the molten coating metal bath and equipment. Furthermore, the excess nitrogen flow will prevent that portion of the ferrous base metal strip **9** within the direct fired furnace **2** and the controlled atmosphere heating furnace **3** from overheating to the extent of damage to or breakage of the strip.

Figure 9.3: Hot Dip Metallic Coating Line

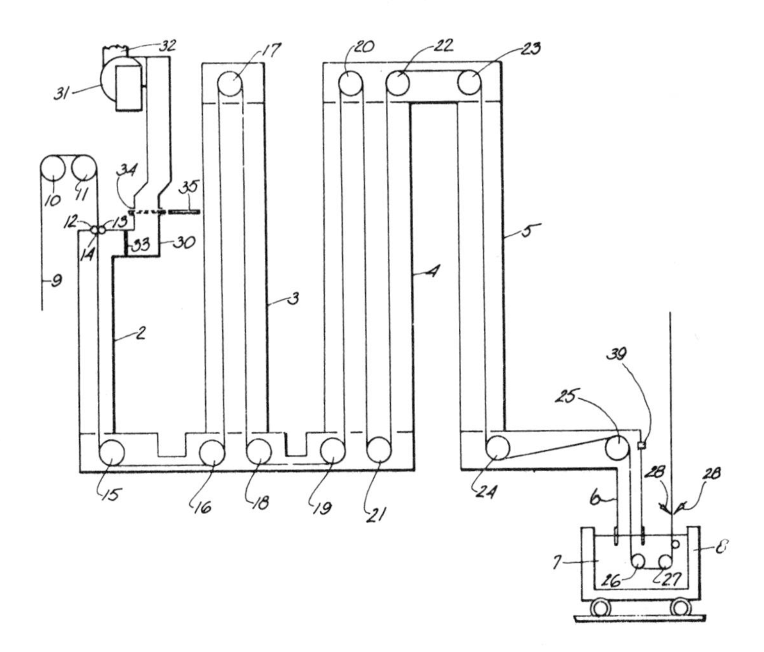

Source: U.S. Patent 4,148,946.

The excess nitrogen flow can be introduced into the strip preparation furnace through existing inlets for the various atmospheres of the parts, the existing inlets (not shown) being provided with appropriate valve means, or it may be introduced through a separate purge valve, or by both. A purge valve may be located at any appropriate position in the strip preparation furnace. For purposes of an exemplary showing, a purge valve is diagrammatically indicated at **39** in Figure 9.3. The purge valve **39** may, for example, be a motor operated valve which is automatically actuated at the time of a line stop. The excess nitrogen flow should be at a sufficient positive pressure to prevent the passage of oxygen from the ambient air into the strip preparation furnace through any openings, seams, or the like therein.

In the operation of the apparatus, at the time of a line stop the door means **35** will be shifted to its extended or closed position to seal off the strip preparation furnace from exhaust fan **31** and air dilution opening **34**. The other end of the strip preparation furnace will, of course, be sealed by virtue of the fact that the free end of snout **6** extends below the surface of molten metal bath **7**.

At the same time, purge valve **39** is activated to introduce an excess of nitrogen into the strip preparation furnace at a positive pressure. Both the door **35** and the purge valve **39** may be set up to be actuated automatically upon the occurrence of a line stop.

Finishing Process for Molten Metallic Coatings

H.F. Graff, J.B. Kohler, M.B. Pierson, N.W. Parks, P.E. Schnedler and R.E. Strait; U.S. Patent 4,137,347; January 30, 1979; assigned to Armco Steel Corporation describe a process concerned primarily with the finishing of a molten metallic coating. Assuming that the surface of the strip to be coated has been properly prepared, the strip is passed into a bath of molten coating metal and withdrawn from the bath in a generally vertical path of travel. As the strip emerges from the bath of molten coating metal, it will withdraw or pull with it a portion of the molten coating metal forming a concave meniscus. According to this process the nonuniform surface layer of coating metal on the strip is cleanly sheared by a laminar flow fluid jet so that the outer portion of the coating metal flows back to the bath, while a desired quantity of coating metal adheres or remains on the strip. The quantity of molten metal adhering to the strip (the thickness of the finished coating) is controlled by varying the velocity of the fluid jet, the angle of impingement of the jet on the coated strip, the height of the point of impingement, and the shape of the nozzle opening producing the jet.

The shape or contour of the nozzle opening will contour the narrow dimension of the fluid jet and hence permit variation of the jet wiping action across the strip width. Thus, it is possible to produce optimum coating weight distribution across the strip width, including the elimination of oxide berries at the strip edges.

Metal-Glass-Fluoride Bearing Material

H.E. Sliney; U.S. Patent 4,136,211; January 23, 1979; assigned to U.S. National Aeronautics and Space Administration has found that low friction surfaces exhibiting self lubrication and oxidation resistance up to and in excess of about 930°C may be formed as a composite of a metallic constituent which lends strength and elasticity to the structure, a glass component which provides oxidation protection to the metal and may also contribute to reduced wear and finally, a fluoride component which provides lubrication. The composite may be fabricated as a porous, sintered metallic matrix which is infiltrated with molten glass and fluoride or, optionally, a suitable substrate with metal, glass, and fluoride components codeposited thereon by plasma spray techniques.

The lubricating material of the process will effectively preclude galling from room temperature to and in excess of 930°C, but is particularly effective over the range 530° to 930°C at which temperature the glass and particularly the fluorides are soft enough to form a smear or glaze of lubricating film on the surface. These composites are easily machined and exhibit plastic response over the temperature range of interest.

The self-lubricating, oxidation resistant bearing composites may be comprised of from 20 to 80% high-temperature resistant metallic component; 0 to 50% silver, gold or alloys thereof; and, 20 to 80% combined lubricating fluorides and oxide inhibiting glass. A preferred composition comprises 60 to 70% of a metallic constituent selected from the group of high temperature iron, cobalt, and nickel alloys; 15 to 20% of a lubricating fluoride salt selected from the group of calcium fluoride, barium fluoride and mixtures thereof; and 15 to 20% of a glass comprised of from 10 to 50% barium oxide, 0 to 30% sodium oxide and/or potassium oxide, 0 to 15% calcium oxide and balance silica.

A most preferred composite according to the process is comprised of 67% of an 80 nickel, 20 chrome alloy, 16½% of barium fluoride-calcium fluoride eutectic, and 16½% of a glass comprised of 20% BaO, 10% K_2O, 10% CaO, and 60% SiO_2. In the case of plasma sprayed composites, the fluoride eutectic may be replaced with the single fluoride CaF_2.

In carrying out the plasma spray method, glass frit is prepared in any manner well known to those skilled in the art: e.g., ball milling a suitable glass composition. The glass frit is sized to particles of 125 μ and is then mixed with the desired metal powder and fluoride lubricant powder.

The underlying substrate surface upon which the bearing composite is to be co-deposited should be grit blasted or otherwise cleaned to remove surface films, foreign materials and the like. The composites may then be sprayed to a thickness from 0.010 to 0.060 cm and subsequently machined back to a working thickness of from about 0.005 to 0.050 cm.

Regardless of the fabrication technique selected, the composite-bearing materials of this process may be surface enriched in the lubricant by a thin fluoride spray film or by subsequent heat treatment in air at from 760° to 900°C for 4 to 24 hours. Such a heat treatment will cause slight exudation and solid state migration of fluorides along the surface and further serves the beneficial purpose of mildly preoxidizing the exposed metal. Additionally, the surfaces will become entirely covered with a combined fluoride-oxide film which is highly desirable to prevent direct metal-to-metal adhesive contact during sliding.

Flaw-Marking Apparatus

J.P. Vild, W.I. Cleary, and D.P. Fox; U.S. Patent 4,127,815; November 28, 1978; assigned to Republic Steel Corporation describe a marking technique which has been developed for marking defective hot workpieces. A conventional flocking or sandblast gun is used. The marking gun used in this process carries a quantity of atomized aluminum powder. The powder has a melting point of approximately 1200°F. A vibrator is connected to the gun to vibrate it on a continuous or substantially continuous basis. This vibration agitates the aluminum powder to facilitate its dispensing. This vibration supplements the material agitation achieved with air flow in the conventional gun and is especially desirable since air flow is quite intermittent occurring only when the flaw is detected.

When a flaw is detected, a signal from the detector actuates the marking gun. A short blast of compressed air expels a small quantity of aluminum powder from the gun and onto the pipe. Since the pipe is above the melting point of the aluminum, the powder is fused onto the surface of the pipe providing highly visible indication of the location of a flaw with a marking material that can withstand the elevated temperatures of the workpiece.

Printing Plate

J. Stroszynski; U.S. Patent 4,098,188; July 4, 1978; assigned to Hoechst AG, Germany describes a printing plate comprising a support bearing either (a) a layer of particles each comprising a hydrophilic metal core and an oliophilic metal casing or (b) a layer of particles each comprising an oliophilic metal core and a hydrophilic metal casing.

Figure 9.4: Printing Plate

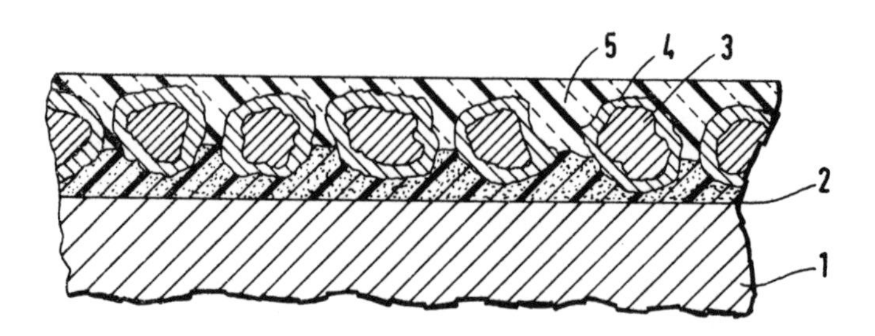

Source: U.S. Patent 4,098,188

Referring to Figure 9.4, a supporting film **1** has applied thereto an adhesive layer **2** and a bimetallic layer cemented thereby to the supporting film **1**. The layer is composed of small particles which have a metallic core **3** and a metallic casing **4**. The supporting film preferably is composed of a dimensionally stable material, for example a metal (aluminum, iron, but preferably steel or zinc) or a plastic (especially polyethylene terephthlalate). Hardenable epoxide resins, for example, are suitable for cementing the bimetallic particles **3**, **4** to the support **1**. As described, of the particles, either the core **3** is hydrophilic and the casing **4** surrounding it oleophilic, or vice versa. As hydrophilic metals may be employed, for example, aluminum, zinc, steel, chromium and nickel, and as oliophilic metals, for example, copper, brass and lead. The printing plate according to the process shown in Figure 9.4 is provided with a light sensitive etch resistant layer **5**.

It has proven particularly advantageous to apply the suspension from a casting die. The combination of particles, adhesive and solvent is advantageously so selected that after drying, the particles are tightly packed at the exposed surface. The process is preferably carried out by multilayer coating, each coating being incomplete in itself, and after a given drying period, the coatings are reduced by means of high pressure, applied in a line to a few layers and, in the limiting case, one layer. It is possible when the cores are deformable to flatten, calibrate and level the surface.

The linear pressure is advantageously produced by a resilient roller and a fixed roller which are pressed positively against one another. Preferably, the fixed roller pressed on the coating. A layer-levelling pressing operation of this type can be repeated as long as the binder conforms to the deformation without damage. Following the coating or covering of the supporting film surface are further operations, for example, repetition of the particle covering or washing, and where necessary, a process rendering the particle layer hydrophilic and sensitizing it.

The production of an image suitable for printing, which is necessary for the production of a printing form from the printing plates according to the process, is most advantageously carried out by applying an etch resistant layer to the particulate covering in the areas necessary for the production of the desired image, by etching the exposed areas of the metal casing covering the metal cores of the particles and subsequently removing the etch resistant layer from the protected areas.

If the casings of the particles are composed of the hydrophilic and the cores of the oleophilic metal, the etch resistant layer must be applied in the areas of the particulate covering which are to be the printing areas in the printing form to be produced. The imagewise application of the etch resistant layer could be carried out manually, e.g., using a brush. Generally, however, the printing plate is designed to be processed into a printing form by a photographic reproduction technique. The etch resistant copying layers used in these processes, and their development into an etched stencil are known.

Dialysis Membrane

Dialysis represents a selective diffusion of dissolved particles by semipermeable membranes. In medicine, the size of a hole for the passage of so-called crystalloids (e.g., electrolytes, glucose, urea, creatine, barbiturates) must be sufficient, while colloids and corpuscular components (protein, fats, blood cells, bacteria and viruses) are not to pass.

In medicine this is accomplished by means of an artificial kidney (hemodialysis) in which the blood and the fluid that is to be purified are separated from one another by a membrane. By diffusion, osmosis and ultrafiltration, an exchange of the components of those parts of the fluid takes place, which can pass through the passages of the separating membrane. The type and number of the parts which participate in an exchange therefore depend on the nature of the membrane, concerning the size of its holes and the distance between the holes.

Since the existence of passages and their effective diameter cannot be predetermined, it is not possible to use a dialyzer suitable for a sickness therapy. A chronic uremia, an acute failure of the kidneys, intoxications from medicaments and similar diseases are caused by bacteria, viruses, poisonous substances of variable molecular size. The as yet unknown poisons of uremia belong, e.g., to the medium molecules which have a molecular weight of about 20,000 and which can be removed only be a long duration of dialysis.

Also ruptures of the membrane are not completely to be excluded and represent a danger for the life of the patient (home dialysis).

B. Schilling; U.S. Patent 4,141,838; February 27, 1979 describes a process to develop a dialysis membrane in such a way, that it will hold back substances of a certain diameter and will guarantee a uniform, quick passage of substances of a smaller size. Furthermore, this membrane should be reusable after use (e.g., in hemodialysis).

The dialysis membrane consists advantageously of a plate or of a foil which is provided with a coating. At the same time, the plate or foil has holes which run approximately perpendicularly in relation to the plane of the dialysis membrane.

Preferably the plate or foil consists of metal, a semiconductor or a dielectric and in a special embodiment, of stainless steel. A membrane may also consist of a fine web (e.g., a steel web). For medicinal purposes it is necessary that this coating consist of a sterilizable material. Preferably, gold, silver or platinum is used. An effect which can be achieved therewith, the decrease of the precipitation of blood protein on the membrane, will also be achieved by silicon, chromium, nickel, vanadium, titanium, glass, etc.

The process has been characterized in that holes are drilled into the plate or foil which are controllable as to size by a coating. Instead of the perforated plate or foil, a steel web can also be used. In that case, the passages are produced preferably by laser beams, electron beams, etc.

In that case, the diameter depends on the task set and it lies in most cases in the μm area. The drilling by means of laser or electron beams represents some of the few possibilities of making passages of such smallness in a reproducible manner. The smallest diameters which are technically reproducible at the present time lie between 1000 and 7000 Å.

The plate or foil is coated by electrolysis and subsequent vaporizing. At the same time average hole diameters of about 50,000 Å are reduced by electrolytic coating, down to about 10,000 Å. After retarding the relatively quick electrolysis process, the desired diameter of the passage of from 20 to 80 Å will be achieved by slow vaporizing on.

The coating may also be applied either by vaporizing on in the vacuum or galvanically. As a result, a narrowing down of the passages of the membrane will be achieved, whereby diameters of a few Å will be achieved, and for another thing, a suitable coating in the case of the use of the membrane for medical dialysis purposes will prevent the precipitation of blood protein on the membrane.

Metallizing of Superhard Man-Made Materials

J.V. Naidich, G.A. Kolesnichenko, N.S. Zjukin, B.D. Kostjuk, S.S. Shaikevich, Y.F. Motsak, V.P. Fedulaev, N.A. Kolchemanov, V.M. Ugarov, V.V. Losev, M.S. Drui, A.A. Lavrinovich, D.F. Shpotakovsky, and S.V. Chizhov; U.S. Patent 4,117,968; October 3, 1978 describe a method for soldering metals to diamond- and boron nitride-base materials whereby a metallizing coating is applied to the soldering surface of the diamond- and boron nitride-base materials, with the materials to be joined being exposed to a temperature sufficient to achieve adequate adhesion of the metallizing coating to the materials, and the metal is joined with the metallized material by use of soldering agents exhibiting ductile properties and having melting points of up to 1000°C.

The metallizing coating is constructed by metals showing high adhesiveness towards the diamond- and boron nitride-base materials, e.g., chromium and/or molybdenum, tantalum, tungsten or titanium taken at the rate of from 3.0 to 65% by wt of the coating, the balance of the coating being accounted for by metals exhibiting improved oxidation resistance, e.g., copper and/or silver, lead, tin, nickel or cobalt.

The application of the coating is effected in two operational steps: first, a coat of a hard metal showing high adhesive activity towards diamond- and boron nitride-base materials is built up to a thickness between 0.0005 and 0.001 mm, and then a coat of a hard metal exhibiting improved oxidation resistance is built up to a thickness between 0.001 and 0.01 mm.

An alternative technique of application of the metallizing coating uses a metal made up of metals highly adhesive towards diamond- and boron nitride-base materials and oxidation-resistant metals, the coating thickness being from 0.01 to 0.5 mm.

The soldering procedure is carried out at the solder melting point for 0.5 to 2 minutes at a rate of heating and cooling of the surfaces being joined, of from 10 to 30 deg/sec in air under a layer of a liquid flux. The liquid flux belongs to the category of fluxes employed for soldering metals by means of metallic spelters. The melting points of such fluxes lie within the range from 500° to 600°C and they further exhibit stability of properties in the temperature range up to 1000°C.

An acceptable flux (Flux No. 1) may have the following composition: 35 wt % boric anhydride, 42 wt % anhydrous potassium fluoride, 23 wt % potassium fluoroborate.

Another flux, No. 2, similar to Flux No. 1 in terms of properties has the following composition: 70 wt % boric acid, 9 wt % calcium fluoride, 21 wt % sodium tetraborate.

Yet a third possible flux, Flux No. 3, has the following composition: 20 wt % boric acid, 80 wt % sodium tetraborate.

A further possibility, Flux No. 4, is a compound of the following composition: 10 wt % sodium fluoride, 8 wt % zinc chloride, 32 wt % lithium chloride, and 50 wt % potassium chloride.

The soldering agent for joining metals with metallized materials on the basis of diamond and cubic boron nitride has the following composition: from 13 to 60 wt % copper, 5 to 15 wt % tin, 5 to 25 wt % titanium, and in accordance with the process, 2 to 15 wt % lead, the balance being constituted by molybdenum and/or tungsten or tantalum.

An alternative, and equally effective, soldering agent has the following composition: from 13 to 60 wt % copper, 5 to 15 wt % tin, 5 to 25 wt % titanium, from 0.5 to 20 wt % solder-insoluble nonmetallic filler, and from 0.3 to 20 wt % molybdenum and/or tungsten or tantalum. The nonmetallic filler is a powder of grain size from 1 to 50 μ.

Example: A polycrystal of a cubic modification of boron nitride such as Elbor of diameter 3.9 mm and height 4.4 mm and a steel holder of diameter 5 mm and height 20 mm are to be soldered together.

The boron nitride polycrystal is premetallized with a chromium layer by electron-beam sputtering of the metal and its deposition to a layer thickness of 0.001 mm, the procedure being effected in a vacuum of 1-5 x 10^{-5} mm Hg. Then the polycrystal is given a protective coat of copper chemically deposited to a thickness of 0.005 mm. The coating components are in the weight ratio of 20 chromium to 80 copper. Then the semicrystal is subjected to baking under a vacuum of 2-5 x 10^{-5} mm Hg at 950°C for 20 min. The metallized polycrystal is placed in a prepared cylindrical bore hole formed in the end face of the holder along the axis thereof, the solder slit being 0.15 mm on a side.

The soldering procedure is effected in air under a layer of Flux No. 1 by high-frequency heating with a soldering agent comprising 20 wt % of tin and 80 wt % of copper at a heating-cooling rate of 20°C per second. The process is carried on at a temperature of 900°C for 1.5 minutes. The solder melts down and by the action of capillary forces flows into the solder slit.

The soldered joint is even, free from cavities and cracks in the cutting element as well as from other soldering flaws; the adhesion of the soldering agent to the polycrystal is satisfactory.

COMPANY INDEX

The company names listed below are given exactly as they appear in the patents, despite name changes, mergers and acquisitions which have, at times, resulted in the revision of a company name.

INVENTOR INDEX

U.S. PATENT NUMBER INDEX

Copies of U.S. patents are easily obtained
from the U.S. Patent Office at 50¢ a copy.

NOTICE

PLATING OF PLASTICS 1979

RECENT DEVELOPMENTS

by Francis A. Domino

Chemical Technology Review No. 138

The adaptability of plastics for so many applications has accelerated development of techniques for plating them with metals. Plated plastics have replaced many automobile parts formerly made of metal, have made major inroads in the data processing and sound recording fields, and have expedited production of relatively inexpensive decorative items.

This text details about 175 processes involving both electro- and electroless plating, the latter comprising catalytic reduction of metal salts to effect the desired finish. Among the procedures described are surface treatments conducive to plating, bath compositions, conditioning agents, sensitizers, stabilizers, activators, and catalysts, all of which facilitate the coating with heavy or precious metals of such substrates as polyolefins, halopolymers, polystyrene, acrylonitrile-butadiene-styrene resins, and polyesters.

The following lists chapter headings, **examples of some** subtitles, and, in parentheses, the number of processes per topic.

1. **MAGNETIC COATINGS, METALIZERS AND PRETREATMENTS (12)**
   Ferromagnetic Memory Layer
   Fired-on Glass Frits
   Grafting on Polymeric Textile
   Metal Salts/Photo- or Chemical Reduction
2. **ABS RESINS & POLYOLEFINS (17)**
   Asbestos-Containing ABS Substrate
   Electrocoating ABS with All-Ni System
   Noble Metal-Acid Solution Pretreatment
   Isotactic Polypropylene with Organic N
   Polymers with Natural Silica Fillers
3. **Al, Ni, Co AND NOBLE METALS (21)**
   Polycarboxylic Acid Activator
   Noble Metal Sensitizer
   Hydrazine-Containing Nickel Bath
   Copper Adhesion Aid for Ni-Cr Film
   Photolytic Use of Gold Complexes
   Chelating Agents and Promoters
   Alkyleneimine and Amine Bath Additives
   S or Se Activator of Silver Bath
4. **GENERAL USE SENSITIZERS, ACTIVATORS, CATALYSTS (29)**
   Novel Metal-Tin Halide Complex
   Colloidal Sensitizer & Redox Treatment
   Chromic Acid and Ceric Ammonium Nitrate
   Photosensitive Palladium Sensitizer
   Autocatalytic Plating
   Monocarbonyl Stabilizer
   Metal-Phosphorus Sulfur Complex
   Alcohol Pretreatment for Persulfate Etch
   Permanganate & Manganate Ion Treatment
   Pretreatment with Sulfur Trioxide
   Sacrificial Anodized Metal Foil Laminate
5. **THERMOPLASTIC AND THERMOSET RESINS (25)**
   Alkaline Alcohol to Pretreat PVC
   Quinoline or Butyrolactone in Preetch Treatment of Polyoxymethylene Polymers
   Electroless Silver Plating of Nylon
   Pd/Ag Pretreating of Polyamides
   Phosphorus-Activated Polyphenylene Oxide
   Resin Pretreated with Fine Metal
   Pre-etch Swelling of Epoxy Substrate
   N Heterocyclic to Precondition Epoxy
   Glycol Ether to Pretreat Thermoset Resin
6. **ELECTRODEPOSITION (10)**
   Sodium Phenolate-Detergent Solution
   Crimped Metal-Coated Synthetic Filament
   Electroplated Through-Hole Circuit Board
7. **ELECTROLESS PLATING PRETREATMENTS (25)**
   Dry Colloidal Catalyst Compositions
   Hydrous Oxide Colloid of Base Metals
   Palladium Salt and Photosensitive Binder
   Undercoat of Polyvinylidene Chloride
   Urea & Tin Salt Catalytic Acid Solution
   Stannous-Cuprous Ion Complex as Primer
   Dye and Chelating Agent as Catalysts
   Composite Particulate Diamond-Metal Coat
8. **ELECTROLESS COPPER PLATING (14)**
   Brucine Additive for Hydrogen Inhibition
   Polyalkylene Oxide to Enhance Ductility
   Ring N Compound to Strengthen Film
   Perfluorocarbon in Nonionic Surfactant
   Dry Replenishment for Copper Ions
   Iodobenzoic Acid Stabilizer
9. **OTHER ELECTROLESS DEPOSITION (15)**
   Noble Metal-Phenol Surface Improver
   Hectorite as a Deposition Catalyst
   Elementary Sulfur Stabilizer
   Pretreating Ni & Co with Reducing Agent
   Thiol Reducer in Noble Metal Plating
   Vacuum or Sputter Deposits of Au or Pd
10. **RELATED APPLICATIONS (6)**
    Dispersion Coating for Flemish Finish
    Plastic Overlay on Metalized Substrate

ISBN 0-8155-0770-4

385 pages

TECHNOLOGY OF METAL POWDERS

Recent Developments 1980

Edited by L.H. Yaverbaum

Chemical Technology Review No. 153

In these energy-conscious times powder metallurgical processes have acquired new significance.

The powder metallurgy approach to the manufacture of metal items, when compared with casting and forging, brings with it serendipitous advantages: there is practically no loss by scrap formation; the temperatures applied are much lower and heat is applied for shorter periods. Most sintering reactions take place below the melting point of the metal or metals being shaped.

Research is being hastened trying to satisfy the demand for tailor-made powders with carefully checked granulometric and morphological characteristics.

This book probably represents the most comprehensive data source available for the production of metal powders, ranging from iron and nickel to precious metals and the superalloys.

The partial table of contents below gives **chapter headings, examples of subtitles** and the number of processes per topic in parentheses.

ISBN 0-8155-0794-1

360 pages

ELECTROPLATING & RELATED METAL FINISHING
POLLUTANT & TOXIC MATERIALS CONTROL 1978

by Marshall Sittig

Pollution Technology Review No. 46

The plating industry makes use of a great many chemical and electrochemical reactions to improve the surface and structural surface properties of metals and other materials. The effluents from a plating plant are highly toxic and corrosive. Regulatory agencies have been paying increasing attention to the discharge of undesirable elements into the environment. Pollution of streams by discharge of rinse water, and air pollution by discharge of fumes from hot electroplating baths have been of particular concern. Leachates from disposal of wastes on land may drain into drinking water supplies.

The EPA on June 26, 1978 issued pretreatment regulations for toxic industrial wastes discharged into public sewer systems. The electroplating industry is first on the list to be regulated under the new rules.

A traditional way of handling the pollution problem has been to use relatively large quantities of water to dilute the pollutants. This is no longer acceptable, and this book presents methods for advanced treatment and detoxification of fumes and wastewater with excellent materials recovery. Recently this industry has turned in the direction of what appears to be the only logical solution to the problem, that is to develop a zero discharge technology and to make the plating operation as nearly as possible a closed ecological system.

This book presents a detailed and up-to-date review of pollution control in metal finishing and is intended to be used as a practical manual by anyone connected with this major industry. A partial and condensed table of contents follows here. The bibliography at the end of the volume cites the important government surveys and federally funded studies which were used as source material.

ISBN 0-8155-0716-X **413 pages**

EXTRACTIVE METALLURGY 1977

Recent Advances

by Edward J. Stevenson

Chemical Technology Review No. 93

The efficient extraction and recovery of metals found in the earth's crust or on the bottom of her oceans has become increasingly important as modern industrial societies consume vast quantities of these materials.

With the increased awareness on the part of both the public and the metals-winning industry of the ecological dangers that can arise from continued customary mining procedures and the increased problems of pollution caused by obsolete refining methods, the search for clean low-energy processes has become paramount.

This book describes over 230 processes from the recent U.S. patent literature. The world-wide search for new and efficient metal recovery methods is nowhere more evident than is shown here by the scope of these processes.

Many of these processes encompass the extraction and isolation of several metals. However, for continuity and clarity, the organization of this book is focused on the major metallic component most commonly sought in a given process. Improved uses of raw materials and of human and energy resources are emphasized as is any step that might lead to a cleaner, less energy-consuming process.

A partial and condensed table of contents follows here. Chapter headings and important subtitles are given. Numbers in parentheses indicate the number of topics covered.

ISBN 0-8155-0668-6

255 pages

RECHARGEABLE BATTERIES 1980
Advances Since 1977

Edited by Robert W. Graham

Energy Technology Review No. 55
Chemical Technology Review No. 160

Rechargeable or secondary batteries represent electrical storage systems of the so-called high energy density type, since they are capable of supplying high-amperage electricity. The high energy capacity and compactness of such batteries renders them particularly suitable as principal or auxiliary sources of electrical energy in both mobile and stationary power plant systems. The latest research goals are emphasized in detail in this book.

Most of the current efforts are concentrated on weight reduction and other refinements, each of which improves a given type of battery. The technology exists and can be engineered rapidly into production procedures when the need for such batteries arises.

The scope of this book is shown by the following partial table of contents which presents **chapter headings and subtitles.**

ISBN 0-8155-0802-2 **452 pages**